NEW GCSE SCIENCE

Marjorie Comley

Additional Science
For Specification Units B2, C2 and P2

AQA

Series Editor: Ken Gadd

Authors: Mary Jones, Louise Petheram and Mike Tingle

D0231609

Student Book

William Collins' dream of knowledge for all began with the publication of his first book in 1819. A self-educated mill worker, he not only enriched millions of lives, but also founded a flourishing publishing house. Today, staying true to this spirit, Collins books are packed with inspiration, innovation and practical expertise. They place you at the centre of a world of possibility and give you exactly what you need to explore it.

Collins. Freedom to teach

Published by Collins
An imprint of HarperCollins*Publishers*
77 – 85 Fulham Palace Road
Hammersmith
London
W6 8JB

Browse the complete Collins catalogue at
www.collinseducation.com

©HarperCollins*Publishers* Limited 2011

10 9 8 7 6 5 4 3 2 1

ISBN-13 978 0 00 741456 7

British Library Cataloguing in Publication Data
A Catalogue record for this publication is available from the British Library

Commissioned by Letitia Luff
Project managed by Hanneke Remsing and 4science
Edited by 4science
Proofread by Life Lines Editorial Services and Vicki Harley
Indexed by Nigel d'Auvergne
Designed by Hart McLeod and Peter Simmonett
New illlustrations by Peters & Zabransky and 4science
Picture research by Caroline Green
Concept design by Anna Plucinska
Cover design by Julie Martin
Production by Kerry Howie
Contributing authors John Beeby, Nicky Thomas and Ed Walsh
'Bad Science' pages based on the work of Ben Goldacre

Printed and bound by L.E.G.O. S.p.A. Italy

Acknowledgements – see page 320

Contents

Biology

Chemistry

Physics

How to use this book

Welcome to Collins New GCSE Science for AQA!

The main content

Each two-page lesson has three levels:

> The first part outlines a basic scientific idea

> The second part builds on the basics and develops the concept

> The third part extends the concept or challenges you to apply it in a new way.

Information that is only relevant to the Higher tier is indicated with 'Higher tier'.

Each section contains a set of level-appropriate questions that allow you to check and apply your knowledge.

Look for:

> 'You will find out' boxes

> Internet search terms (at the bottom of every page)

> 'Did you know' and 'Remember' boxes

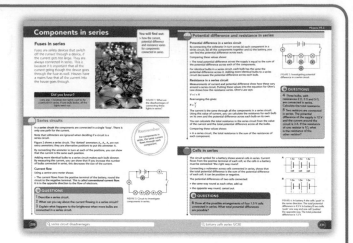

Units and sections

Each Unit is divided into two sections, allowing you to easily prepare for Route 1 or Route 2 assessment.

Link the science you will learn with your existing scientific knowledge at the start of each section.

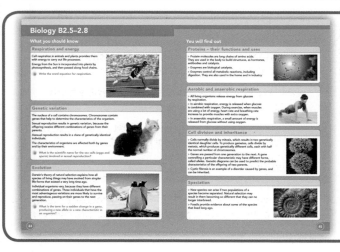

Checklists

Each section contains a checklist.

Summarise the key ideas that you have learned so far and see what you need to know to progress.

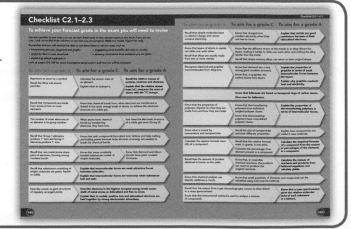

Exam-style questions

Every section contains practice exam-style questions for both Foundation and Higher tiers, labelled with the Assessment Objectives that they address.

Familiarise yourself with all the types of question that you might be asked.

Worked examples

Detailed worked examples with examiner comments show you how you can raise your grade. Here you will find tips on how to use accurate scientific vocabulary, avoid common exam errors, improve your Quality of Written Communication (QWC), and more.

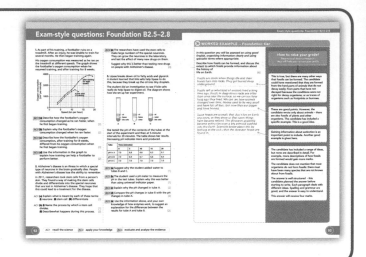

Preparing for assessment

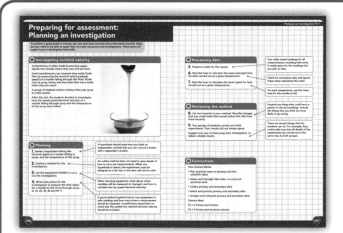

Each Unit contains Preparing for assessment activities. These will help build the essential skills that you will need to succeed in your practical investigations and Controlled Assessment, and tackle the Assessment Objectives.

Each type of Preparing for assessment activity builds different skills.

> Applying your knowledge: Look at a familiar scientific concept in a new context.

> Planning an investigation: Plan an investigation using handy tips to guide you along the way.

> Analysing and evaluating data: Process data and draw conclusions from evidence. Use the hints to help you to achieve top marks.

Bad Science

Based on *Bad Science* by Ben Goldacre, these activities give you the chance to be a 'science detective' and evaluate the scientific claims that you hear everyday in the media.

Assessment skills

A dedicated section at the end of the book will guide you through your practical and written exams with advice on: the language used in exam papers; how best to approach a written exam; how to plan, carry out and evaluate an experiment; how to use maths to evaluate data, and much more.

Biology B2.1–2.4

What you should know

Cells, tissues, organs and organ systems

Animals and plants are made up of cells.

Cells are specialised to carry out different functions.

Cells are organised into tissues, organs and organ systems, each of which carries out a particular function.

 State two differences between a typical animal cell and a typical plant cell.

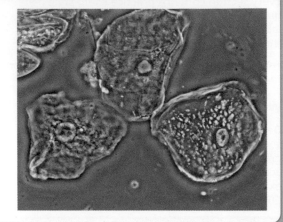

Photosynthesis

Plants remove carbon dioxide from the air, to make food.

In photosynthesis, carbon dioxide combines with water to produce glucose and oxygen.

● Write the word equation for photosynthesis.

Environment

Each kind of living organism is adapted to live in particular conditions.

Changes in the environment can affect the distribution of living organisms.

 Give one example of an animal that is adapted to live in an arctic environment, and explain two ways in which its features help it to survive there.

You will find out

Cells and cell transport

> Different types of cells have different structures. This enables cells to carry out their specific functions.

> Oxygen and other dissolved substances move into and out of cells by diffusion.

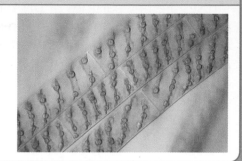

Tissues, organs and organ systems

> In animals and plants, cells are grouped into tissues.

> Different organs are made up of aggregations of tissues performing specific functions.

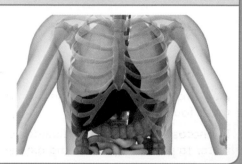

Photosynthesis

> Chlorophyll in plant cells absorbs energy from sunlight. This energy is used to convert carbon dioxide and water to glucose and oxygen.

> Factors, including light, temperature and carbon dioxide concentration, affect the rate of photosynthesis.

> Glucose is used to supply energy, to make cellulose for cell walls and to make proteins.

Organisms and their environment

> The distribution of different species of organisms in the environment is affected by physical factors. These factors include temperature, nutrients, light, water, oxygen and carbon dioxide.

> Quantitative data about the distribution of organisms can be collected using quadrats and transects.

Animal and plant cells

Hijacked

This cell has been attacked by viruses. Viruses are tiny particles of DNA and protein. They are not made from cells. They cannot reproduce on their own. Instead, they invade cells and hijack them, making the cell produce multiple copies of the virus. The new viruses burst out of the cell, ready to find more cells to invade. Not surprisingly, the cell is killed by the viral attack.

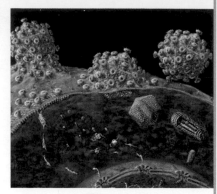

FIGURE 1: Viruses are bursting out of this cell. Can the cell survive?

You will find out:

> most human and animal cells have a nucleus, cytoplasm, mitochondria, ribosomes and a cell membrane

> plant and algal cells also have a cell wall, and may have chloroplasts and a permanent vacuole

> the functions of all of these cell parts

Animal and plant cells

All living organisms are made of cells.

Every one of the cells in your body has been produced from the one single cell – the zygote – that began your life. The resulting cells divided over and over and over again to make your body.

Animal and plant cells have many similarities, but also some differences.

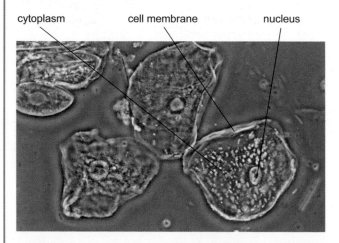

cytoplasm   cell membrane   nucleus

FIGURE 3: A photograph of human cheek cells taken through a microscope.

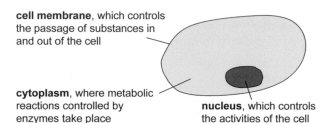

cell membrane, which controls the passage of substances in and out of the cell

cytoplasm, where metabolic reactions controlled by enzymes take place

nucleus, which controls the activities of the cell

FIGURE 2: An animal cell.

cell membrane, which controls the passage of substances in and out of the cell

cell wall, which strengthens the cell

chloroplast, which absorbs light energy to make food

vacuole, filled with cell sap

cytoplasm, where metabolic reactions controlled by enzymes take place

nucleus, which controls the activities of the cell

FIGURE 4: A plant cell.

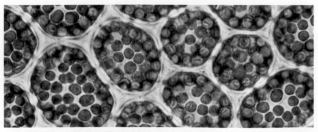

FIGURE 5: A photograph of moss cells taken through a microscope. Each cell has a nucleus, but they are hidden under the chloroplasts.

QUESTIONS

1 List three structures that are found in animal cells and also in plant cells.

2 List three structures that may be found in plant cells but not in animal cells.

Q cell structure … plant animal cells

Cell organelles

The different parts of a cell are called **organelles**. Each organelle has a particular function.

With a really good microscope, you can see more detail in cells. Figures 6 and 7 show some very small organelles that you cannot usually see with the type of microscope you use in school.

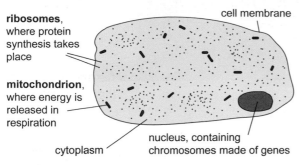

ribosomes, where protein synthesis takes place

cell membrane

mitochondrion, where energy is released in respiration

cytoplasm

nucleus, containing chromosomes made of genes

FIGURE 6: A more detailed look at an animal cell.

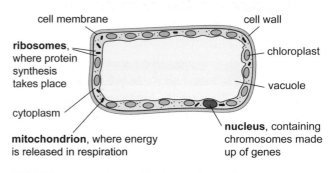

cell membrane

cell wall

ribosomes, where protein synthesis takes place

chloroplast

vacuole

cytoplasm

mitochondrion, where energy is released in respiration

nucleus, containing chromosomes made up of genes

FIGURE 7: A more detailed look at a plant cell.

QUESTIONS

3 Copy and complete this table. You will need to add several more rows.

Structure	Present in animal cell?	Present in plant cell?	Function
nucleus	yes	yes	controls the activity of the cell

Remember
All cells have cell membranes. Plant cells and algal cells also have cell walls.

Electron microscopes

Research laboratories use electron microscopes. Scientists send beams of electrons through the specimen. The electrons hit a screen or a sheet of photographic film where they make an image.

Electron microscopes can magnify an object by two million times. This is considerably more than the two thousand times with a light microscope.

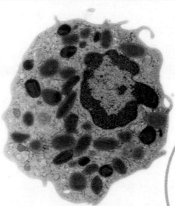

FIGURE 8: This image of a white blood cell was made using an electron microscope.

QUESTIONS

4 In Figure 8, the blue object in the cell is the nucleus. What do you think the red objects are? What other parts of the cell can you identify?

Remember
Although the nucleus controls most things in a cell, it is the cell membrane that controls what goes into and out of the cell.

🔍 cell organelle functions GCSE

Microbial cells

A different world

The world is teeming with living things that are too small to see. You can find microorganisms like these by looking at a drop of pond water under a microscope.

You will find out:

> bacteria are made of a single cell with cytoplasm, a cell membrane and cell wall, but no nucleus
> yeast is a single-celled fungus, with a nucleus, cytoplasm, cell membrane and cell wall

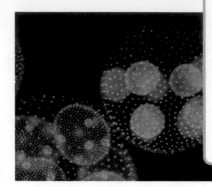

FIGURE 1: These microscopic balls are organisms called *Volvox*. How do you think they feed? (There is a big clue in the photo.)

Microorganisms

Microorganisms are living things that are too small to see with the naked eye.

Microorganisms include yeast, bacteria and some kinds of algae.

Most microorganisms are made of a single cell.

Yeast

Yeast is a single-celled fungus. Fungi include mushrooms, moulds and toadstools. The cells of a fungus have cell walls. However, the cell walls are not made of cellulose like the cell walls of plant cells.

Fungi do not contain chloroplasts. Fungi cannot photosynthesise. Yeast often feeds on sugars from fruits.

Algae

Algae are simple, plant-like organisms. Seaweeds are algae. Seaweeds are made of many cells. However, many algae are single-celled, or are made of a single long row of cells joined together.

The cells of algae are similar to those of plants.

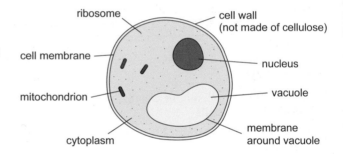

FIGURE 2: The structure of a yeast cell.

ribosome — cell wall (not made of cellulose)
cell membrane — nucleus
mitochondrion — vacuole
cytoplasm — membrane around vacuole

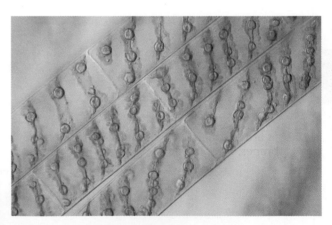

FIGURE 3: This is an alga called *Spirogyra*, seen through a microscope. The green spirals are its chloroplasts.

FIGURE 4: The green scum on this pond is made up thousands of *Spirogyra* strands all tangled together. It is often called blanket weed.

🔍 microorganisms GCSE ... yeast cells

QUESTIONS

1 What is a microorganism?

2 Give three examples of microorganisms.

3 List three ways in which a yeast cell is similar to a plant cell.

4 List two ways in which a yeast cell differs from a plant cell.

5 How do you think that **Spirogyra** feed?

Remember

One alga, lots of algae. Alga is singular, and algae is plural.

Bacteria

The cells of **bacteria** are very different from animal cells, plant cells, algal cells and yeast cells. The big difference is that bacteria do not have a nucleus.

In animal cells, plant cells, algal cells and yeast cells, the nucleus contains the genes. In a bacterium, the genes are in the cytoplasm.

QUESTIONS

6 List three similarities between a bacterial cell and a yeast cell.

7 List two differences between a bacterial cell and an animal cell.

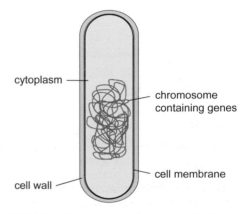

FIGURE 5: The structure of a bacterial cell.

Viruses

Some kinds of bacteria make you ill. They are **pathogens**. But there is another kind of pathogen – **viruses**.

Viruses are not made of cells at all. They do not have cell membranes, cytoplasm or a nucleus.

A virus is hundreds of times smaller than a cell. Most viruses are made of a sphere of protein, with some DNA inside it. The virus cannot do anything unless it gets inside a living cell. It reproduces inside the cell, using the cell's organelles to help.

QUESTIONS

8 Discuss whether or not viruses should be classified as living organisms.

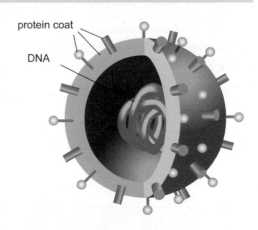

FIGURE 6: The structure of a virus.

Diffusion

You will find out:

> diffusion is a result of the random movement of particles, so that they spread out evenly

> gases, and substances in solution, can diffuse

> oxygen, required for respiration, moves into cells by diffusion

Finding a female by smell

This male moth's feathery antennae are the equivalent of your nose – but millions of times more sensitive. The antennae contain smell receptors, picking up scents that tell the moth there is a female around. Female moths emit scents called pheromones. The scent molecules from the female moth spread through the air by diffusion.

FIGURE 1: Can you suggest why the moth's antennae are so huge and feathery?

Moving particles

Diffusion

Scents are substances that you can smell. A scent consists of molecules of gases that spread out in the air.

Molecules naturally spread out from a place where there are a lot of them into a place where there are fewer. They spread from a high concentration to a lower concentration. This is **diffusion**.

The greater the difference in concentration, the faster diffusion happens.

Particles are always moving. Any particles that can move around freely can diffuse. This includes particles in gases and liquids.

> *Diffusion is the spreading of the particles of a gas, or of any substance in solution, resulting in a net movement from a region where they are of a higher concentration, into a region where they are in a lower concentration.*

Cells and diffusion

Most cells need oxygen so that they can respire. Oxygen diffuses into the cells from a higher concentration outside to a lower concentration inside.

The concentration of oxygen inside a cell is because the cell keeps on using it up, for respiration.

The oxygen diffuses through the cell membrane on its way into the cell.

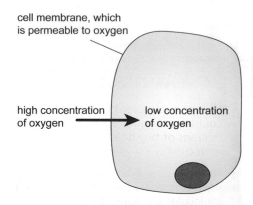

cell membrane, which is permeable to oxygen

high concentration of oxygen → low concentration of oxygen

FIGURE 2: The red arrow shows oxygen moving into a cell. What is the name for the way that the oxygen moves?

QUESTIONS

1 Explain why oxygen diffuses into a cell.

2 Which part of the cell does the oxygen have to pass through, to get inside the cell?

3 Do you think that molecules in solids can diffuse? Explain your answer.

Did you know?

A male moth can sense a female if just five molecules of scent per second reach his antennae.

Q cell membrane diffusion

Exchanging gases

All cells respire. Respiration normally uses oxygen and makes carbon dioxide.

Diffusion and temperature

The faster that particles move around, the faster they will diffuse. So, diffusion tends to happen faster at high temperatures than at low temperatures.

Aromatherapists make use of this fact. They place scented oil in a container and warm it over a candle or little spirit lamp. The warm temperature helps the scent to evaporate and spread out into the air.

FIGURE 3: Aromatherapists make use of diffusion.

QUESTIONS

4 Explain how and why the carbon dioxide made in respiration will leave the cell.

5 If a cell is respiring faster than usual, then oxygen moves into it faster. Use what you know about diffusion to explain why this happens.

A model cell

Visking tubing looks a bit like sticky tape without the sticky part. If you could see it under a very powerful microscope, you would be able to see millions of tiny holes in it. These holes are so small that they can stop big molecules, such as starch molecules, from getting through.

Figure 4 shows some Visking tubing containing starch solution. Starch solution is a mixture of starch molecules and water molecules. The starch molecules are huge compared with the tiny water molecules.

The tubing is in a beaker containing iodine solution. This is a mixture of iodine molecules and water molecules. Iodine molecules can pass through the holes in the Visking tubing.

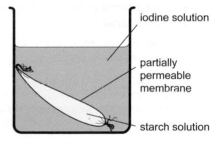

iodine solution

partially permeable membrane

starch solution

FIGURE 4: An experiment with a model cell, to demonstrate diffusion.

QUESTIONS

6 Which molecules can diffuse freely through the Visking tubing?

7 When iodine mixes with starch, a blue–black colour is produced. Where will a blue–black colour be produced in this experiment? Explain why.

8 It takes a few minutes to see a deep blue–black colour. Suggest two ways in which you could speed up the experiment. Explain why each of your suggestions would work.

Specialised cells

You will find out:
> how different cells are specialised for a particular function

Life in a small world

Parameciums are very small organisms. A paramecium is made of only one cell. It lives in ponds, in a world where most of what it experiences is too small for us to see. Parameciums feed on other organisms even smaller than themselves, such as bacteria.

Did you know?

There are about 210 different cell types in your body.

FIGURE 1: Paramecium, magnified about 400 times. It is covered with tiny hairs called cilia. The cilia push against the water, like oars, to make the paramecium move.

 ## Different cells for different functions

In a single-celled organism such as a paramecium, the cell has to do everything. In a multicelled organism, such as you or a plant, cells can share out different tasks between them. Each cell can become **specialised** – really good at doing just one or two particular things.

You began your life as a little ball of unspecialised cells. As the cells divided, over and over again, the ball became an embryo. Different cells began to take on special tasks. This is **differentiation**.

Sperm cells and egg cells obviously have very different roles (functions) from all the other cells in the body. The diagrams show how their structures are adapted to suit their functions.

Cells are so tiny that they are measured in a unit called a micrometre, symbol μm. There are 1000 micrometres in one millimetre.

vesicle containing enzymes, to digest a way into the egg

nucleus containing one set of chromosomes

cell membrane

cytoplasm

mitochondria, to provide energy for swimming

tail for swimming

30 μm

FIGURE 2: A sperm cell.

QUESTIONS

1 Describe three similarities between a sperm cell and an 'ordinary' animal cell.

2 What is the function of (a) a sperm cell (b) an egg cell?

3 (i) Copy this table and add two more features of sperm cells.

Feature	How it helps the sperm with its function
only one set of chromosomes in the nucleus	so there will be two sets when it joins with the egg nucleus

(ii) Construct a similar table about egg cells.

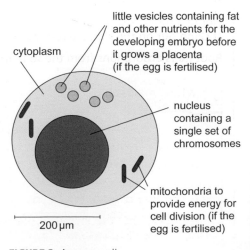

little vesicles containing fat and other nutrients for the developing embryo before it grows a placenta (if the egg is fertilised)

cytoplasm

nucleus containing a single set of chromosomes

mitochondria to provide energy for cell division (if the egg is fertilised)

200 μm

FIGURE 3: An egg cell.

cellular differentiation

More examples of specialised cells

The human body contains hundreds of different kinds of specialised cells. Red blood cells, goblet cells and ciliated cells are just three of them.

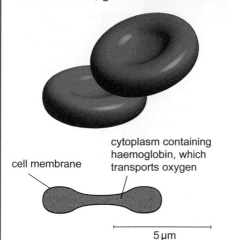

cytoplasm containing haemoglobin, which transports oxygen

cell membrane

5 μm

This is what the cell would look like if cut in half. The shape is good for taking in and letting out oxygen.
Notice that the red blood cell does not have a nucleus.

FIGURE 4: A red blood cell.

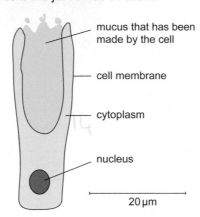

mucus that has been made by the cell

cell membrane

cytoplasm

nucleus

20 μm

Goblet cells are found in the lining of the alimentary canal, and in the tubes leading down to the lungs. They make mucus, which helps food to slide easily through the alimentary canal, and helps to stop bacteria getting down into your lungs.

FIGURE 5: A goblet cell.

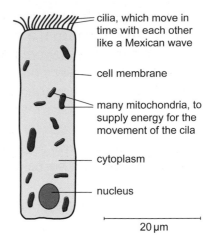

cilia, which move in time with each other like a Mexican wave

cell membrane

many mitochondria, to supply energy for the movement of the cila

cytoplasm

nucleus

20 μm

Ciliated cells are found alongside goblet cells in the trachea and other tubes leading into the lungs. The sweeping movements of their cilia move mucus up the tubes and into the back of the throat, where you swallow it.

FIGURE 6: A ciliated cell.

 QUESTIONS

4 State two features that all of the cells in Figures 4, 5 and 6 have in common.

5 (i) Which of the cells, described on these two pages, move around?
(ii) How are their structures and shapes suited to this?

 Specialised plant cells

Plants also have specialised cells.

Figure 7 shows a slice cut downwards through a root. The little things sticking out on the side are **root hairs**. Each one is part of a single cell.

Root hairs absorb ions from the soil, often by diffusion.

 QUESTIONS

6 Describe how the shape of root hair cells differs from the shapes of other cells in the root.

7 If you had not been told that this was part of a root, how could you have worked out that the cells shown in Figure 7 were plant cells and not animal cells?

8 Suggest how the shape of the root hair cells helps them to carry out their function.

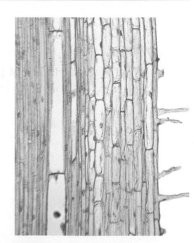

FIGURE 7: A longitudinal (lengthways) section through a root, seen through a microscope.

erythrocytes ... root hair cells

Tissues

Growing new skin

The photo shows a sheet of skin that has been grown from a person's own cells. The skin cells were put into a dish containing a nutrient medium. The cells divided over and over again, and stayed stuck together to form this big sheet. It is going to be used by a surgeon to replace some of the person's skin that was damaged by burns.

> **Did you know?**
>
> The total surface area of your skin is about two square metres.

You will find out:
> a tissue is a group of cells with similar structure and function
> how some tissues in animals are specialised to carry out their functions

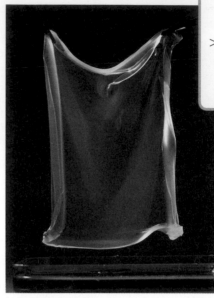

FIGURE 1: This sheet of skin is an example of a tissue.

What is a tissue?

In everyday life, a tissue is a piece of thin, soft paper. Some tissues in living organisms are rather like that, but there are also many other sorts of tissues.

In an animal or a plant, a tissue is a group of similar cells that all work together to carry out a particular function. As they are all specialised to carry out the same function, they all have similar structures.

Figures 2, 3 and 4 show examples of tissues that are found in a human body.

muscle cell

The cells are long and thin.

FIGURE 2: The cells in muscular tissue can make themselves shorter. This is called contraction.

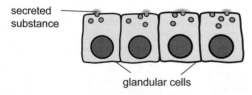

secreted substance

glandular cells

FIGURE 3: The cells in glandular tissue make and release useful substances. This is called secretion.

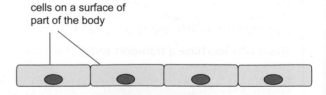

cells on a surface of part of the body

FIGURE 4: Epithelial tissue covers surfaces in the body. The tissue protects the cells underneath it.

QUESTIONS

1 Write down a definition of a tissue.
2 List three examples of tissues in the human body.

Structure and function of tissues

The cells that make up tissues are adapted for their particular roles. Figures 5, 6 and 7 show examples.

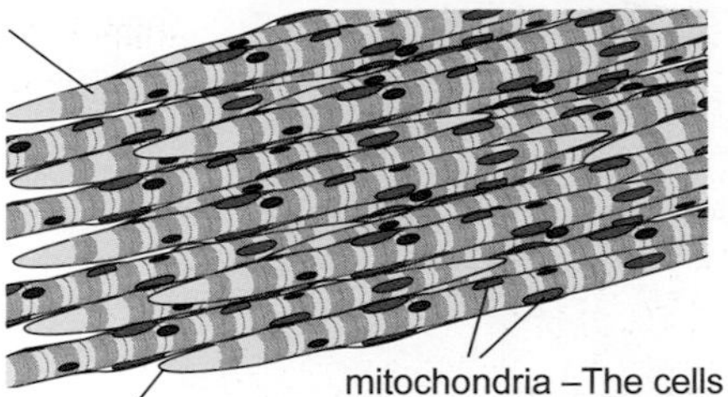

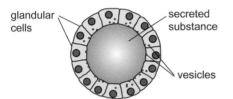

muscle cell

The cells are long and thin.

mitochondria –The cells use energy to make themselves get shorter (contraction). The many mitochondria in the cells provide the energy for contraction.

FIGURE 5: Muscular tissue is specialised to produce movement.

glandular cells

secreted substance

vesicles

FIGURE 6: Glandular tissue is specialised to secrete useful substances.

The glandular cells contain many small vesicles of useful substances that the cell has made, such as enzymes or hormones. The substances are released outside the cell. This is called secretion.

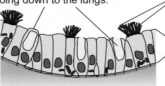

goblet cells – These make mucus, which traps particles of dust or bacteria in the air going down to the lungs.

cilia – These sweep the mucus upwards, keeping the dust and bacteria out of the lungs.

epithelial tissue – These cells make a single layer lining the tubes leading down to the lungs.

FIGURE 7: This is the epithelial tissue that covers the tubes leading down to the lungs.

QUESTIONS

3 Explain why the cells in muscle tissue have large numbers of mitochondria.

4 Cells in glandular tissue usually have large numbers of ribosomes. Suggest why.

5 When someone smokes, the smoke destroys the cilia in the tubes leading down to the lungs. Explain why that increases the chances of getting infections in the lungs.

Single-celled and multicellular organisms

There is a big advantage to being a multicellular organism, rather than a single-celled one. You can have many different cells, each specialised for a particular function, rather than one cell having to do everything.

However, being big does cause problems. Here is one example: think of a cell deep inside your body – how is it going to get oxygen for respiration? For a cell living on its own, it would be easy – it could just take in oxygen by diffusion from the air or water around it. For oxygen to diffuse all the way from the air into the centre of your body, it would take far too long.

This is why you have a transport system – a blood system that carries substances to all the different tissues in the body. You also have special surfaces that are adapted for allowing things to move in and out of the body quickly. For example, inside the lungs there are millions of tiny spaces in contact with the air, where oxygen can quickly diffuse into your blood.

QUESTIONS

6 Describe one advantage and one disadvantage of being a multicellular organism rather than a single-celled organism.

7 Suggest how lung tissue is adapted for its function.

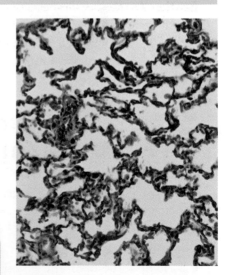

FIGURE 8: This is lung tissue, seen through a microscope. It is nearly all spaces, called air sacs, which are filled with air. There are tiny blood vessels in the walls of these air sacs.

Animal tissues and organs

You will find out:

> an organ, such as the stomach, is made of different tissues

> a system, such as the digestive system, is made of different organs

Inside the digestive system

This is an X-ray of part of someone's digestive system, inside the abdomen – from just above the waist downwards. The blue tubes are the large intestine. The dark things inside are the remains of food, on its way out of the body.

Did you know?

The total length of the human digestive system can be up to nine metres.

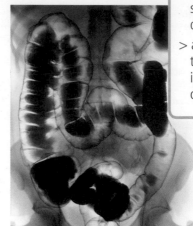

FIGURE 1: The colon is an example of an organ. It is part of the digestive system.

Organs

Your body contains many different organs. The eye, the stomach and the thigh muscle are all examples of organs.

Organs are made of tissues. Each organ contains many different tissues.

For example, the wall of the stomach contains:

> **glandular tissue**, which secretes **digestive juices** to break down the food

> **muscular tissue**, which contracts and relaxes to produce churning movements and mix up the contents of the stomach

> **epithelial tissue**, which covers the outside and the inside of the stomach wall.

Systems

A system is a group of organs that performs a particular function. The function of the digestive system, for example, is:

> to break down the food you eat

> then to allow the broken down food to seep into your blood system.

Each body system contains many different organs. Figure 3 shows some of the organs in the digestive system.

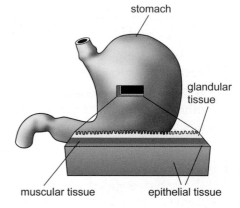

FIGURE 2: The tissues that make up the wall of the stomach.

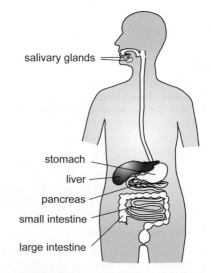

FIGURE 3: Some of the organs in the human digestive system.

QUESTIONS

1 Arrange these in order, smallest first: organ, tissue, cell, system.

2 Give two examples of organs that are part of your digestive system.

3 Name three different tissues that make up the stomach. What is the function of each of these tissues?

4 The heart and the aorta (the main artery) are two examples of organs. They are part of a body system. Which one?

Q organ GCSE ... body system

Functions of the digestive organs

Figure 4 shows the functions of some of the organs in the digestive system.

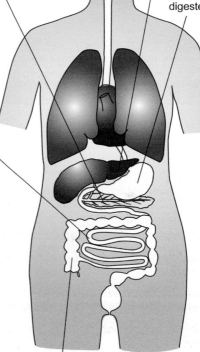

salivary glands, which secrete a digestive juice called saliva – Saliva contains enzymes that break down starch in the food.

liver, which secretes bile to aid digestion of fats

pancreas, which secretes pancreatic juice – This contains more enzymes that break down starch, proteins and fats in the food.

stomach, where proteins are digested

small intestine, where the enzymes from the pancreas work – After they have digested the food and broken it down to small, soluble molecules, these seep through the wall of the intestine into the blood, along with most of the water in the food. This is called absorption.

FIGURE 4: The functions of organs in the digestive system.

large intestine – Undigested food passes through on its way to the outside world as faeces. As food moves through here, more water is absorbed from it.

QUESTIONS

5 Name two glands in the digestive system. What does each of these glands secrete?

6 Glands contain glandular tissue. Name one other organ in the digestive system that contains glandular tissue.

7 Name two organs in which food is digested (broken down) to small molecules.

8 In which organ do the small molecules of food move through the digestive system wall and into the blood? What is the name for this process?

Protecting the digestive system

Many of the organs in the digestive system secrete enzymes, which break down large molecules in the food to small ones. They could also break down molecules in the cells in the walls of the organs, if these were not protected. This is especially important in the stomach, where glandular tissue in the walls secretes not only enzymes but also hydrochloric acid.

The whole of the digestive system is therefore lined with protective epithelial tissue. This secretes large quantities of mucus, which forms a slimy layer all over the inner surface of the digestive organs. The mucus is a barrier between the acid and enzymes, and the cells making up the walls of the organs.

The mucus has another function, too. It is very slippery, so it makes it easier for the food to slide through the various organs of the digestive system as it makes its way from the mouth to the anus.

QUESTIONS

9 Name one type of cell that you would expect to find in the epithelium lining the digestive system.

10 Mucus is secreted by the epithelial tissue in another body system besides the digestive system. What is this system, and what is the function of the mucus there?

Plant tissues and organs

You will find out:
> plant organs include stems, roots and leaves
> plant tissues include epidermal tissue, mesophyll, xylem and phloem
> about the functions of each of these plant tissues

Eating organs

Much of the food that we eat comes from plant organs. Potatoes, for example, are swollen underground stems, which are plant organs. Leaves are organs. And so are fruits.

Did you know?

The largest organ in the world is probably the trunk of a giant redwood tree.

FIGURE 1: How many different plant organs are there in this display?

Plant organs

You have seen that an organ is a structure with a particular function. It is made up of different tissues. It is not only animals that have organs. Plants have organs, too.

Figure 2 shows three different organs in a plant.

FIGURE 3: The stems of this cucumber plant have an unusual function. What is it? How do you think this helps the plant to survive?

QUESTIONS

1 (i) List three organs found in plants.

(ii) For each organ you have listed, state one function.

FIGURE 2: The major organs of a plant.

leaves – The leaves are where the plant makes its food. This is where photosynthesis takes place. Leaves are usually broad and flat, so they can get plenty of sunlight – the energy source that drives photosynthesis.

stem – The stem holds the flowers and leaves up in the air. This allows the leaves to get plenty of light, so that they can photosynthesise. It also allows the flowers to attract insects, so that they can be pollinated.

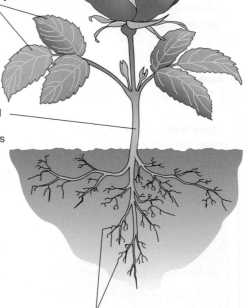

roots – The roots anchor the plant firmly into the soil. Roots absorb water from the soil, and the water then moves through the root and stem into all the other parts of the plant. Roots also absorb mineral ions from the soil, which the plant needs for healthy growth.

plant organs and their functions

Plant tissues

Figure 4 shows the tissues that are present in a plant leaf.

It is not only leaves that are covered with **epidermis**. The whole plant has a layer of epidermis all over its surface. This helps to protect the underlying cells.

> In the leaf, the epidermis helps to stop water vapour escaping from the cells inside the leaf.

> In the root, some of the cells in the epidermis have long, thin extensions called root hairs. These are specialised to absorb water and mineral ions from the soil.

Most of the cells in a leaf are **mesophyll** cells. Mesophyll means 'middle of the leaf'. This is where photosynthesis takes place.

Xylem and **phloem** tubes run through the entire plant. Wood is made of xylem. Xylem tissue is really unusual because its cells are completely dead. They are hollow tubes surrounded by a strong, tough waterproof material. In the living plant, water flows through these dead tubes.

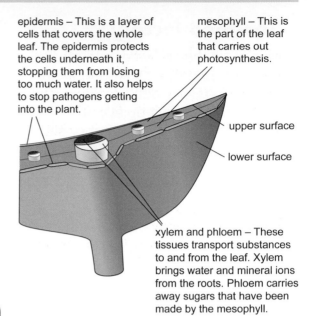

epidermis – This is a layer of cells that covers the whole leaf. The epidermis protects the cells underneath it, stopping them from losing too much water. It also helps to stop pathogens getting into the plant.

mesophyll – This is the part of the leaf that carries out photosynthesis.

upper surface

lower surface

xylem and phloem – These tissues transport substances to and from the leaf. Xylem brings water and mineral ions from the roots. Phloem carries away sugars that have been made by the mesophyll.

FIGURE 4: The tissues in a leaf.

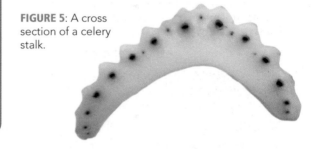

FIGURE 5: A cross section of a celery stalk.

QUESTIONS

2 (i) Give three examples of tissues in a plant.

(ii) For each tissue that you have named, describe its functions in a leaf.

3 A celery stalk had been standing in a coloured dye before the slice shown in Figure 5 was cut. The stalk took up the dye, which travelled up its xylem tissue. Make a drawing of the celery stalk and label the epidermis and the xylem.

Leaf epidermis

Figure 6 shows micrographs of the epidermis on the upper surface and the lower surface of a leaf. The epidermal cells fit tightly together. The epidermal tissue on the lower surface of the leaf has little holes called stomata. These allow gases to diffuse into and out of the leaf.

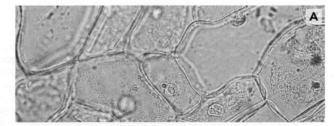

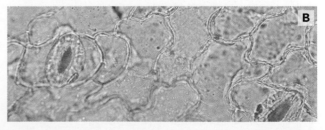

FIGURE 6: The upper (A) and lower (B) epidermal tissue of a leaf.

QUESTIONS

4 For the epidermis of a leaf, suggest:

(a) how the tight fitting of cells helps the epidermal tissue to carry out its function

(b) which gases diffuse through the stomata

(c) why the stomata are mostly on the lower surface, rather than the upper surface.

Photosynthesis

You will find out:

> plants and algae use energy from light to combine carbon dioxide with water to make sugar (glucose) and oxygen

> energy from the light is captured by a green pigment called chlorophyll

The food factory

Plants use the air, water and other substances in the soil around them to make food. They don't make much fuss about it. They do it quietly and steadily, so we do not notice what they are doing. However, if it were not for plants, there would be no animals in the world, because there would be neither food to eat nor any oxygen to breathe.

Did you know?

Only 1 in every 2500 molecules in the air is carbon dioxide. Yet this is where all our food comes from – thanks to plants.

FIGURE 1: This is a food factory. All of these plants are making food.

Making food and oxygen

Making food

Plants make food by **photosynthesis**. They take carbon dioxide from the air and water from the soil. In their leaves, they use energy from sunlight to make the carbon dioxide and water react together. Some of the energy in sunlight is absorbed by a green pigment called chlorophyll.

This is the word equation for the reaction.

energy from sunlight
carbon dioxide + water ⟶ glucose + oxygen

The first kind of food that the plant makes is **glucose**. Glucose is a sugar. It is a type of carbohydrate.

This is where all food comes from. Everything that you eat can trace its history back to a plant.

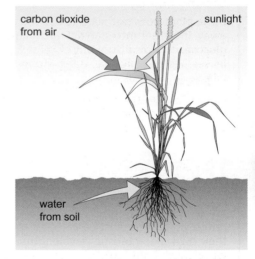

FIGURE 2: How plant leaves obtain what they need for photosynthesis.

QUESTIONS

1 Write down the word equation for photosynthesis, without looking it up. Then check to see if you have it right.

2 (i) Name the two raw materials needed for photosynthesis.

(ii) For each one, say where the plant leaf obtains it.

3 Why is light needed for photosynthesis?

4 Think back to what you know about plant cells. Which part of a plant cell contains chlorophyll?

Q glucose GCSE

Making oxygen

Millions of years ago, before there were any organisms that could photosynthesise, there was hardly any oxygen in the air.

Oxygen was first made when bacteria evolved that could photosynthesise. Gradually, over millions of years, the amount of oxygen in the air built up. Now more than 20% of the air is oxygen. It has all been made by bacteria, algae and plants.

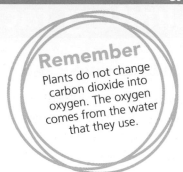

Remember

Plants do not change carbon dioxide into oxygen. The oxygen comes from the water that they use.

Energy for living things

Photosynthesis uses energy. This energy usually comes from the Sun, although some indoor plants use energy from artificial lighting.

In photosynthesis, energy in electromagnetic radiation from the Sun is transferred to plants. They absorb energy and store it in glucose molecules.

This is where all of your energy comes from. It comes from the food that you eat. Somewhere – it might have been in Canada or just down the road from you – electromagnetic radiation transfers energy from the Sun to plants that absorb it and, after photosynthesis, store it as glucose. Some of the energy you are using today might have begun its travels as a sunbeam in Canada.

Storing energy

If the plant does not need to use the glucose it has made straight away, it turns it into starch. Starch molecules are big and, unlike glucose, cannot diffuse out of cells. So, starch is stored in the plant's leaves. Just like glucose, starch stores energy that the plant can use whenever it needs it.

Figure 3 shows the results of testing a variegated (green and white) leaf for starch. The leaf has had iodine solution added to it. Iodine solution goes blue–black if starch is present.

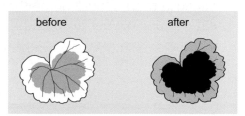

before   after

FIGURE 3: Testing a variegated leaf for starch.

QUESTIONS

5 What happens to the energy from sunlight that is used in photosynthesis?

6 In the leaf in Figure 3, which parts contained starch? What does this tell you about chlorophyll and photosynthesis?

Chlorophyll and light absorption

White light is a continuous spectrum of colours, from red to violet. Different parts of the spectrum have their own characteristic wavelengths. Short wavelength light looks blue and long wavelength light looks red.

Figure 4 shows the proportion of the different colours of white light that are absorbed by chlorophyll.

QUESTIONS

7 Which wavelength of light is chlorophyll best at absorbing?

8 The colours of light that are not well absorbed are reflected. Explain why chlorophyll looks green.

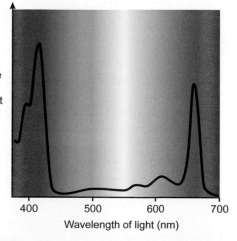

Proportion of the different colours of white light that are absorbed by chlorophyll

Wavelength of light (nm)

FIGURE 4: The colours of light that are absorbed by chlorophyll.

Limiting factors

Just survive

In winter, many plants shut down. It is too cold for them to photosynthesise. The days are short and there is not much sunlight. If all the water is frozen, they cannot get water to their leaves. They may as well just drop all their leaves and wait for spring.

FIGURE 1: There's not much a tree can do in winter except wait until spring arrives.

Did you know?

As carbon dioxide concentrations in the atmosphere increase, some plants may be able to grow faster.

> **You will find out:**
>
> > how the rate of photosynthesis can be limited by light, temperature and the availability of carbon dioxide

---

 ## Speeding up photosynthesis

How fast can a plant make food? It depends on how fast it can get the raw materials it needs, and how much energy it can absorb.

Light intensity

The brighter the light, the more energy the plant is receiving and the faster it can photosynthesise.

Carbon dioxide concentration

Carbon dioxide is one of the raw materials for photosynthesis. If you give a plant more carbon dioxide, it will probably be able to photosynthesise faster.

Temperature

Most plants cannot photosynthesise at all when it is really cold. They photosynthesise faster in spring and summer, when the days are warmer.

QUESTIONS

1 List three factors that limit the rate of photosynthesis.

2 Some gardeners grow plants in a greenhouse, where they burn paraffin in a heater. Suggest two ways that this could help the plants to photosynthesise faster. (Think about which gases are produced when a fossil fuel burns.)

 ## Limits to the speed of photosynthesis

Light intensity

The graph in Figure 2 shows how fast a plant photosynthesises at different light intensities.

When the light intensity is 0, there is no photosynthesis. As the light intensity increases, the rate of photosynthesis also increases.

In region A of the graph, light intensity is limiting the rate of photosynthesis. You can tell this, because the graph shows that if the plant is given more light, then it photosynthesises faster. Light is a **limiting factor** for photosynthesis.

FIGURE 2: How light intensity affects the rate of photosynthesis.

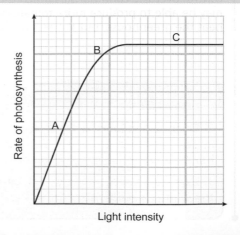

Q factors affecting photosynthesis

However, there comes a point when the rate of photosynthesis does not increase any more, even when the plant is obtaining more light. It is already photosynthesising as quickly as it can. Perhaps it does not have enough chlorophyll to absorb light any faster. Perhaps it does not have enough carbon dioxide. Perhaps the temperature is too low.

Carbon dioxide concentration

The graph in Figure 3 shows how fast a plant photosynthesises at different carbon dioxide concentrations.

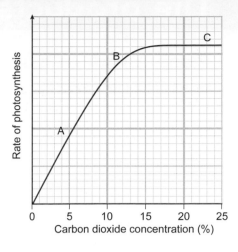

FIGURE 3: How carbon dioxide concentration affects the rate of photosynthesis.

⊙ QUESTIONS

3 Look at Figure 2.

(i) What is the limiting factor in region A of the graph?

(ii) Suggest what could be the limiting factor in region C.

4 Look at Figure 3.

(i) What is the limiting factor in region A of the graph? How can you tell?

(ii) At what point on the graph does another limiting factor begin to take effect? How can you tell?

(iii) Suggest what this other limiting factor might be.

5 The concentration of carbon dioxide in the air is approximately 0.04%. Using the information in the graph, suggest whether a gardener could make the plants in his greenhouse grow faster by giving them extra carbon dioxide. Explain your answer.

Greenhouses

Growing crops in a greenhouse gives the grower a lot of control over the conditions in which the plants live. A grower may be able to produce more tomatoes more quickly if they heat the greenhouse, but the cost of the fuel might outweigh the increase in what they are paid for the tomatoes.

⊙ QUESTIONS

6 Despite the long distance they have to travel, imported strawberries are often cheaper to buy than home grown ones in springtime. Suggest why.

7 Explain why tomatoes do not grow well in Britain in the winter, even if greenhouses are heated.

FIGURE 4: Early strawberries may be grown in greenhouses.

 greenhouse plant growth

The products of photosynthesis

Plants produce fuels

Many plants make oils, which are energy stores for the plant. We often steal the oils to use for ourselves, to make or cook food. We even use them as fuel in vehicles.

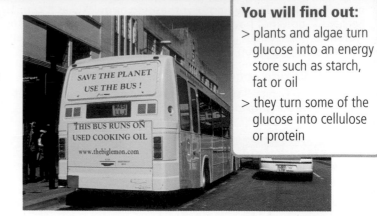

FIGURE 1: This bus in Brighton is helping to reduce the use of fossil fuels.

You will find out:

> plants and algae turn glucose into an energy store such as starch, fat or oil

> they turn some of the glucose into cellulose or protein

How plants use their food

Just like us, a plant needs food for two reasons – for energy and for building its structure.

Releasing energy

When a plant cell needs energy, it breaks down some of the glucose that it made in photosynthesis. The breaking-down process is **respiration**. Respiration releases some of the energy that was stored in the glucose.

Making new substances

Glucose is a carbohydrate. It is a sugar. Plants can make many different substances from glucose. These include:

> **starch**, which is also a carbohydrate. Starch is good for storing energy.

> **cellulose**, which is another carbohydrate. Cellulose is used for making cell walls.

> **fats and oils**, which are not carbohydrates. They are good energy stores.

> **proteins**, which are not carbohydrates either. They are needed for making new cytoplasm, new cell membranes and enzymes.

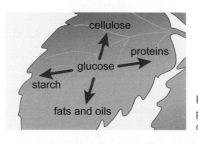

FIGURE 2: What plants can make out of glucose.

QUESTIONS

1 Where did the energy in the glucose come from, originally?

2 What is the name of the process that releases energy from glucose?

3 Name (a) two different substances that plants use for storing energy (b) two others that they use for building new cells.

Mineral salts

Glucose is made from carbon, hydrogen and oxygen atoms. Plants make other substances from glucose. These are shown in Table 1, together with the types of atoms from which they are made.

TABLE 1: Glucose products.

Substance	Atoms the substance is made from			
	Carbon	Hydrogen	Oxygen	Nitrogen
carbohydrates	✓	✓	✓	
fats and oils	✓	✓	✓	
proteins	✓	✓	✓	✓

🔍 cellular respiration plants

To convert glucose into proteins, plants need nitrogen.

The air is almost 80% nitrogen, so you would think that it would not be a problem. Unfortunately, this nitrogen is so unreactive that plants cannot do anything with it. It just diffuses in and out of their leaves.

Plants have to get their nitrogen in the form of mineral ions. They normally use nitrate ions, NO_3^-. They absorb them from the soil, through their roots.

healthy potato plant

potato plant without nitrate

FIGURE 3: The plant on the left is healthy. The one on the right is short of nitrate ions.

QUESTIONS

4 Which three elements are combined in a glucose molecule?

5 Which other element is needed in order to change glucose into proteins?

6 Explain why the plant in Figure 3 that is short of nitrate ions has not been able to grow tall.

Choosing the storage product

Plants have choices when it comes to storing energy.

Most plants store at least some energy as starch. Starch, like glucose, is a carbohydrate. Starch is a polymer made up of many glucose molecules linked together in a long chain. Unlike glucose, starch is insoluble. It forms grains inside cells, instead of dissolving and getting mixed up with everything else inside the cell.

In some ways, fats and oils are even better for energy storage than starch. This is because they can store more energy per gram. So, if you want lightweight energy storage, fats and oils are the best bet.

The downside of fats and oils is that it is a bit more difficult to make them and to break them down. Plants tend to use them when weight is a factor. For example, many seeds contain oils as energy stores for the embryo plant inside the seed. This makes it easier for the lightweight seed to be made and carried a long way from the parent plant, so it can grow in a new place.

FIGURE 4: Oil-rich seeds and nuts.

QUESTIONS

7 Explain why plants store energy in the form of starch, rather than as glucose.

8 Give one advantage and one disadvantage of storing energy as starch rather than as fats or oils.

Preparing for assessment: Applying your knowledge

To achieve a good grade in science, you not only have to know and understand scientific ideas, but you need to be able to apply them to other situations and investigations. These tasks will support you in developing these skills.

✷ Wild woodlands

There is a wild-looking woodland near Shelley's house. Throughout most of the winter, it looks dull and grey, because the trees have all lost their leaves. Then, in February, before the leaves have reappeared, it is full of snowdrops, and later in the spring the ground is a sea of bluebell flowers.

The woodland is a nature reserve, so Shelley was surprised (and a bit upset) to see people chopping down some of the trees, one winter's day. She stepped gingerly over the muddy ground and fallen branches to ask them what they were doing. "We are coppicing the hazel trees", said one of the workers. "Don't worry – we just cut the branches off, close to the ground. They will all grow back again, looking just as beautiful as before. If we don't do this, they will get old and overgrown, and the conditions will not be right for all the different kinds of plants, mammals, birds and insects that live in the wood."

Shelley was told that the hazel trees in one part of the wood are coppiced about once every seven or eight years. Each year, a different part is coppiced.

This lets more light onto the woodland floor in those areas, while the un-coppiced parts stay shady and sheltered.

Shelley also learned that there were dormice living in the wood. Dormice love eating hazel nuts. Someone had put boxes on some of the trees for the dormice to make nests in, and the coppicing team were taking care to avoid damaging the nest boxes. They were moving the boxes to new trees, if necessary.

The team wanted more helpers, so Shelley went along at the weekend to join in. She discovered that some of the wood from the trees would be used to burn as fuel, providing heating for the local wildlife trust's offices nearby. "Trees harvest energy from sunlight and use it to process carbon from the air, turning the energy and carbon into fuel for themselves" said the team leader. "We can take parts of the trees and use them as fuel for ourselves, without harming the trees or the other wildlife in the wood."

 Task 1

Snowdrops and bluebells take the opportunity to grow very early in the year, before the leaves are fully out on the trees above them. This means that they receive plenty of light.

Why do plants need light to grow?

Task 2

Unlike the snowdrops and bluebells, hazel trees do not produce their leaves until spring is well under way.

Suggest the advantages of coming into leaf later in the spring.

Task 3

Wood contains energy and can be burned as a fuel to provide heating.

Explain where the energy came from and how it came to be stored in the tree.

Task 4

Dormice are not very common. In England, they are mostly found in the southeast. Dormice like to live in large woodlands where they have plenty of space to move around without having to leave the trees.

(a) Suggest how coppicing the woodland provides a range of different habitats in the wood.

(b) Suggest how this could allow a wide range of different species, including dormice, to live in the wood.

 Maximise your grade

	Answer includes showing that you...
E	recall that plants need light for photosynthesis.
	recall factors that may limit the rate of photosynthesis.
	recall physical factors that may affect the distribution of organisms.
	can explain that low temperatures slow down photosynthesis.
C	know that, as a result of photosynthesis, energy carried by light is transferred to and stored in substances such as glucose and cellulose.
	understand that the distribution of a particular species in a habitat is affected by environmental conditions.
A	understand that different conditions suit different species, so a variety of different conditions in a wood means that many species can live there.

31

Distribution of organisms

You will find out:

> how the distribution of organisms may be affected by temperature, light, and the availability of nutrients, water, oxygen and carbon dioxide

Penguin distribution

Many different species of penguins live in the Antarctic. Conditions are very similar in the Arctic, but no penguins live there. Penguins evolved in the southern hemisphere and have never managed to reach the Arctic regions.

Did you know?

Arctic terns spend the northern summer in the Arctic and then migrate to the Antarctic for the southern summer, a round trip of almost 71 000 km. This means they get more daylight each year than any other living thing on Earth.

FIGURE 1: Can you suggest why penguins from the Antarctic have never managed to reach the Arctic?

Factors affecting distribution

Organisms can only live in environments for which they are adapted. There are many physical factors that affect their distribution.

Temperature

Many species can live only in a particular temperature range. For example, many tropical plants are killed if the temperature falls below 0 °C.

Availability of nutrients

Plants need mineral ions from the soil. Very wet soils are usually short of nitrate ions, so only certain plants can live there. Animals can live only where their food is found.

Amount of light

Plants and algae must have light for photosynthesis. They cannot grow in really dark places, such as inside caves or deep in the oceans.

FIGURE 2: This house plant was accidentally left outside on a frosty night. It probably will not recover.

Availability of water

All organisms need water. Species that live in deserts have special adaptations that help them to obtain and conserve water.

Availability of oxygen and carbon dioxide

Most organisms need oxygen, for respiration. They cannot live where oxygen is in short supply, such as on very high mountains or in deep, dark lakes. Plants also need carbon dioxide, for photosynthesis.

◉ QUESTIONS

1 Give three examples of physical factors that affect organisms.

2 Explain why not many species of plants or animals can live in a desert.

3 Suggest why the availability of light affects plants more than animals.

FIGURE 3: This plant is adapted to its desert conditions. Why might it not thrive in other environments?

🔍 plant distribution factors

Seaweed distribution

Egg wrack is a seaweed – a multicellular alga that grows on rocky shores. It gets its name because it has egg-shaped bladders that help its fronds to float in the water when the tide comes in.

The parts of the shore where the egg wrack grows are covered by water twice a day when the tide comes in.

Egg wrack cannot grow very far up the shore because it would be uncovered for a long time when the tide goes out. It would get too hot and too dry in summer.

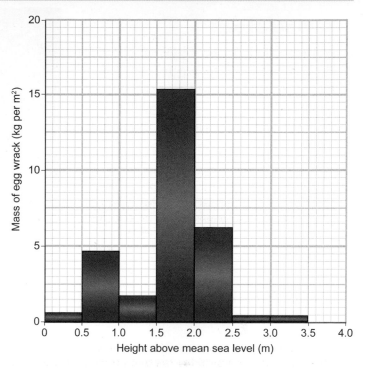

FIGURE 5: The distribution of egg wrack on a rocky shore.

FIGURE 4: Egg wrack at low tide.

QUESTIONS

4 State the physical factors that limit the distribution of egg wrack at the top of a rocky shore.

5 Some seaweeds are able to live higher up the shore than egg wrack can. Suggest what adaptations they might have, to allow them to survive there.

Interaction of physical and biotic factors

Physical factors are not the only ones that affect the distribution of organisms. They are also affected by biotic factors – factors involving other organisms, including competition and predation.

If egg wrack grows low down on the shore, it is covered by water for most of the time. This means that sea-dwelling herbivorous animals can graze on it for a much longer part of the day. The egg wrack also has to compete for space and light with other seaweeds that are better adapted for growing in deep water.

QUESTIONS

6 Competition with other seaweeds and grazing by herbivorous animals prevent egg wrack growing lower down the shore. Outline a laboratory-based experiment to find out which is the more important factor.

Using quadrats to sample organisms

Taking stock

It is really important to know what lives where, if we want to take care of our environment. Knowing what species are present, and how many organisms of each species there are, helps to prevent species being lost.

FIGURE 1: Surveying plants on Dartmoor.

What lives where?

Imagine that you have been given the task of finding out (a) which species live in a grassy field and (b) how many individuals of each species are present. You cannot possibly count every individual organism in the field. You will need to **sample** the area.

You could identify and count the organisms in a small, square area called a **quadrat**. Quadrats can be any size, but they often have sides 0.5 metres long. Quadrat frames can be wire or wood.

You will need to count the organisms in several quadrats. Often, it is best to place the quadrats randomly. Figure 2 shows one way of doing this.

Once you have placed your quadrat, you need to identify each species inside it.

Then you can either:

> count the numbers of each one, or

> estimate the percentage of the area inside the quadrat that each species occupies.

You should repeat this process many times. Work out the mean (average) number of, or area covered by, each species.

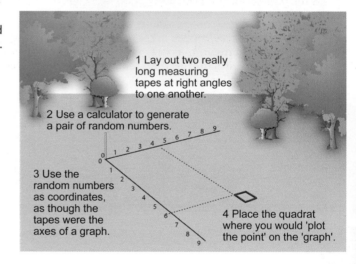

1 Lay out two really long measuring tapes at right angles to one another.

2 Use a calculator to generate a pair of random numbers.

3 Use the random numbers as coordinates, as though the tapes were the axes of a graph.

4 Place the quadrat where you would 'plot the point' on the 'graph'.

FIGURE 2: How to place quadrats randomly.

FIGURE 3: Two ways of collecting data from a quadrat.

(a) counting individuals within a quadrat

Results:
species 1   *
species 2   6
species 3   1

* too numerous to count

(b) estimating percentage cover in a quadrat

Results:
species 1   70%
species 2   20%
species 3   10%
bare ground 15%

species 1

species 2

species 3

bare ground

Transects

Sometimes, you might spot that the habitat steadily changes from one kind to another. For example, your field might have a pond in it, and you want to know if the species of plants change as you get closer to the pond. In that case, you could put one of your long tapes with one end in the pond and the other end out in the field. This is a transect. You can place your transects at intervals – say every two metres – along the tape.

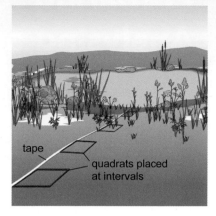

FIGURE 4: Using a transect.

QUESTIONS

1 What is a quadrat?

2 Explain why it is not always easy to count the number of individual organisms inside a quadrat.

Validity and sampling

Ensuring validity

It is important to use many quadrats, and to place these quadrats randomly, in order to get valid results. This means that the results really do tell you about the distribution of the organisms in the whole field.

For example, if you choose where to put the quadrats in the field, you might unconsciously choose areas that do not have stinging nettles in them. This could mean that the results of your survey might show lower numbers of stinging nettles than really are present in the field. The results would not be valid.

Sample size

If you counted the organisms in only two or three quadrats, you are very unlikely to get a representative sample of the whole field. On the other hand, if you covered the whole field with quadrats, it would take weeks and weeks to collect the results. The trick is to find a compromise – as many quadrats as you can manage to survey in the time available. In any case, you should always do at least ten.

QUESTIONS

3 Explain why it is important to place quadrats randomly, when sampling the organisms in a habitat.

4 Explain why it is important to survey the organisms in many quadrats, not just two or three.

Choosing quadrat size

The larger the quadrat:

> the more likely you are to have a representative sample of all the organisms in the area inside it

> the more difficult it will be to count them all, or to make a fair estimate of the percentage of the ground covered by each species.

One way of finding the optimum size of quadrat is to try out different sizes, and then count how many different species you find in each size. Figure 5 shows one set of results obtained in a trial.

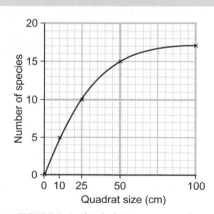

FIGURE 5: In this habitat, one quadrat with 50 cm sides includes most of the species present.

QUESTIONS

5 Use the information in Figure 5 to suggest the best size of quadrat to use in this habitat. Explain your answer.

Preparing for assessment: Analysing and interpreting data

To achieve a good grade in science, you not only have to know and understand scientific ideas, but you need to be able to apply them to other situations and investigations. These tasks will support you in developing these skills.

Buttercups and soil moisture

Hypothesis

Ewan cycles to school on a path across a meadow. In spring, the meadow is full of buttercups.

Ewan noticed that there were two different kinds of buttercup. One kind seemed to grow in the drier parts of the meadow, and Ewan identified this one as the bulbous buttercup. The other kind seemed to grow in wetter places, and he identified this as the creeping buttercup.

Ewan decided to collect data to test this hypothesis: Creeping buttercups grow in wetter places than bulbous buttercups.

Method

He laid out a 40 metre long tape to form a transect across part of the meadow, where the ground changed from being quite dry to very wet.

He put down 0.5 metre square quadrats every 4 metres along the tape.

He estimated the percentage of each quadrat that was covered with each kind of buttercup. He made his estimates to the nearest 10%.

Ewan also measured the percentage of water in the soil, using a soil moisture meter and a data logger.

He did this twice in each quadrat, in two slightly different places.

Results

Ewan's results are shown in the Table.

Quadrat	1	2	3	4	5	6	7	8	9	10
Water content in the soil – test 1 (%)	33.6	32.8	38.2	41.5	49.3	54.6	56.1	66.2	67.9	71.7
Water content in the soil – test 2 (%)	32.9	34.1	38.1	43.9	49.5	56.2	56.4	38.4	70.3	72.8
Cover of bulbous buttercup (%)	60	40	60	50	40	20	10	0	0	0
Cover of creeping buttercup (%)	0	0	20	10	0	30	50	60	80	70

☀ Processing data

1. Calculate the mean percentage water content of the soil in each quadrat. Take care – there is an anomalous result that you will need to deal with.

2. Think about possible ways in which you could use Ewan's results to draw a graph or chart showing the relationship between the percentage cover of the two kinds of buttercups and the percentage water content of the soil.

Choose one way, and draw a graph or chart.

> Remember that the mean should have the same number of decimal places as the numbers you used to calculate it.

> There is no 'right' way to display these results. For a graph, what would go on each axis? Would you draw a line graph, bar chart or scattergram? If you know about histograms or kite diagrams, you might choose to use one of these instead.

☀ Analysing and interpreting data

3. Comment on these choices that Ewan made, and how they might affect the validity or repeatability of his investigation:

a. using a soil moisture meter and data logger to measure the soil water content

b. measuring the soil water content twice in each quadrat, in slightly different places

c. using ten quadrats

d. estimating percentage cover to the nearest 10%

4. Ewan's friend said that the difference between the two measurements of soil water content in each quadrat meant that the soil moisture meter was not accurate or precise. Ewan disagreed. Who was right, and why?

5. Suggest the most important probable source of error in Ewan's investigation. Was this a random error or a systematic error?

6. Describe the relationship between soil moisture content and the distribution of each of the two species of buttercup.

7. Assess the extent to which Ewan's data supported the hypothesis.

8. Suggest what Ewan could do next, in order to obtain extra evidence to increase the confidence with which he can draw a conclusion.

> Validity means that Ewan measured the appropriate variables to test his hypothesis, and that he controlled other important variables. Repeatability means that, if he did the same investigation again, he would expect to get similar results.

> Accuracy means how close the measurement is to the true value. Precision means that the measurements are close together; it does not say how close they are to the true value. Think about where Ewan made the two measurements in each quadrat.

> Remember that you cannot prove that a hypothesis is correct with just a few experiments. You can find that it is supported. However, one investigation could disprove it.

> Make sure that your suggestions would allow Ewan to find more evidence in relation to the same hypothesis, not a different one.

☀ Connections

How Science Works

- Collect primary and secondary data

- Select and process primary and secondary data

- Analyse and interpret primary and secondary data

- Use scientific models and evidence to develop hypotheses, arguments and explanations

Science ideas

B2.4.1   Distribution of organisms

Checklist B2.1–2.4

To achieve your forecast grade in the exam you will need to revise

Use this checklist to see what you can do now. Refer back to the relevant topics in this book if you are not sure. Look across the three columns to see how you can progress.

Remember that you will need to be able to use these ideas in various ways, such as:

> interpreting pictures, diagrams and graphs
> suggesting some benefits and risks to society
> applying ideas to new situations
> drawing conclusions from evidence you are given.
> explaining ethical implications

Look at pages 278–299 for more information about exams and how you will be assessed.

To aim for a grade E	To aim for a grade C	To aim for a grade A
Know that all living things are made of cells.	Describe the main differences between animal and plant cells.	
Know that different parts of the cell have different functions.	Describe the function of the different parts (organelles) of a cell.	
Know that dissolved substances move into and out of cells by diffusion.	Describe how oxygen needed for respiration passes through cell membranes by diffusion.	Explain that, during diffusion, particles move from regions of high concentration to regions of low concentration.
Recall that most microorganisms are single-celled organisms.	Describe the main features of bacterial and yeast cells.	Explain the key difference between bacterial and all other types of cell.
Recall that living organisms can be single celled or multicellular.	Explain that, in multicellular organisms, cells may differentiate and become specialised.	Explain an advantage and a disadvantage of being a multicellular, as opposed to single-celled, organism.
Recall the names of at least one animal tissue, organ and system. Label the organs of the digestive system on a diagram.	Describe the function of a number of animal tissues, organs and systems.	

To aim for a grade E   To aim for a grade C   To aim for a grade A

Recall the main plant organs: stems, roots and leaves.	Give some examples of plant tissues, including their function.	
Know that, in multicellular organisms, different cells are designed to carry out different functions.	Describe an example of an animal cell that has differentiated and become specialised.	Explain the concept of cell specialisation using a number of examples from plants and animals.
Recall that green plants and algae use energy from sunlight to make food by photosynthesis.	Use a word equation to summarise the process of photosynthesis.	Explain the stages of photosynthesis.
Know that the rate of photosynthesis is dependent on varying environmental conditions.	Describe the three factors that limit the rate of photosynthesis and how they affect the process.	Explain how the three limiting factors interact to affect photosynthesis.
Recall that the main product of photosynthesis is glucose and that glucose is used in respiration.	List the range of substances, together with their uses, that plants can make from glucose.	Explain that plants need nitrate ions in order to make protein. Explain where plants obtain these ions.
Recall that organisms need to be adapted to the environment in which they live.	Describe the range of factors that affect the distribution of organisms.	
Know why samples are taken, rather than counting every individual organism, when studying distribution.	Describe how and why you would use quadrats and transects when investigating the distribution of organisms.	Explain how investigators ensure that data collected on the distribution of organisms is both valid and reproducible.

1. The diagrams show a bacterium and a yeast cell.

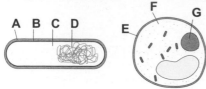

AO1 **(a)** Give the **letters** of the part on **each** cell that controls what enters and leaves the cell. [2]

AO2 **(b)** Describe how the parts D and G differ. [2]

AO2 **(c)** Describe **one** way in which both of these two cells differ from an animal cell. [1]

2. A student investigated the distribution of a green alga called *Pleurococcus* on tree trunks.

She drew a square with sides of 10 cm on a piece of flexible, transparent plastic. She drew lines to divide it into 100 small squares, each with sides of 1 cm.

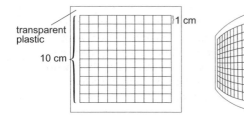

The student placed the quadrat on the south side of a tree trunk, one metre up from the ground. She counted the number of squares in which *Pleurococcus* was present. She repeated this on the north side of the tree trunk.

She then repeated the counts on nine more trees. Her results are shown in the table.

Tree		1	2	3	4	5	6	7	8	9	10
Squares with alga	South side	38	38	12	29	16	49	4	38	27	13
	North side	69	69	91	54	49	87	43	94	86	81

AO2 **(a)** Explain why the piece of plastic needed to be bendy and transparent. [2]

AO2 **(b)** Suggest why it was a good idea to divide the 10 cm square into 100 squares, rather than counting the *Pleurococcus* in the whole 10 cm square. [2]

AO2 **(c)** The student always placed the quadrat the same height above the ground. State **two** other variables that the student should have controlled in her investigation. [2]

AO3 **(d)** Describe any patterns that you can see in the student's results. [1]

AO3 **(e)** Suggest an explanation for the pattern that you have described in (d). [2]

3. (a) Tomatoes are often grown in greenhouses. In the UK, growers may use paraffin heaters to maintain a suitable environment inside the greenhouse for the tomato plants.

AO2 **(i)** Describe **two** ways in which the use of paraffin heaters could increase the rate of photosynthesis of the tomato plants. [2]

AO2 **(ii)** Explain why increasing the rate of photosynthesis of the plants will also increase the quantity of tomatoes that are produced. [2]

(b) In some parts of the world, freshwater is in short supply. The water that is used to irrigate crop plants may be salty.

An investigation was carried out to see if adding a product called PEO granules to the water surrounding a plant's roots would increase their growth rate, when grown in salty water. The graph shows the results.

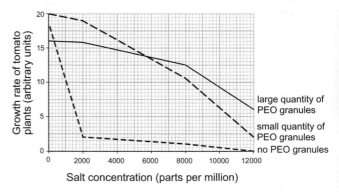

AO3 **(i)** In this experiment, two quantities were varied. What were these two quantities? [2]

AO2 **(ii)** Suggest two variables that should have been kept constant during this experiment. [2]

AO3 **(iii)** Describe the effect of salt concentration on the the tomato plants, when no PEO granules were used. [2]

AO3 **(iv)** A grower uses water with a salt concentration of 8000 parts per million to water his tomato plants. PEO granules are very expensive.

What do you recommend he should use for growing his tomatoes? Explain your answer. [3]

AO1 recall the science   AO2 apply your knowledge   AO3 evaluate and analyse the evidence

✳ WORKED EXAMPLE – Foundation tier

The diagram shows a transverse section through a leaf.

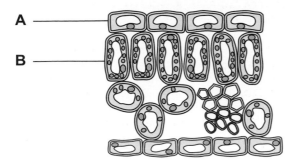

A

B

(a) Structures **A** and **B** are tissues.

(i) Explain the meaning of the term, tissue. [2]

A group of cells

(ii) Suggest how the structure of the cells in tissue A adapt them for their functions. [2]

They make a waxy layer on top of themselves to stop the leaf drying out.

(b) Carbon dioxide, needed for photosynthesis, moves into the leaf by diffusion.

Suggest and explain two ways in which the structure of the leaf allows carbon dioxide to diffuse quickly to the cells that require it. [4]

First way

The leaf is very thin

This allows carbon dioxide to diffuse quickly.

Second way

It has holes in the bottom.

To allow carbon dioxide to go in from the air

How to raise your grade!
Take note of these comments – they will help you to raise your grade.

This is correct, but it is not the full definition. It receives one mark. To obtain the second mark, the candidate needs to add that the cells in a tissue have similar structure and function.

It is important to learn the definitions that are given in the specification.

This is a good answer, and receives two marks. The answer says what the adaptation is (making the waxy layer) and how that helps them with their function (stopping the leaf drying out).

The candidate receives three marks for this.

There is a mark for saying that the leaf is thin, but the explanation just repeats the question. The candidate should explain how this helps carbon dioxide to diffuse rapidly to the cells inside the leaf – for example, by saying that this means there is only a short distance for carbon dioxide to diffuse from the air to the photosynthesising cells inside the leaf.

There is another mark for saying that the leaf has holes in the bottom, and another for saying that these allow carbon dioxide to enter the leaf from the air.

1. Phytoplankton are tiny, plant-like organisms that float in the sea. An experiment was carried out to investigate how temperature affects the rate of photosynthesis of phytoplankton that live in the Antarctic Ocean. The table shows the results.

Temperature (°C)	-1	3	7	11	15	19	23	27
Rate of photosynthesis (arbitrary units)	100	143	198	120	70	42	16	3

AO3 **(a)** State the temperature at which the rate of photosynthesis was highest. [1]

AO3 **(b)** During the Antarctic spring, the mean temperature of the sea water is -0.8 °C. Use the data in the table to suggest a value for the rate of photosynthesis during spring. [1]

AO2 **(c)** Suggest **two** factors, other than temperature, that could **reduce** the rate of photosynthesis of these phytoplankton. [2]

AO3 **(d)** Use the data in the table, and your own knowledge, to suggest how global warming might affect the rate of photosynthesis of phytoplankton in the Antarctic Ocean. [4]

2. In the 1970s, the effluent from a large oil refinery was discharged onto a rocky shore. A study was carried out to see how this affected the distribution of limpets on the shore.

A transect was placed along the shore at low tide, running along the water's edge. The centre of the transect line was at the point where the effluent ran down the beach. The density of the limpets was measured at six points along the transect.

More transects were then placed, parallel to the first one, at various positions higher up the shore.

The kite diagram shows the results. The width of each shaded bar represents the density of limpets at that point.

Bird's eye view of rocky shore

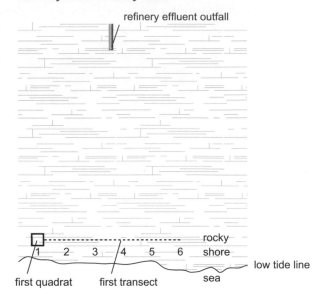

Results of transects

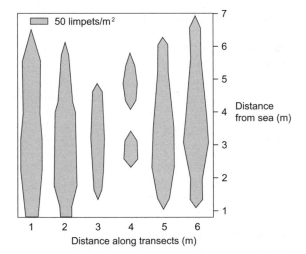

AO2 **(a)** Suggest how the density of the limpets, at different points along the transect, could be measured. [3]

AO3 **(b)** Describe the effect of the oil refinery effluent on the distribution of limpets on this shore. [3]

AO3 **(c)** Suggest why no limpets were found further than 7 metres away from the low tide sea level. [3]

AO1 recall the science   AO2 apply your knowledge   AO3 evaluate and analyse the evidence

✳ WORKED EXAMPLE – Higher tier

Roots absorb water, dissolved ions and oxygen from the soil. The photograph shows a young root of a mung bean.

(a) Name the process by which oxygen moves into the root cells. [1]

The process by which oxygen moves into the root cells is diffusion. The oxygen moves down its concentration gradient.

(b) Suggest how the structure of the root, shown in the photograph, helps this process to take place quickly. [3]

The root has lots of tiny root hairs. These increase the surface area. Because there is a large surface area, this means the root hairs are in contact with a lot of air spaces in the soil. A lot of oxygen can therefore move into the root at the same time. The oxygen can easily diffuse through the cell wall and cell membrane of the root hair cells.

(c) In some circumstances, the cells in the root respire more quickly than usual.

Explain how an increase in the rate of respiration in the root cells will affect the rate of movement of oxygen into the root. [3]

When the root cells are respiring quickly, they use up a lot of oxygen. This reduces the concentration of oxygen inside the root cells. There will therefore be a steeper diffusion gradient from the high concentration of oxygen in the air spaces in the soil, to the low concentration of oxygen inside the root cells. Diffusion takes place more quickly when there is a bigger difference in concentration, so oxygen will diffuse into the root cells faster.

How to raise your grade!
Take note of these comments – they will help you to raise your grade.

The candidate receives the mark 'diffusion'. However, they have written much too much.

It is important to look carefully at what the question is asking you to do. Here, you are asked to 'Name' something, so all that was needed was a one-word answer – 'diffusion'. The candidate has wasted a lot of time in writing unnecessary words.

This is an excellent answer, and it receives all three marks. The candidate clearly states that the root hairs increase the surface area. The next two sentences explain clearly how this helps a lot of oxygen to diffuse quickly into the root. The last sentence also relates the structure of the cell to the process of uptake of oxygen by diffusion.

This is another excellent answer that receives all three marks. Each step is clearly described and explained.

It is worth planning the answer if you are trying to *explain* something. Explaining is more difficult than describing, because you need to say *why* or how something happens, not just *what* happens.

Biology B2.5–2.8

What you should know

Respiration and energy

Cell respiration in animals and plants provides them with energy to carry out life processes.

Energy from the Sun is incorporated into plants by photosynthesis, and then passed along food chains.

 Write the word equation for respiration.

Genetic variation

The nucleus of a cell contains chromosomes. Chromosomes contain genes that help to determine the characteristics of the organism.

Sexual reproduction results in genetic variation, because the offspring receive different combinations of genes from their parents.

Asexual reproduction results in a clone of genetically identical individuals.

The characteristics of organisms are affected both by genes and by their environment.

 What is the scientific name for the sex cells (eggs and sperm) involved in sexual reproduction?

Evolution

Darwin's theory of natural selection explains how all species of living things may have evolved from simpler life-forms that existed a very long time ago.

Individual organisms vary, because they have different combinations of genes. Those individuals that have the most advantageous variations are more likely to survive and reproduce, passing on their genes to the next generation.

What is the term for a sudden change in a gene, producing a new allele or a new characteristic in an organism?

You will find out

Proteins – their functions and uses

> Protein molecules are long chains of amino acids. They are used in the body to build structures, as hormones, antibodies and catalysts.

> Enzymes are biological catalysts.

> Enzymes control all metabolic reactions, including digestion. They are also used in the home and in industry.

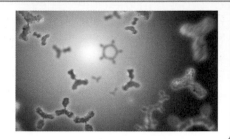

Aerobic and anaerobic respiration

> All living organisms release energy from glucose by respiration.

> In aerobic respiration, energy is released when glucose is combined with oxygen. During exercise, when muscles are using a lot of energy, heart rate and breathing rate increase to provide muscles with extra oxygen.

> In anaerobic respiration, a small amount of energy is released from glucose without using oxygen.

Cell division and inheritance

> Cells normally divide by mitosis, which results in two genetically identical daughter cells. To produce gametes, cells divide by meiosis, which produces genetically different cells, each with half the normal number of chromosomes.

> Genes are passed from one generation to the next. A gene controlling a particular characteristic may have different forms, called alleles. Genetic diagrams can be used to predict the probable characteristics of the offspring of two parents.

> Cystic fibrosis is an example of a disorder caused by genes, and can be inherited.

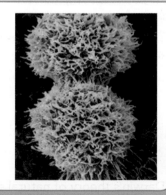

Speciation

> New species can arise if two populations of a species become separated. Natural selection may result in them becoming so different that they can no longer interbreed.

> Fossils provide evidence about some of the species that lived long ago.

Proteins

Body building

Body builders eat huge amounts of protein in their diet. They exercise to make their muscles develop. The muscles are mostly made of protein.

> You will find out:
> - protein molecules are long chains of smaller molecules called amino acids
> - some proteins are structural components of tissues and others act as hormones, antibodies and enzymes (catalysts)

FIGURE 1: A body builder may eat more protein in a day than you eat in a week.

What are proteins?

Proteins have very large molecules. Each protein molecule is made of a very long chain of hundreds of smaller molecules linked together. These small molecules are **amino acids**.

There are 20 different amino acids. They can be linked in any order. Even a tiny change in the order makes a different protein.

Proteins in the body

> Every **cell** contains a lot of protein. The cell membrane, nucleus and cytoplasm all contain protein.

> The cells in **muscular tissue** contain especially large amounts of protein. This gives the muscles their structure and strength and brings about movement.

> Some of your **hormones** are proteins.

> The **antibodies** that destroy bacteria are proteins.

> **Enzymes** are all proteins.

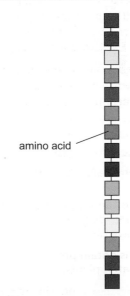

amino acid

FIGURE 2: The structure of part of a protein molecule.

QUESTIONS

1 Which of these makes up the smallest percentage of your body mass: protein, fat or carbohydrate?

2 Name the smaller molecules that link together to form a protein molecule.

3 Suggest why meat is a good source of protein in the diet.

Q amino acids and proteins GCSE

Protein shapes

Protein molecules can fold up into different shapes. The shape of a protein molecule affects the way that it behaves.

Some protein molecules make long, thin fibres. The proteins in muscles do this. Fibrous proteins help in forming structures in the body.

Other protein molecules curl up into a ball. These are often involved in the chemical reactions in the body. This is what enzymes, antibodies and hormones do. The ball often has a dent in it that is a perfect fit for one other kind of molecule.

For example, an antibody molecule might have a dent that perfectly fits a particular molecule on a particular bacterium. The antibody molecule can stick to that molecule and then kill the bacterium.

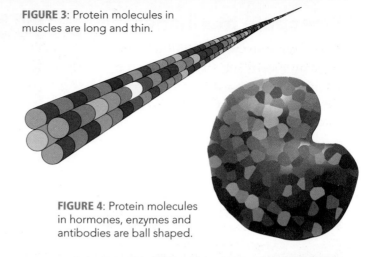

FIGURE 3: Protein molecules in muscles are long and thin.

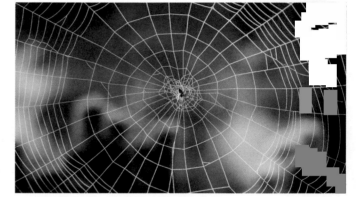

FIGURE 4: Protein molecules in hormones, enzymes and antibodies are ball shaped.

FIGURE 5: Spider silk is made up of fibres of protein. It is immensely strong.

QUESTIONS

4 Keratin is the protein in hair. Suggest whether a keratin molecule stays in a long, thin shape or whether it folds up into a ball. Explain your answer.

5 Explain why one particular kind of antibody will work against one kind of bacterium but not another.

Making proteins in the body

Animals cannot make proteins unless they have amino acids as a starting point. Plants, however, can make proteins out of carbohydrates, so long as they have nitrate ions.

Humans obtain proteins by eating plants, or by eating other animals that have eaten plants.

> During digestion, the long protein molecules in food are broken down into individual amino acids.

> These are then absorbed through the wall of the small intestine and are carried all over the body in the blood.

> Each cell takes up the amino acids that it needs from the blood.

> Inside the cell, on ribosomes, amino acids are linked together to make the particular kinds of proteins that the cell requires.

The sequence of amino acids in the protein is very important. Different sequences make different protein molecules with different shapes. The genes in the nucleus provide instructions about exactly which amino acids to link together, and in which order.

QUESTIONS

6 Draw a flow diagram to explain how a nitrogen atom, in a nitrate ion in the soil, could become part of a protein in your muscle cells.

7 Suggest how slightly different versions of the gene for keratin could produce straight hair or wavy hair.

🔍 protein molecule shape

Enzymes

Pineapple marinades

Pineapple contains a protein-digesting enzyme called bromelain. If you put meat in a marinade containing fresh pineapple, it becomes more tender as the tough protein fibres in the meat are broken down.

FIGURE 1: Adding pineapple to kebabs will make the meat more tender.

> **You will find out:**
> > enzymes are proteins that act as catalysts
> > temperature affects an enzyme's shape
> > different enzymes work best at different pH values

Speeding up chemical reactions

Biological catalysts

There are hundreds of different chemical reactions going on in our bodies, all the time. Each one is controlled by an **enzyme**.

Enzymes are biological **catalysts**. All enzymes:

> are protein molecules

> control one specific chemical reaction

> make the reaction happen quickly.

Enzymes, like a catalyst for any chemical reaction, are not used up.

How enzymes work

An enzyme molecule is a long chain of amino acids, folded into a ball. There is a dent in the ball into which another molecule can fit. This dent is the **active site** of the enzyme.

The molecule that fits into the enzyme is called its **substrate**. The enzyme makes the substrate react, changing into a new substance.

High **temperature** changes the shape of the enzyme molecule, so the enzyme stops working.

The **pH** also affects the shape of an enzyme. Different enzymes work best at different pH values.

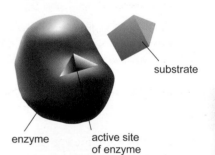

substrate

enzyme — active site of enzyme

FIGURE 2: An enzyme's substrate fits perfectly into its active site.

TABLE 1: Examples of pH at which enzymes work best.

Name of enzyme	Where it is found	pH at which it works best
amylase	saliva, in the mouth and in the small intestine	7.5
pepsin (a protease)	the stomach	2.0
lipase	the small intestine	7.5

QUESTIONS

1 What are enzymes and what do they do?

2 Explain why an enzyme has to be a special shape.

3 Name an enzyme that works best in acidic conditions.

🔍 lock and key theory enzymes

Factors affecting enzyme activity

Most of the enzymes in the body work best at about 37 °C, which is normal body temperature. This is their **optimum temperature**.

At temperatures above the optimum, the enzyme begins to uncurl. It starts to lose its shape. Once the active site has lost its shape, the substrate no longer fits. When the enzyme is permanently changed in this way, it is said to be **denatured**.

Enzymes are also sensitive to pH. If the pH is a long way from the enzyme's optimum pH, then the enzyme denatures.

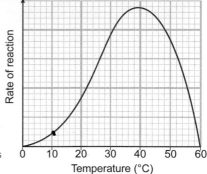

FIGURE 3: How temperature affects enzyme activity.

QUESTIONS

4 Give the optimum temperature for the enzymes in the human body.

5 What happens to an enzyme when it is denatured? Suggest one factor that can cause this.

6 Look at Figure 3. Describe what happens to an enzyme-catalysed reaction as the temperature increases.

7 Look at Figure 4. Describe how the pepsin curve is different from the amylase curve.

FIGURE 4: How pH affects enzyme activity.

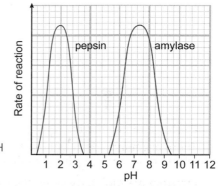

How temperature affects enzymes

As temperature increases, molecules move more quickly. When the temperature is low, the enzyme and its substrate are both moving slowly. They do not bump into each other very often. When they do collide, there is not very much energy involved, so the substrate may not react. This is why enzyme-controlled reactions happen slowly at low temperatures.

Like many reactions, enzyme-controlled reactions tend to double their rate for a 10 °C rise in temperature.

Catalase

Catalase is an enzyme that catalyses the breakdown of hydrogen peroxide.

hydrogen peroxide → water + oxygen

In an experiment, some catalase was put into eight separate test tubes and equal volumes of hydrogen peroxide in eight other test tubes. Pairs of test tubes (one with catalase and one with hydrogen peroxide) were kept at a different temperature from other pairs.

For each pair, the hydrogen peroxide was poured into the test tube containing catalase. The oxygen released was collected for one minute.

Table 2 shows the results obtained.

Remember
Enzymes are not 'killed' at high temperatures. Enzymes are molecules. You cannot kill a molecule.

QUESTIONS

8 (i) Draw a graph of the results in Table 2.

(ii) Is the rate of reaction shown to double for each increase of 10 °C?

(iii) Explain what happens at temperatures above 37 °C.

TABLE 2: Volumes of oxygen produced in one minute.

Temperature (°C)	0	10	20	30	40	50	60	70
Volume of oxygen (cm³)	2	4	9	17	32	15	1	0

Enzymes and digestion

You will find out:
> digestive enzymes work inside the gut, but outside cells
> what amylase, protease and lipase enzymes do and where they work
> acid in the stomach helps stomach enzymes to work well

Hard to swallow

This X-ray shows a key in a small girl's stomach. She swallowed it by accident. The digestive system is just a long, hollow tube, so – unless the key jabs into the cells that line the digestive system – it will not do any harm as it travels through and exits at the other end.

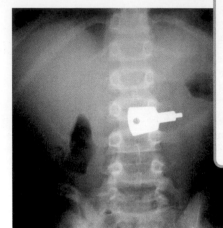

FIGURE 1: A swallowed key in a child's stomach.

Digestion

When you swallow food, it is not really *inside* your body. This is because your gut (digestive system) is a long, hollow tube. It is not until your food gets through the walls of this tube that it is truly inside.

> Molecules from your food can get through the gut wall only if they are small.

> The small molecules can seep through the gut wall and diffuse into your blood.

> The blood carries them all over your body.

This means that large molecules in the food in your gut must be broken down into small ones. This is **digestion**. Most digestion is done by enzymes.

Digestive enzymes

Digestive enzymes are made by cells in the tissue that lines the inside of the gut. The cells that make them are often in **glands**.

The enzymes pass out of these cells and go into the space inside the gut. They become mixed up with the food in the gut.

Digestive enzymes are like scissors. They cut large food molecules into smaller bits. For example, the enzyme amylase cuts starch molecules into much smaller sugar molecules.

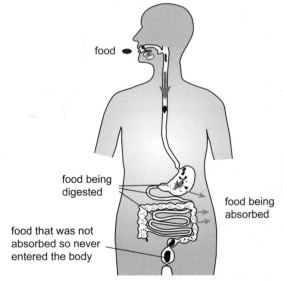

food

food being digested

food being absorbed

food that was not absorbed so never entered the body

FIGURE 2: Food inside your gut is not really inside your body.

QUESTIONS

1 What do food molecules have to do, in order to pass into your blood?

2 Where are digestive enzymes made?

3 Explain how amylase digests starch molecules.

starch – a long molecule made of many glucose molecules joined together

sugar molecules – small enough to pass through the gut wall and into the blood

amylase

FIGURE 3: How amylase digests starch.

Types of enzymes

There are three main groups of digestive enzymes.

> **Amylase** is produced in the salivary glands and in the pancreas. Amylase breaks down starch into sugars in the mouth and small intestine.

> **Protease** enzymes are produced in the stomach, the pancreas and the small intestine. Protease breaks down proteins into amino acids in the stomach and small intestine.

> **Lipase** enzymes are produced in the pancreas and the small intestine. Lipase breaks down lipids (fats and oils) into fatty acids and glycerol in the small intestine.

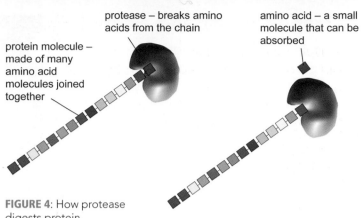

protease – breaks amino acids from the chain

amino acid – a small molecule that can be absorbed

protein molecule – made of many amino acid molecules joined together

FIGURE 4: How protease digests protein.

QUESTIONS

4 Complete this table.

Enzyme	Where it is made	Where it works	What it does
amylase			
	stomach, pancreas and small intestine		breaks protein molecules into amino acids
		small intestine	

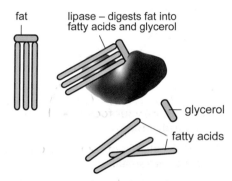

fat

lipase – digests fat into fatty acids and glycerol

glycerol

fatty acids

FIGURE 5: How lipase digests fats.

Acid and alkali

Acid in the stomach

The protease enzyme in the stomach is **pepsin**. The glands in the stomach wall also produce **hydrochloric acid**. This acid:

> kills many of the bacteria that you eat

> provides a suitable pH for pepsin, which can work only in acid conditions.

Bile

The liver produces **bile**, which is stored in the gall bladder. When food arrives in the small intestine, bile flows along the bile duct and mixes with the food.

Bile contains sodium hydrogencarbonate, dissolved in water. This is an alkali. It neutralises the acid from the stomach when it passes into the small intestine. The enzymes in the small intestine need a slightly alkaline pH to work most effectively.

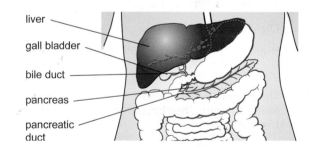

FIGURE 6: Bile is produced in the liver.

liver

gall bladder

bile duct

pancreas

pancreatic duct

QUESTIONS

5 When you chew food, it mixes with amylase in your mouth. The amylase starts to digest any starch in the food, but as soon as the food arrives in your stomach the amylase stops working. Explain why this happens.

Enzymes at home

You will find out:

> some microorganisms produce enzymes that pass out of their cells
> how enzymes are used in biological detergents
> biological detergents work at lower temperatures than other types of detergents

Grapes, yeast and alcohol

People have been making wine for about 8000 years. The process involves fermentation. Grapes have a white 'bloom' on their skin, which is a type of yeast. This makes an enzyme that changes the sugar in grape juice into alcohol.

FIGURE 1: The white powdery covering on grapes is yeast.

Did you know?

If everyone washed clothes at 30 °C instead of 40 °C, that could reduce the UK's CO_2 output by 200 000 000 kilograms every year.

Enzymes from microorganisms

Some microorganisms, such as yeast and bacteria, make enzymes that pass out of their cells. If these microorganisms are grown in large quantities, the enzymes that they produce can be collected. There are many different uses to which these enzymes can be put.

Biological detergents

Washing powders contain detergents, which help greasy dirt to mix with water, so that it can be washed away.

Some detergents also contain enzymes. These include:

> proteases, which digest proteins

> lipases, which digest fats.

Some stains – for example blood stains – on clothes cannot be removed using ordinary detergents. The enzymes help to break down the stains into substances that dissolve in water. The stains can then wash away.

The enzymes in detergents often work best at about 30 °C, whereas 'ordinary' detergents generally work best at higher temperatures.

FIGURE 2: Biological washing powders contain enzymes.

QUESTIONS

1 What do protease enzymes do?

2 Explain how protease enzymes can help to remove stains from clothes.

Q enzymes biological detergents

Removing protein stains

Many different chemicals stain clothes. These include proteins (such as haemoglobin in blood) and lipids (such as egg yolk).

When proteins get onto clothes, they may cling onto the fibres of the clothes and are very difficult to remove.

If you get blood onto a white shirt, the best thing to do is to wash it off straight away, using cold water. Haemoglobin is a soluble protein, so it usually washes off quite easily.

However, if you use hot water, the high temperature will denature the haemoglobin. This makes the haemoglobin insoluble and it sticks permanently to the cloth. Even the best ordinary detergents will not be able to remove it.

Biological detergents contain protease enzymes that can break down the haemoglobin molecules. The proteases in the detergent change the haemoglobin to small, soluble amino acid molecules. These can easily be washed away.

FIGURE 3: Why is it best to wash a fresh blood stain with cold, not hot, water?

QUESTIONS

3 What gives a blood stain its red colour?

4 Explain how a biological detergent removes a blood stain.

5 Using your knowledge of enzymes, explain why biological detergents should normally be used at temperatures below 40 °C.

Skin complaints

In the 1970s, doctors began to notice many more people visiting their surgeries with sore skin on their hands.

Some particularly observant doctors questioned their patients to determine what they had been doing with their hands. They began to suspect that the new biological detergents might be related to these skin problems.

The doctors thought that perhaps the proteases in these detergents could break down the keratin in skin. Keratin is a protein found in the thin layer of dead cells covering the outer surface of the skin. If the keratin is broken down, this outer layer flakes away, leaving unprotected living skin cells underneath.

Research has failed to confirm this idea. However, many people still report that their skin is sensitive to biological detergents.

QUESTIONS

6 A biological detergent carton has this advice on the label:

Not suitable for wool or silk garments. Not recommended for hand washing. Do not wash above 50 °C.

Explain each of these pieces of advice. (You may need to find out what wool and silk are made of.)

7 Suggest how to investigate and test the hypothesis that biological detergents cause skin problems.

Enzymes in industry

You will find out:

> proteases are used to pre-digest some baby foods

> carbohydrases are used to convert starch into sugar syrup

> how isomerase helps to make sweet foods sweeter with less energy content

Tenderising meat

One way to make tough meat tender is to beat it with a mallet. The mallet smashes up the cells in the meat and releases the enzymes inside them. The proteases start to digest the tough fibrous proteins. This makes the meat tender and easier to eat.

Did you know?

Some Amazonian tribes chew roots and spit them out to ferment into alcoholic drinks. The enzymes in their saliva break down the starch in the roots to sugar.

FIGURE 1: Why does this make meat tender?

Using enzymes for food manufacture

Enzymes have many uses in industry. Enzymes can make reactions happen at low temperatures and pressures. Without enzymes, a lot of money would have to be spent on increasing temperatures and pressures to make chemicals react together.

Enzymes are especially useful in the food manufacturing industry.

Baby foods

In young babies, the digestive system is not fully developed. Some baby food manufacturers add proteases (protein-digesting enzymes) to their products. This makes it easier for the baby to absorb nutrients from the food.

Sugar syrup

A concentrated solution of sugar is called a **syrup**. Sugar syrup is used in making sweets and sports drinks.

Starch solution is easy to make by cooking potatoes or maize and mixing them with water. The starch can then be changed into sugar syrup by adding **carbohydrase** enzymes.

Carbohydrases are enzymes that convert starch to sugar. Amylase is an example of a carbohydrase enzyme.

FIGURE 3: Honey naturally contains fructose. Where do the honeybees get this sugar from?

FIGURE 2: Many manufactured baby foods have been pre-digested using enzymes.

Slimming foods

Foods for slimmers often contain a sugar called fructose instead of glucose. You need less fructose than glucose or sucrose to get the same sweetness.

Fructose is made from glucose using an enzyme called **isomerase**.

Digesting baby foods

Babies need the same kinds of nutrients as everyone else. They particularly need plenty of protein in their diet, because they are growing fast.

Although babies can digest proteins themselves, their enzymes may not manage to digest absolutely all the protein in their food. This is why baby food manufacturers sometimes add proteases to the food. These enzymes break down large protein molecules into amino acids.

When the baby eats this pre-digested food, it can absorb the amino acids. As the baby grows, its own digestive system develops and produces all the enzymes it needs.

In countries where mothers cannot obtain, or cannot afford, manufactured baby foods, mothers often pre-chew food before giving it to their babies.

Making soft-centre chocolates

Soft-centred chocolates are made with the help of enzymes.

This is how they are made:

1. A mixture of sucrose, flavouring, colouring and a little water are mixed together to make a paste.

2. A very small amount of an enzyme called sucrase is added to the paste.

3. The paste is moulded into appropriate shapes and left to set.

4. Liquid chocolate is poured over them and left to set.

The chocolates are then warmed up a little, so that the enzyme begins to work on the sucrose inside them.

The following chemical reaction happens inside the chocolate case:

sucrose → glucose + fructose

The mixture of glucose and fructose makes a thick, soft syrup inside the chocolates.

FIGURE 4: Which enzyme is used to help to make the soft centres in chocolates?

🔍 protease baby food

Preparing for assessment: Planning an investigation

To achieve a good grade in science, you not only have to know and understand scientific ideas, but you need to be able to apply them to other situations and investigations. These tasks will support you in developing these skills.

✷ Investigating the effect of pH on the catalytic activity of catalase

Coralie had learnt that the contents of a human stomach are so acidic that they can burn through clothes. Yet she also knew that enzymes can work in those conditions.

Coralie had done experiments with catalase, an enzyme which catalyses the breakdown of hydrogen peroxide to water and oxygen.

hydrogen peroxide → water + oxygen

She decided to investigate how pH affects the ability of catalase to catalyse this reaction.

First, Coralie made a catalase solution, by crushing up a potato, mixing it with water and filtering the mixture. The filtrate contained catalase from the cells in the potato. Then she divided the solution between five small beakers, and added a different buffer solution to each one (buffers keep a solution at a particular pH).

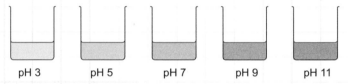

| pH 3 | pH 5 | pH 7 | pH 9 | pH 11 |

Coralie's teacher suggested two methods that she could use to measure the rate of the breakdown of the hydrogen peroxide.

Method 1

> Put some hydrogen peroxide into a test tube.

> Cut out lots of small circles of filter paper using a hole punch.

> Using forceps, dip one circle of paper into the solution in one of the beakers.

> Push the paper circle to the bottom of the tube of hydrogen peroxide.

> Measure how long it takes for bubbles of oxygen to collect on the paper and lift it to the surface. The faster the reaction, the shorter the time this will take.

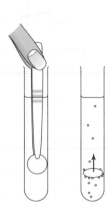

Method 2

Put some hydrogen peroxide into a small glass container. Place it on a sensitive top pan balance attached to a data logger. Add catalase solution to the container. Measure the decrease in mass as oxygen is produced and lost to the air.

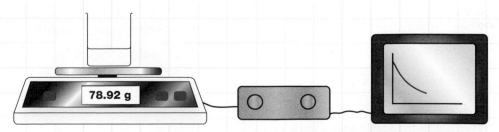

78.92 g

✳ Planning

1. Suggest a hypothesis that Coralie could test.

2. Which of the two methods would you choose? Explain your choice.

> Your answer should discuss the advantages and disadvantages of each method. Finally, make your choice based on these.

3. What would be the independent variable in your investigation? What would be the dependent variable?

> Try to think of the variables that are most likely to make your results less valid if you do not control them.

4. State three variables that you would keep constant in your experiment. For each one, explain carefully how you would control this variable.

5. Predict the results that you would expect to obtain. Explain your prediction.

> You may like to look up the optimum pH of catalase. Otherwise, use your knowledge of other enzymes to suggest an answer.

6. Assess the risks involved with the method that you have chosen. Explain what you would do to reduce these risks.

> Look up each of the reagents used in this reaction to see if they are hazardous.

✳ Processing data

7. Construct a results table in which you could record your results.

> A good results table is easy for someone else to look at and understand. You may want to try out two or three designs before choosing the best one. Remember to put units in the headings of columns and rows.

✳ Reviewing the investigation

8. Say briefly how you would decide whether or not your investigation had supported the proposed hypothesis.

> Think back to the hypothesis and what results you would have expected had it been correct.

9. Explain what you would do to evaluate the reliability (repeatability and reproducibility) of your results.

> Repeatability means that, if you did your experiment again, you would get the same results. 'Reproducibility' means that, if someone else did the experiment, using different materials and equipment, they would get the same results as you.

✳ Connections

How Science Works

- Plan practical ways to develop and test scientific ideas

- Assess and manage risks when carrying out practical work

- Collect primary and secondary data

Science ideas

B2.5.2 Enzymes

Aerobic respiration

Cyanide poisoning

Cyanide is a poison that can be fatal even in very small quantities. Cyanide kills because it stops an enzyme from working. This enzyme is involved in respiration. If respiration stops in an organism's cells, it dies.

Did you know?

On 20 July 1944, there was a plot to assassinate Hitler. Field Marshall Rommel was implicated and committed suicide by taking a cyanide pill.

> **You will find out:**
> > respiration releases energy from glucose
> > respiration happens inside every living cell
> > in aerobic respiration, oxygen is combined with glucose

FIGURE 1: Cyanide has been used in mines to extract gold and silver. The waste kills all living organisms.

Getting energy

Your body uses energy all the time. Without energy, cells quickly die.

All of your energy comes from the food that you eat.

The energy is locked up inside the food molecules – food is an energy store. To release the energy, cells have to break these molecules apart. This is normally done using **glucose** molecules.

This is **respiration**. Every living cell respires all the time. If the cell stops respiring, it dies.

Most of the time, cells break down the glucose molecules by combining them with oxygen. This is called **aerobic respiration**, because it uses oxygen from the air.

This is the word equation for respiration:

glucose + oxygen → carbon dioxide + water (+ energy)

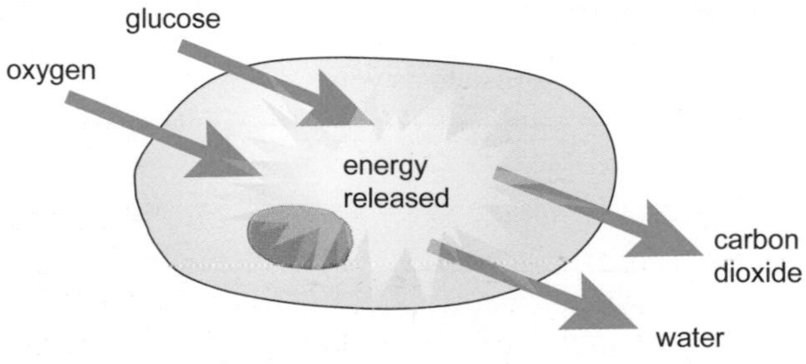

Heart and lungs

Aerobic respiration uses oxygen and produces carbon dioxide. Your body obtains oxygen from the air around you. The oxygen enters your blood in your lungs. It is transported in the blood to all your body cells. The carbon dioxide that the cells make is carried back, in the blood, to the lungs.

FIGURE 3: Why do you breathe faster when you exercise?

glucose

oxygen

energy released

carbon dioxide

water

FIGURE 2: Summarising respiration. What two substances do cells need for aerobic respiration? What two substances do they produce?

Remember
The energy for cells to use is already inside glucose molecules. Respiration releases this energy.

Q aerobic respiration process

Breathing in and out helps to move fresh air into your lungs and to push out stale air. The fresh air contains a lot of oxygen. The stale air contains less oxygen and a lot more carbon dioxide.

The heart pumps to push the blood around your body. It keeps every cell well supplied with the oxygen and glucose that they need, and quickly takes away the carbon dioxide.

QUESTIONS

1 Explain why your cells need oxygen.

2 (i) Where does oxygen go into your blood?

(ii) How is the oxygen transported to the cells that need it?

3 Explain why your heart beats faster when your muscles are working hard.

Mitochondria

Mitochondria are tiny structures found in animal and plant cells. It is inside mitochondria that aerobic respiration takes place. Mitochondria contain all the enzymes that are needed to make the reactions of respiration happen quickly.

Cells that need a lot of energy usually have large numbers of mitochondria. Sperm cells, for example, have many mitochondria packed tightly into the part between their tail and their nucleus. Muscle cells are also packed full of mitochondria.

QUESTIONS

4 In which types of cells are mitochondria found?

5 (i) Name two types of cells that have particularly large numbers of mitochondria

(ii) Explain why they need so many.

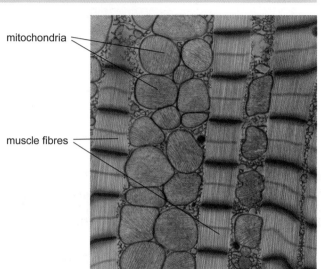

FIGURE 4: This is a piece of heart muscle. Why are there so many mitochondria packed in?

Respiration and photosynthesis

If you compare the respiration equation with the photosynthesis equation (page 24), you can see that they are mirror images of each other.

Photosynthesis locks up energy from sunlight inside glucose molecules. Respiration releases this energy so that cells can use it.

When there is sunlight, mesophyll cells in plant leaves are photosynthesising and respiring. Both reactions are happening at once – photosynthesis in their chloroplasts and respiration in their mitochondria. In bright light, photosynthesis happens a lot faster than respiration.

At night, photosynthesis stops, but the cells carry on respiring.

QUESTIONS

6 Why do plants first make glucose by photosynthesis and then break it down again by respiration?

7 Suggest what happens during daylight (a) to the carbon dioxide that is produced in the mitochondria in mesophyll cells (b) to the oxygen that is produced in the chloroplasts in mesophyll cells.

8 Which gas do plants release from their leaves during the night?

Using energy

Battling the cold

A man doing hard manual work in a normal environment needs about 15 000 kJ per day, but a man hauling a sledge in the freezing temperatures near the poles needs around 27 300 kJ per day. Most of this extra energy is used to produce heat to keep the body temperature constant.

Did you know?

Your body probably uses around 10 000 kJ per day. That is enough to boil more than 300 litres of water.

FIGURE 1: Scott's expedition to the South Pole, 1912.

Why do cells need energy?

Energy is constantly being released from food molecules inside your cells, by respiration. Figure 2 shows what the body uses the energy for.

QUESTIONS

1 (i) List three ways in which the human body uses energy.

(ii) Which of these also happen in plants?

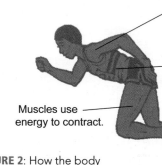

Cells use energy to keep the body temperature constant.

Cells use energy to build up large molecules – such as proteins – from small ones.

Muscles use energy to contract.

FIGURE 2: How the body uses energy.

Using the energy

Building large molecules

It takes a lot of energy to make small molecules join together to make long chains. Cells use energy to make:

> amino acid molecules link together to make protein molecules

> sugar molecules link together to make starch or cellulose molecules.

Muscle contraction

Muscles use energy to make themselves shorter. Muscles usually keep a store of fuel inside their cells, so that they can start working even if the blood is not bringing them very much glucose.

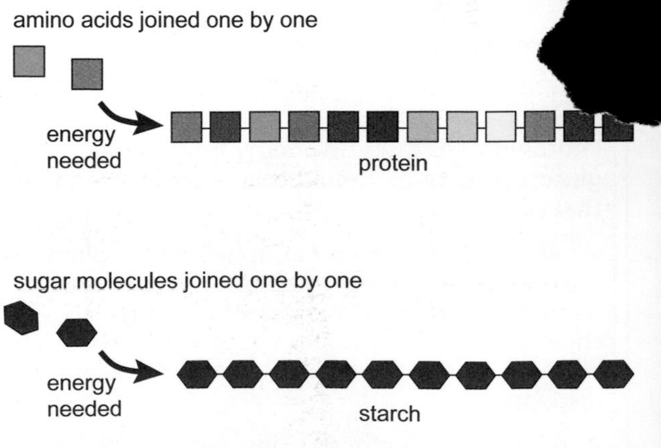

amino acids joined one by one

energy needed

protein

sugar molecules joined one by one

energy needed

starch

FIGURE 3: Molecules linking together. Which of these processes takes place only in plant cells, not in animal cells?

Q cell protein production GCSE

The storage substance is **glycogen**. It is very similar to starch. A glycogen molecule is made of many glucose molecules linked together. When a muscle needs energy, it breaks down the glycogen, producing glucose for use in respiration.

Maintaining a steady body temperature

Mammals and birds keep their body temperature around 37 °C, no matter what the temperature around them. If their environment is colder than they are, then heat is lost from their body. Respiration in cells releases energy, which increases body temperature. The body regulates this, so that the amount of heat produced balances the amount of heat that is lost.

Making amino acids

Plants can make amino acids from glucose. Animals cannot do this. They obtain amino acids by eating ready-made proteins.

In a plant, sugars made in photosynthesis can be converted into amino acids by combining them with nitrate ions. The nitrate ions are taken up from the soil, into the roots.

Making amino acids from sugar and nitrate ions requires energy. Once the plant has made amino acids, it can combine them to make protein molecules.

FIGURE 4: On this thermogram of penguins, the yellow areas are warm and the blue areas are cold. Where is the heat coming from?

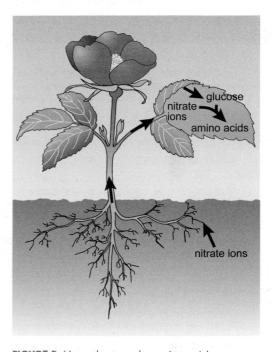

FIGURE 5: How plants make amino acids.

⬤ QUESTIONS

2 Distance runners often eat meals containing a lot of carbohydrate (such as pasta) over three or four days before the race. This is called carbohydrate loading. How does this help muscles to work well during a race?

3 Describe one way in which animals, but not plants, use energy.

4 Describe one way in which plants, but not animals, use energy.

Energy for swimming

Sperm cells are tiny, but they use a huge amount of energy.

When a sperm cell is released at the top of a woman's vagina, it still has to cover a distance of up to 15 centimetres before it reaches its goal. Not surprisingly, sperm cells are packed full of mitochondria. They do not have any food reserves such as glycogen. Instead, they use sugars from the fluid in which they were ejected from the male.

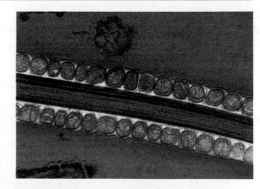

FIGURE 6: This photograph, taken using a powerful microscope, shows part of a sperm tail (running across the picture). The yellow circles with black lines in them are mitochondria.

⬤ QUESTIONS

5 Explain why sperm cells contain so many mitochondria.

6 Suggest why sperm cells do not contain glycogen stores.

Anaerobic respiration

You will find out:

> anaerobic respiration is used by muscles to get a little extra energy when oxygen is in short supply

> lactic acid, produced by anaerobic respiration, builds up in muscles and makes them fatigued

> anaerobic respiration produces an oxygen debt that must be repaid

Ten seconds to fame

Feet, in running shoes, are carefully placed into starting blocks. Fingers are meticulously positioned just behind the starting line. Senses are straining for the signal to start. Mind and body are totally focused on the race. BANG! Many months of training are now put to the test.

FIGURE 1: Usain Bolt won the 100 m at the 2008 Olympics in 9.69 s.

Anaerobic respiration

When someone sprints in a 100 metre race, they are asking their muscles to give every last scrap of energy that they can. All of this energy is released in the muscle cells by respiration.

Respiration needs sugar and oxygen. Muscles rely on oxygen being brought to them by the blood.

> During a sprint race, the muscles are using energy so quickly that they run out of oxygen.

> Even though the heart is working as hard as it can to pump more blood – carrying oxygen – to them, it still is not enough.

> The muscles therefore resort to releasing energy from glucose without using oxygen.

> This does not release anywhere near as much energy as usual, but even a small amount extra is going to help.

The normal type of respiration, using oxygen, is called aerobic respiration. This 'last resort' type of respiration, not using oxygen, is called **anaerobic respiration**. It is incomplete breakdown of glucose to lactic acid and not to carbon dioxide and water.

glucose → lactic acid (+ a little energy)

Anaerobic respiration releases far less energy than aerobic respiration (see Table 1).

Anaerobic respiration produces **lactic acid**. The lactic acid builds up in the muscles. As blood flows through the muscles, it removes the lactic acid, but it takes time to remove all of it.

TABLE 1: Energy released by respiration.

Type of respiration	Energy released (kJ) per gram of glucose
aerobic	16.1
anaerobic	0.8

Did you know?

The record for holding your breath under water is over eleven minutes. Definitely not something to try at home.

 QUESTIONS

1 How does anaerobic respiration differ from aerobic respiration?

2 How does the amount of energy released during anaerobic respiration compare with that during aerobic respiration?

3 What substance builds up in your muscles when they respire anaerobically?

Q anaerobic AND aerobic respiration

Lactic acid and oxygen debt (Higher tier)

Lactic acid is a mild poison. It makes muscles feel tired and can cause cramps. Muscles cannot work properly when lactic acid builds up in them. They stop contracting efficiently.

After you have stopped running (or whatever strenuous exercise you were doing), you have to get rid of this lactic acid. This needs extra oxygen. It is as though you have 'borrowed' some energy without 'paying for it' with oxygen. The extra oxygen is called the **oxygen debt**.

When you stop exercising, you continue to breathe heavily. This takes in the extra oxygen that you need to pay off the oxygen debt.

When your muscles are respiring anaerobically, they are still respiring aerobically as well. They use up every scrap of oxygen they can in aerobic respiration, and then just top up their energy needs using anaerobic respiration.

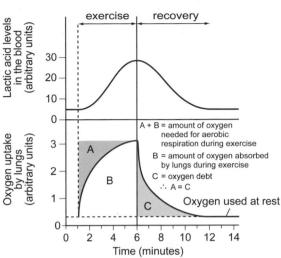

 QUESTIONS

4 Explain why a person continues to breathe deeply after strenuous exercise has stopped.

5 Explain the relationships between areas A, B and C in Figure 2.

FIGURE 2: Anaerobic respiration, lactic acid and oxygen debt.

Maximum oxygen uptake

When you exercise, your muscles use more oxygen. To obtain this extra oxygen you breathe more deeply and more quickly. Your heart beats faster so that oxygenated blood is pumped to your active muscles more quickly.

The maximum volume of oxygen your body is able to use per minute is called **VO$_2$ max**. The fitter you are, the higher the value of your VO$_2$ max.

The bar chart in Figure 3 compares VO$_2$ max values for athletes specialising in different sports.

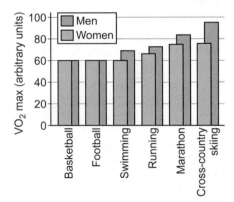

FIGURE 3: VO$_2$ max for different sports people.

QUESTIONS

These questions relate to Figure 3.

6 In which sports do athletes show the lowest and the highest VO$_2$ max values?

7 Explain where you would place netball and tennis in the bar chart.

8 Which athletes would be able to pay off an oxygen debt most quickly? Explain why.

9 Suggest how an athlete could develop a higher VO$_2$ max.

Remember

No carbon dioxide is made in anaerobic respiration.

🔍 oxygen debt lactic acid ... increase VO2 max

Cell division – mitosis

Happening all the time

As you are reading this paragraph, millions of cells all over your body are dividing. Not all cells divide at the same pace. Those lining your small intestine are the fastest – once every two to three days. By contrast, your nerve cells never divide. Between these two extremes are the cells of the rest of your body. For example, some cells in your skin divide every week or so.

Did you know?

The cells of an elephant are the same size as the cells of a mouse. The elephant just has more of them.

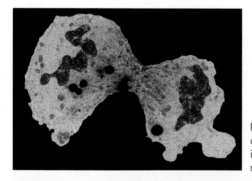

FIGURE 1: The lining of the small intestine. Why might these cells need to divide so often?

Cell division

Why do cells need to divide?

Human bodies are made of billions and billions of cells.

When you grow:

> cells divide to make more cells

> each small new cell grows to its full size

> then it may divide again.

There is a limit to how large a cell can grow and stay alive. If cells become too big, then oxygen and carbon dioxide cannot diffuse into and out of them fast enough.

QUESTIONS

1 Why is mitosis necessary?

2 Explain why cells do not keep on growing.

Mitosis

More cells are made when existing cells divide into two. The normal kind of cell division is **mitosis**.

Mitosis is very important because it provides cells:

> for growth

> to replace dead or damaged cells.

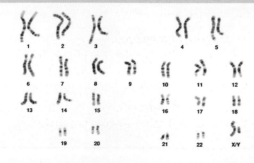

FIGURE 2: A cell dividing into two by mitosis.

Mitosis and chromosomes

Normal body cells have two sets of chromosomes. There are 46 chromosomes in each cell – two sets of 23. We therefore have a pair of each kind of chromosome.

FIGURE 3: This picture has been made by cutting out images of each chromosome from other photos, and arranging them in their pairs. Can you see one pair where the two chromosomes do not match?

mitosis … cell life cycle

Each chromosome consists of a long chain of genes. The genes contain instructions for inherited features.

Before a cell divides by mitosis, it first makes perfect copies of each chromosome. When the cell divides by mitosis it makes two new cells. The chromosomes are shared out equally between each of these new cells.

> The new cells therefore have exactly the same number and kind of chromosomes as the parent cell.

> They are **genetically identical** to the parent cell and to each other.

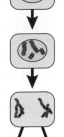

1 Before mitosis begins, each chromosome is copied exactly. The two copies stay attached to one another.

2 During mitosis, the two copies of each chromosome move apart.

QUESTIONS

3 A human body cell has 46 chromosomes. How many chromosomes will there be in each of the new cells, if this cell divides by mitosis?

4 Suggest why it is important that the new cells made by mitosis are genetically identical to the parent cell.

3 When the cell divides, each new cell has two complete sets of chromosomes. The two new cells are genetically identical.

FIGURE 4: Before and during mitosis.

How mitosis happens

This is a series of photomicrographs (photographs taken through a microscope) that show the stages of mitosis.

FIGURE 5: The process as a cell divides by mitosis.

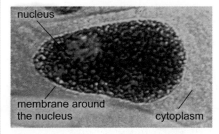

A The cell prepares to divide. You cannot see the chromosomes at this stage.

B Now the chromosomes have become visible, as mitosis gets under way.

C The chromosomes are arranged in the middle of the cell.

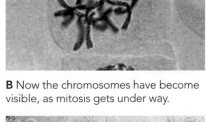

D The identical copies of each chromosome split apart and move to opposite ends of the cell.

E The chromosomes now form into two nuclei.

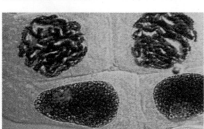

F New cell membranes form and two new cells have been made.

QUESTIONS

5 Something very important is happening to the chromosomes at stage A. What is it?

6 In stage C, all the chromosomes collect up in the middle of the cell. Suggest how this helps mitosis to work properly.

Q mitosis stages GCSE ... chromosomes definition

Cell division – meiosis

You will find out:

> a special kind of cell division, meiosis, is used for making gametes

> the cell divides twice to produce four cells, each with a single set of chromosomes

Early ideas about sperms

The Dutchman Antonie van Leeuwenhoek lived between 1632 and 1723. A very keen amateur scientist, he built his own microscopes. He was the first person to look at human sperms. He suggested that each sperm contained a little person, which would grow into maturity after fertilisation.

FIGURE 1: Sperm cells do not contain a tiny person. What do they contain?

Making gametes

Gametes (sex cells) have only half the normal number of chromosomes. This is so that, when they fuse at fertilisation, the new cell ends up with the normal number again.

In humans, gametes are made by a special kind of cell division called **meiosis**. **Sperm cells** are made in the **testes**. **Egg cells** are made in the **ovaries**.

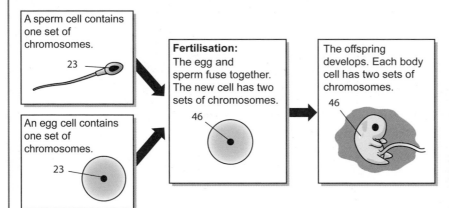

A sperm cell contains one set of chromosomes.

23

An egg cell contains one set of chromosomes.

23

Fertilisation: The egg and sperm fuse together. The new cell has two sets of chromosomes.

46

The offspring develops. Each body cell has two sets of chromosomes.

46

FIGURE 2: Chromosome numbers in gametes and other cells.

QUESTIONS

1 Name the kind of cell division that is used to make gametes.

2 Why would mitosis not be suitable for making gametes?

3 Name two organs in humans where gametes are made.

Meiosis and chromosomes (Higher tier)

Before meiosis happens, the chromosomes in the cell that is going to divide are copied. This is just the same as in mitosis.

Then the cell divides. Unlike mitosis, it divides twice. This means that four cells are produced – not two, as happens in mitosis.

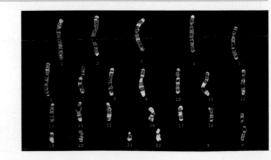

FIGURE 3: The chromosomes from an egg cell. Compare this with Figure 3 on page 64. How do they differ? Why?

🔍 meiosis stages ... gametes higher meiosis

Each of the new cells gets only half the original number of chromosomes.

For example, in a normal human body cell there are 46 chromosomes – two sets of 23. The gametes each have only 23 chromosomes – a single set.

Figure 4 shows what happens when a cell divides by meiosis.

1 Before meiosis begins, each chromosome is copied exactly. The two copies stay attached to one another.

2 Each chromosome finds its partner from the other set.

3 The chromosomes separate from their partners, and the cell divides.

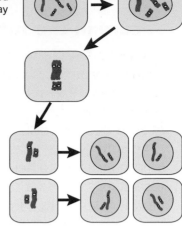

4 Each cell divides again, this time separating the two copies of each chromosome.

5 Four new cells have been produced, each with a single set of chromosomes.

FIGURE 4: Before and during meiosis.

QUESTIONS

4 How many sets of chromosomes are there in (a) a sperm cell (b) a fertilised egg?

5 Suggest which type of cell division – mitosis or meiosis – is used when a zygote divides to produce all of the cells in the body of an embryo. Explain your decision.

Comparing mitosis and meiosis

There are important differences between mitosis and meiosis (see Table 1).

TABLE 1: Comparing mitosis and meiosis

Mitosis	Meiosis
The cell divides once.	The cell divides twice.
Two cells are made.	Four cells are made.
The new cells have the same number of chromosomes as the original cell.	The new cells have half the number of chromosomes as the original cell.
This is how new body cells are made.	This is how gametes are made.
It happens in all parts of the body.	It happens only in the testes and ovaries.

QUESTIONS

6 Would it be possible for a cell with a single set of chromosomes to divide by (a) mitosis (b) meiosis? Explain your answer.

7 Occasionally, a cell in a woman's ovaries may fail to separate all the chromosomes properly when it divides by meiosis. One of her egg cells ends up with a complete, single set of chromosomes (which is right) plus one extra copy of chromosome 21 (which is not right).

(i) If this egg cell is fertilised by a normal sperm, how many chromosomes will the zygote have?

(ii) A person who develops from this zygote has Down's syndrome. How many chromosomes will there be in each of their body cells?

Did you know?

Sometimes a human sperm or egg cell ends up with two sets of chromosomes. This can result in a zygote with three sets of chromosomes. It will almost always die after a few days.

Stem cells

A cure in the future

Michael J. Fox is one of tens of thousands of people with an illness called Parkinson's disease. Although there are drugs that help to treat the disease, there is no cure. Michael has founded an organisation to help to fund researchers to find a cure. Hopes are that stem cells might prove to be that cure.

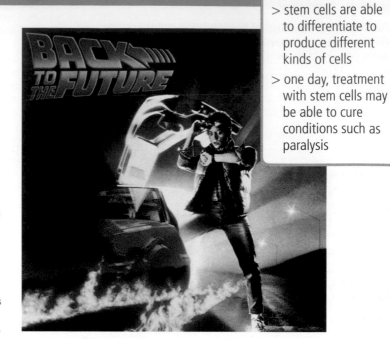

FIGURE 1: The actor Michael J. Fox now has Parkinson's disease, a condition in which sufferers have difficulty with muscle control.

What are stem cells?

Most cells in your body are specialised. This means they are **differentiated** for one particular function.

In animals, most cells differentiate soon after they are made. In plants, many cells can differentiate at any time during their lives.

Once a cell has differentiated, it cannot change its function. For example, a heart muscle cell is specialised to contract and help the heart to beat. It cannot change and become a brain cell.

However, some of your cells have not differentiated. These are **stem cells**. They can turn into different types of cell.

Stem cells divide by mitosis. Some of the new cells stay as stem cells, so there are always some stem cells left. The rest of the new cells differentiate into another type of cell.

FIGURE 2: These are stem cells taken from an embryo. Each cell is able to change into any kind of specialised cell found in the human body.

QUESTIONS

1 (i) Name three types of cells in the human body that have differentiated for a particular function.

(ii) What is the function of each?

2 When does cell differentiation happen in (a) animals (b) plants?

3 What is special about stem cells?

Did you know?

In 2010, Italian surgeons successfully restored sight to people whose eyes had been damaged in accidents, using stem cells.

Q stem cells GCSE

Embryo and adult stem cells

In in-vitro fertilisation (IVF), the fertilised egg divides over and over again to produce an embryo. For the first few days, all of these cells stay as stem cells. Each one has the potential to develop into any kind of cell in the human body. Stem cells can be obtained by taking one of these cells, without damaging the embryo.

Most body tissues in an adult human also contain stem cells. Unfortunately, in most tissues, there are so few stem cells that they are almost impossible to find. However, in the marrow of many bones there are lots of stem cells.

Figure 3 shows some of the cells that can be produced from these stem cells.

Using stem cells

Although scientists have known about stem cells for some time, it is only recently that they have been able to find ways of using them.

For example, imagine that someone has had an accident and snapped their spinal cord. No nerve impulses can now reach the lower part of their body, so they are paralysed. If stem cells could be put into their spinal cord, maybe these would be able to produce new nerve cells. Perhaps the person might be able to walk again.

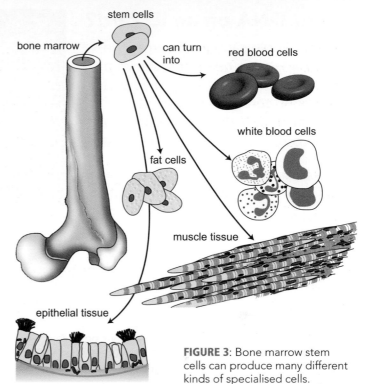

FIGURE 3: Bone marrow stem cells can produce many different kinds of specialised cells.

QUESTIONS

4 (i) Explain how embryo stem cells are obtained.

(ii) Discuss the arguments for and against this approach and explain whether or not you support it.

5 Explain how stem cells might one day be used to help a paralysed person to walk again.

Stem cells for the future

Many people with incurable diseases are ready to try anything to help them, even when doctors have no cure. This has led to the sale of 'stem cell cures' over the internet. Most of these have never been tested, and do not work. Sometimes people are prepared to travel to China or elsewhere, to receive 'treatment' for their disease. Unscrupulous doctors, or even people who are not qualified doctors, can sell their so-called treatments for huge amounts of money.

It is important that patients are protected from these untested treatments. On the other hand, it is also important that trials are carried out as they *could* lead to success. Finding the balance is quite tricky.

QUESTIONS

6 Discuss how trials of new and so far untested stem cell treatments should be regulated. How can a balance be achieved between keeping research moving forward, yet making sure that vulnerable people are not tricked into allowing potentially dangerous things to be done to their bodies?

Genes, alleles and DNA

Your DNA on an ID card?

Most people carry ID of some kind. We sometimes need to prove who we are. It has been suggested that everyone could have an ID card with their own DNA fingerprint on it. Each DNA fingerprint is completely unique.

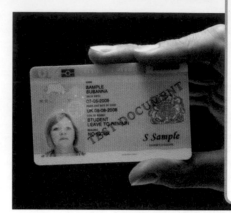

FIGURE 1: Can you think of any disadvantages of having an ID card that contains your DNA profile?

> ### You will find out:
>
> > a gene is a small section of DNA that helps to control a particular characteristic
> > there are different forms of most genes, called alleles
> > each person has unique DNA, which can be used to identify them

Genes

Genes and alleles

You have 46 chromosomes in each of your body cells – two sets of 23. Chromosomes are made of **DNA** (deoxyribonucleic acid). Each of these chromosomes contains hundreds of genes.

Each gene is a short length of DNA that gives instructions to the cell. Different genes give instructions for different characteristics. For example, there are genes for making the pigments that give your hair its colour. There are genes for making haemoglobin.

Most genes come in several different forms, called **alleles**. For example, a gene for hair colour might have one allele that gives black hair, another that gives brown hair, another that gives red hair and so on.

Genes and reproduction

In asexual reproduction, a cell divides by mitosis. It produces two new cells that have exactly the same chromosomes – and therefore exactly the same alleles – as the original cell.

In sexual reproduction, two different gametes fuse together. These gametes may contain different alleles. The zygote that is produced has a mixture of different alleles from both the parents.

> You have two sets of chromosomes in each cell.

> There is a gene for the same characteristic at the same place on the same chromosome in each set.

> The two genes might be the same allele or different alleles.

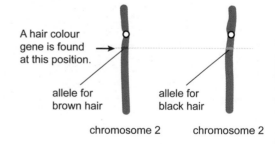

A hair colour gene is found at this position.

allele for brown hair | allele for black hair

chromosome 2 | chromosome 2

FIGURE 2: Genes have different forms, called alleles.

QUESTIONS

1 What is (a) a gene (b) an allele?

2 Explain why organisms produced by asexual reproduction contain the same alleles as their parent.

3 Explain why organisms produced by sexual reproduction contain a different mixture of alleles from either of their parents.

Q DNA structure GCSE ... asexual sexual reproduction

DNA fingerprinting

Everyone has DNA that is unique. Everybody's DNA is slightly different, unless they have an identical twin. Because of this, anyone can be identified from a sample of their DNA. Samples of DNA can be found in body cells in fluids such as saliva, blood and semen, or in hairs or bones. Forensic scientists (scientists who study objects related to crimes) often use DNA to identify a body, or to identify a person who was at the scene of the crime.

To make a DNA fingerprint, a sample of DNA is cut up into little pieces. The DNA fingerprint looks like a series of stripes. Each stripe represents a little piece of DNA. Each person's pattern of stripes is unique.

Figure 4 shows how DNA fingerprints can be used to identify who is the father of a child. All of the child's DNA has come from its father and its mother. So, every stripe in the child's DNA fingerprint must be present in either its mother's or its father's DNA fingerprint.

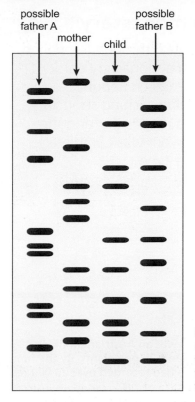

FIGURE 4: All bands in the child will either come from the real mother or the real father. Who is the father of the child?

FIGURE 3: DNA fingerprinting has made it possible to identify a person who has left only one or two hairs behind at a crime scene.

QUESTIONS

4 What is a DNA fingerprint?

5 Describe an advantage of using DNA fingerprints rather than ordinary fingerprints.

6 Explain how DNA fingerprints can help to identify the father of a child.

Analysing your DNA

There is a lot of scientific research going on into human DNA. Scientists have been able to link certain alleles with an increased risk of getting some diseases. For example, one particular length of DNA might be found to be more common in people who develop Parkinson's disease than in people who do not. This does not mean that if you have that DNA, you will develop the disease – it just means that you are slightly more likely to.

There are now several commercial companies, mostly in the USA, who will analyse your DNA (for quite a price) and tell you which diseases you might develop later in your life.

QUESTIONS

7 Discuss the arguments for and against people having their DNA analysed by commercial companies and offer your opinion.

Mendel

Science in a monastery

A major scientific breakthrough happened in the late 1800s in a monastery garden in Brno, now in the Czech Republic. No one was expecting exciting science to be produced in a monastery, and the importance of the work was completely unrecognised at the time.

FIGURE 1: This is the monastery where the science of genetics saw its first big breakthrough.

 ## Gregor Mendel

Gregor Johann Mendel was an Austrian monk. In 1856 he started doing experiments on inheritance in pea plants. By the time he did his last experiment, in 1868, he had recorded over 20 000 results.

Mendel's experiments

Mendel noticed that some of the pea plants growing in the monastery garden had purple flowers, while others had white flowers.

He decided to set up an experiment that involved crossing different pea plants – getting them to breed together. To do this, he took pollen containing male gametes from one flower and put it into another flower, so the male gametes could fertilise its female gametes.

Mendel pollinated purple flowers with pollen from white flowers. The resulting seeds were collected and planted. Figure 3 shows his results.

All the offspring had purple flowers. Mendel then tried breeding these together. Figure 4 shows his results.

FIGURE 2: Pea plants can have either purple flowers like these, or white flowers.

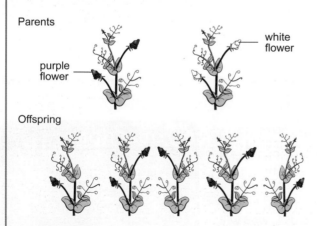

FIGURE 3: The result of crossing purple flowered pea plants with white flowered pea plants.

FIGURE 4: The result of crossing the offspring from the first cross with each other.

Q Mendel pea plant experiment

Mendel decided that these results meant that there were 'factors' for flower colour that were passed down from parents to offspring. Some 'factors' could be hidden in a particular generation, even though they must be there.

QUESTIONS

1 When Mendel did his experiments, no one knew about DNA, chromosomes, genes or alleles. What would Mendel's 'factors' be called today?

2 (i) Which 'factor' was 'hidden' in the purple flowered pea plants that resulted from Mendel's first cross?

(ii) Explain your answer.

Mendel's findings

Interpreting the results

Mendel decided that each plant must have two 'factors' for flower colour. In the original purple flowers, each factor was for purple colour. In the original white flowers, each factor was for white colour.

In the first set of offspring (the ones that all had purple flowers), each plant must have had a purple 'factor' from one parent and a white 'factor' from the other. Mendel thought that the purple factors were 'stronger' than the white factors, so all the flowers were purple. Because none of these plants had pale purple flowers, Mendel decided that each factor must be separately inherited. The factors did not 'blend' together.

The results in Figure 4 suggested to Mendel that, even though the white factors did not show up in the previous generation of plants, they must still be there. When these plants were crossed, some of the offspring inherited a white factor from both parents, so they had white flowers.

Recognising Mendel's work

Mendel presented his findings to the local natural history society in 1865. At that time no one knew anything about how cells divided. They did not know that chromosomes and genes existed, let alone DNA. It was not until 1900 – long after Mendel had died – that other scientists rediscovered his work.

Today, we know that Mendel's 'factors' are alleles of genes, made of DNA. The different varieties of his factors – for example the purple factors and white factors – are different alleles of genes. They are indeed inherited separately.

FIGURE 5: Mendel did not live to see his name become famous.

QUESTIONS

3 Suggest why no one recognised the importance of Mendel's work until after his death.

4 Explain the results of Mendel's first cross, shown in Figure 3, in terms of genes and alleles.

5 Explain the results of Mendel's second cross, shown in Figure 4, in terms of genes and alleles.

Luck or good judgement?

Mendel was lucky with his experiments. By chance – or was it good judgement? – he picked on characteristics that never do 'blend' together.

There are many other features that do 'blend' together. For example, with some other kinds of flowers, if you cross red ones with white ones, you obtain pink ones.

QUESTIONS

6 Imagine that Mendel had crossed purple flowers with white flowers and obtained pale purple flowers. How might this have affected his conclusions?

How genes affect characteristics

The sneezing gene

Many people sneeze when they look at the Sun, or a bright sky. It is called photic sneezing. The cause is an allele of a gene that you inherit from your parents. You only need one copy of this allele to make you a photic sneezer.

FIGURE 1: It is estimated that about one-third of the population has a photic sneezing allele.

Sex and alleles

How sex is determined

The nucleus of each body cell contains two sets of chromosomes. One chromosome in each set is a sex chromosome. Thus, each body cell contains two **sex chromosomes**.

There are two kinds of sex chromosomes, **X** and **Y**. In a female, both sex chromosomes are X. In a male, one is X and one is Y.

Female   XX

Male   XY

Alleles come in pairs

Many other characteristics are determined by a person's genes. Different forms of a gene are called **alleles**.

Example

In rabbits, there might be a gene for hair colour: one allele of the gene may give black fur, another allele may give white fur.

In a rabbit's cells, there are two complete sets of chromosomes. This means that there are two copies of each gene.

If you call the allele for black fur **B**, and the allele for white fur **b**, then there are three possible combinations: BB, Bb or bb.

Table 1 shows the fur colours that each combination of alleles gives.

You will notice, in Table 1, that one B and one b allele produce fur that is just as black as two B alleles. This is because the B allele is **dominant**. The b allele is **recessive**. When a dominant and a recessive allele are together, only the dominant one has an effect.

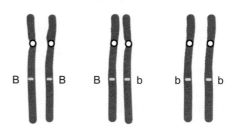

FIGURE 2: The three possible combinations of alleles for fur colour in a rabbit's cells.

FIGURE 3: A rabbit with black fur could have the alleles BB or Bb. A rabbit with white fur can only have the alleles bb.

TABLE 1: Allele combinations.

Allele combination	Fur colour
BB	black
Bb	black
bb	white

QUESTIONS

1 Explain why an animal or plant has two copies of each gene in its cells.

2 In guinea pigs, smooth coat is controlled by an allele H. Curly coat is controlled by an allele h. Allele H is dominant, and allele h is recessive.

(i) Write down the three possible combinations of alleles that a guinea pig could have.

(ii) For each of these combinations, say what kind of coat the guinea pig would have.

Some important terms used in genetics (Higher tier)

The combination of alleles that an organism has in its cells is called its **genotype**. The effect of those alleles on its characteristic is called its **phenotype**. So, you can say that the phenotype of a rabbit with the genotype BB is to have black fur.

> If the two alleles that an organism has are the same, the genotype is **homozygous**.

> If the two alleles are different, it is **heterozygous**.

QUESTIONS

3 Construct a table similar to Table 1. Change the headings of the two columns to Genotype and Phenotype. Then write 'heterozygous' or 'homozygous' next to each genotype.

4 Look back to question 2. What is the phenotype of a guinea pig that is heterozygous for this gene?

Intersex conditions

Most people are either clearly male or clearly female. However, it is estimated that as many as 1 in 100 people have some characteristics of both sexes. These are sometimes called intersex conditions.

If things go wrong when gametes are made, then a zygote could end up with a combination such as XXY or XYY or even just a single X chromosome and nothing else. Usually, if at least one Y chromosome is present, the person develops as a male, because it is the Y chromosome that determines maleness.

In other cases, someone might have a perfectly normal XY combination of chromosomes. However, their body cells do not respond correctly to hormones called androgens, produced as a result of having a Y chromosome. This is androgen insensitivity syndrome. A person with this condition develops as a female.

FIGURE 4: Crystals of androgen, photographed at x10 magnification.

QUESTIONS

5 Discuss the arguments for and against allowing a person with androgen insensitivity syndrome to compete in sports events as a female.

Inheriting chromosomes and genes

The sex ratio

Genetics tells us to expect approximately equal numbers of boys and girls to be born. However, in reality, slightly more boys are born than girls. The ratio is often somewhere around 106 boys to 100 girls.

Did you know?

In 2007, a woman gave birth to quintuplets in an Oxford hospital – and they were all girls.

FIGURE 1: How many of these babies do you think are girls and how many are boys?

How sex is inherited

Genetic diagrams for gender

Gametes have only *one* copy of each chromosome, rather than two.

A woman can make only one kind of gamete. All her egg cells will contain an X chromosome. A man makes two kinds of sperm cells. Half of them will contain an X chromosome and half will contain a Y chromosome.

There is an equal chance of a sperm with an X chromosome, or one with a Y chromosome, fertilising an egg. This can be shown in a **genetic diagram** – a standard way of setting out each of the steps involved as chromosomes or genes pass from the parents to their offspring.

The genetic diagram in Figure 2 shows the chances of a baby being a boy or a girl. The chances are equal.

Parents	XX female	XY male
Gametes	Ⓧ	Ⓧ and Ⓨ

Offspring

	Ⓧ	Ⓨ
Ⓧ	XX	XY

FIGURE 2: A genetic diagram for gender.

Genetic diagrams for alleles of genes

Genetic diagrams are also used to show how alleles of genes are inherited.

A male rabbit with alleles BB for fur colour will produce sperm cells that each contain one B allele. However, a male rabbit with alleles Bb will produce two different kinds of sperm cells. Half of them will have allele B and the other half will have allele b.

Imagine that this male rabbit mates with a white female rabbit. She has alleles bb, so each of her eggs contains a b allele. Any of the sperm cells might fertilise any of her egg cells.

This genetic diagram shows you to expect about half of the baby rabbits to have the alleles Bb and have black fur. The other half would be bb and would have white fur.

A complete genetic diagram includes the whole sequence of descriptions, not just the square showing the gametes and offspring.

Parents	male rabbit with black hair	female rabbit with white hair
	Bb	bb
Gametes	Ⓑ and Ⓑ	all Ⓑ

Offspring

	Ⓑ	Ⓑ
all Ⓑ	Bb black	bb white

FIGURE 3: Rabbit fur colour, example 1.

Chances

It is important to remember that a genetic diagram only shows chances.

Look at the diagram in Figure 3. If the female gave birth to 7 babies and 5 of them were white and 2 were black, it is not really a surprise. After all, if you tossed a penny 7 times, it would not be impossible to get 5 heads and 2 tails.

Q genetic diagram

QUESTIONS

1 Look back at Figure 3 on page 64. (i) Is the person these chromosomes came from a male or a female?

(ii) Explain how you can tell.

2 Which parent's gamete determines the sex of a child – the mother or the father? Explain your answer.

Remember
It is important to construct a complete genetic diagram or you will not receive full marks.

Probability in genetics (Higher tier)

It is usually a good idea to complete a genetic diagram by summarising the approximate chances of getting each of the different genotypes and phenotypes. For the rabbits in Figure 3, the genetic diagram could end like this:

expected genotype ratio   1Bb : 1bb

expected phenotype ratio   1 black : 1 white

Here is another example. What if both the male and female rabbit have the genotype Bb? Once again, a genetic diagram, as shown in Figure 4, will explain what might happen.

Parents	male rabbit with black hair	female rabbit with black hair
	Bb	Bb
Gametes	(B) and (b)	(B) and (b)

Offspring

	(B)	(b)
(B)	BB black	Bb black
(b)	Bb black	bb white

FIGURE 4: Rabbit fur colour, example 2.

 QUESTIONS

3 Look back at question 2. A guinea pig with a curly coat is crossed with a guinea pig with one H and one h allele.

(i) Construct a complete genetic diagram to predict what kind of offspring they will have.

(ii) State the ratios of the different genotypes and phenotypes you would expect to get.

Test crosses

You have probably realised by now that, if an organism has the characteristic that is controlled by a recessive allele, then you can immediately tell what alleles it has just by looking at it. However, if it has a characteristic that is controlled by a dominant allele, then there are two possible combinations of alleles that it might have. The only way to find out is by doing a breeding experiment. This is a **test cross**.

 QUESTIONS

4 Dalmatian dogs can have either black spots or liver-coloured spots. The allele for black spots, A, is dominant, and the allele for liver spots, a, is recessive.

A breeder has a male dog with black spots. She wants to find out what alleles he has.

(i) What sort of female dog should she breed the male dog with, to help find out?

(ii) Draw two genetic diagrams to explain the two possible outcomes of the cross.

(iii) Explain how the results of the cross would tell the breeder which alleles the black spotted male dog has.

How genes work

Rosalind Franklin – the Nobel Prize winner that never was

Rosalind Franklin was born in 1920. Given unusual opportunity for a woman at that time, she became an outstanding scientist. Her work was crucial for providing new information that allowed the much more famous male scientists Crick, Watson and Wilkins to go on and suggest a structure for DNA. They all won Nobel Prizes, but Rosalind did not. She died of cancer when she was only 37.

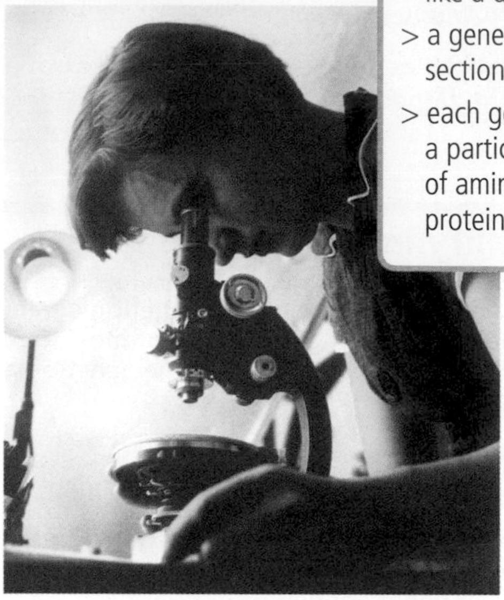

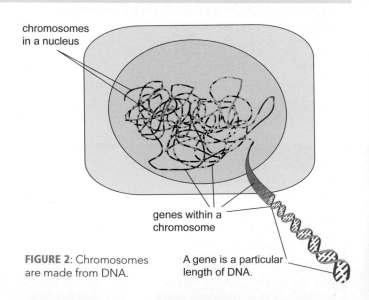

FIGURE 1: Rosalind Franklin.

You will find out:

> chromosomes are made of DNA, whose molecules are shaped like a double helix

> a gene is a small section of DNA

> each gene codes for a particular sequence of amino acids in a protein

DNA and codes

DNA

Each of the chromosomes in the nucleus of a cell is made of a very long molecule of **deoxyribonucleic acid, DNA** for short. The DNA molecules are twisted into a 'double helix' shape.

Chromosomes are divided into sections called **genes**. There are many genes on each chromosome. Each gene is a length of DNA that codes for a particular characteristic.

The genetic code

Genes store information as a type of code, called the genetic code. It is a bit like an instruction manual. It provides instructions to the cell about which proteins it should make.

Some of these proteins are enzymes.

> The enzymes control all the chemical reactions that go on in the cytoplasm of body cells.

> Each enzyme controls a different chemical reaction. There are thousands of different chemical reactions, so cells need thousands of different enzymes.

> Each enzyme has to be coded for by a gene. This explains why we have thousands of genes.

chromosomes in a nucleus

genes within a chromosome

FIGURE 2: Chromosomes are made from DNA.

A gene is a particular length of DNA.

QUESTIONS

1 What chemical are chromosomes made from?

2 Write a sentence that shows you understand the meanings of chromosome, gene and DNA.

3 What is the genetic code?

DNA and protein synthesis (Higher tier)

A protein molecule is made up of many amino acid molecules, all linked in a long chain. There are twenty different amino acids commonly found in proteins. Even a difference of just one amino acid in a chain of hundreds will make a different protein with different properties.

A gene is a length of DNA that carries the code for making one protein. Each strand of a DNA molecule is made up of a sequence of four bases, shown in different colours on Figure 2.

The sequence of bases in the DNA determines the sequence of amino acids in the protein that is made.

Did you know?

There are some lengths of DNA that are almost identical in all living organisms so far tested.

QUESTIONS

4 Explain, in your own words, how a DNA molecule determines which proteins are built up in a cell.

5 Explain how this directly affects the chemical reactions (metabolic reactions) that take place in the cell.

Mutations

Each DNA molecule is carefully copied before a cell divides by mitosis, so that the new cells can each have a copy. The cell machinery for copying DNA is very, very good. Only rarely are mistakes made.

However, mistakes do happen. The new DNA molecules are sometimes slightly different from the originals. This is a **mutation**, and it changes the code. The different DNA has become a different allele of the gene.

One small difference like this can make a massive difference to how the cell works. Sickle cell anaemia is an example.

Everyone has a protein called haemoglobin in their red blood cells. Most of us have genes that code for normal haemoglobin. Many people, though, have a slightly different version of the gene. It differs by only one base in the DNA, out of hundreds, but it is enough to cause a serious condition called sickle cell anaemia.

The faulty gene codes for a faulty version of haemoglobin. In conditions where oxygen is in short supply (such as when doing a lot of exercise), it makes the red blood cells go so badly out of shape that they clump together and get stuck in blood capillaries.

FIGURE 3: Sickled red blood cells. How and why does the shape of these red blood cells differ from normal?

QUESTIONS

6 What is a mutation?

7 Explain how a mutation in a DNA molecule can affect the structure of a protein.

8 Suggest whether sickle cell anaemia can be passed on from a parent to their children. Explain your answer.

Genetic disorders

You will find out:
> how polydactyly and cystic fibrosis are inherited
> how to interpret family trees

Inbreeding and inherited diseases

People who breed cats for showing often breed quite closely related cats together. They try to produce kittens with exactly the right features. This increases the chances of bringing two rare recessive alleles together. If these recessive alleles cause a disease, then a kitten that inherits two of them will have that disease. Abyssinian cats, for example, occasionally inherit a disease that causes blindness.

FIGURE 1: How could cat breeders reduce the risk of kittens inheriting diseases caused by rare recessive alleles?

Did you know?

Marriages between human cousins have quite a high chance of bringing together two harmful recessive alleles.

Inherited disorders

An inherited disorder is something that is not right with the body and that is caused by faulty genes. As their name suggests, inherited disorders are inherited from a person's parents.

There are quite a lot of different inherited disorders that people can have. Polydactyly and cystic fibrosis are two examples.

Polydactyly

Polydactyly means 'having many fingers'. It is a condition in which a person has more than five fingers on their hands, or more than five toes on their feet.

Polydactyly is caused by a dominant allele. This means that you only need to inherit one allele in order to have this condition. The family tree, in Figure 3, shows how polydactyly was inherited in one family.

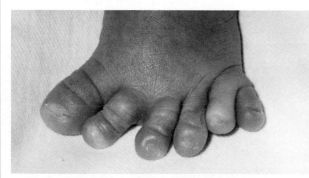

FIGURE 2: Polydactyly usually does not cause a person any problems.

Cystic fibrosis

Cystic fibrosis is a disorder of cell membranes. It affects the lungs, and also the pancreas (a gland that makes digestive enzymes). It is a serious condition. People with this disorder must have regular treatment all their lives.

Cystic fibrosis is caused by a recessive allele. A person with the disorder will have two recessive alleles.

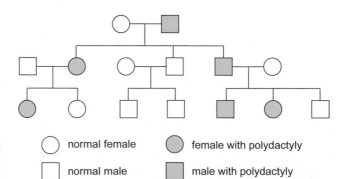

○ normal female	● female with polydactyly
□ normal male	■ male with polydactyly

FIGURE 3: Example of a family tree for polydactyly.

QUESTIONS

1 Explain why both of the parents of a person with cystic fibrosis must have an allele for the disorder.

2 Explain why an inherited disorder cannot be cured.

Q genetic disorder inheritance

Inheriting cystic fibrosis

The family tree, in Figure 4, shows the members of a family who had, or did not have, cystic fibrosis.

Parents A and B must have been very surprised and upset when their third child was born with cystic fibrosis. They must have wondered how this could have happened when they were both perfectly healthy. They did not know that they were both **carriers** of the recessive allele for this disorder.

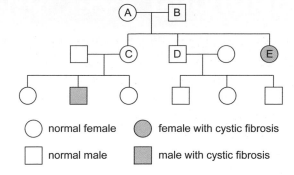

○ normal female   ● female with cystic fibrosis

□ normal male   ▨ male with cystic fibrosis

FIGURE 4: Example of a family tree for cystic fibrosis.

QUESTIONS

3 Suggest why a person with two different alleles for a disorder caused by a recessive allele is said to be a 'carrier'.

4 This question refers to Figure 4.

(i) Using the symbol for the recessive allele and F for the dominant allele, write down the genotype of person E.

(ii) Draw a complete genetic diagram to show how person E inherited cystic fibrosis from her parents A and B.

(iii) Each time A and B have a child, what is the probability that it will have cystic fibrosis?

(iv) What must be the combination of alleles of person C and her partner?

FIGURE 5: What would a genetic counsellor tell parents A and B about the chances of their next child having cystic fibrosis?

Embryo screening

If an embryo is produced by IVF, the embryo remains in a dish for a few days before it is put into the mother's uterus. This gives an opportunity to test the embryo's DNA, to find out if it has alleles for a serious disorder.

It is expensive to do these tests, so most parents do not have them done. However, parents who know they are carriers might decide to have their next child by IVF. They could then choose an embryo that did not have even one copy of this allele, let alone two copies.

QUESTIONS

5 Suggest at least two reasons why many parents who have IVF do not choose to have their embryos screened.

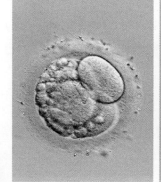
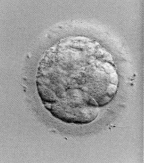

FIGURE 6: These are tiny human embryos that have been produced by IVF. Doctors and parents have to choose which one is to be put into the mother's uterus and which to discard.

Preparing for assessment: Applying your knowledge

To achieve a good grade in science, you not only have to know and understand scientific ideas, but you need to be able to apply them to other situations and investigations. These tasks will support you in developing these skills.

✳ Inherited characteristics

Will (14) has a brother, Andy (11), and a sister, Ellie (16). All three are blonde, though Ellie has brown eyes, unlike her brothers' blue eyes. Two are slim, but Will has a slightly stockier build.

They all inherited characteristics from their parents, of course. The nuclei of the sperm and egg cells that formed them contained chromosomes. Each chromosome contains genes, which determine their characteristics.

Some of these characteristics are easy to see. Their mother has brown eyes and their father has blue eyes (not that sons always inherit characteristics from fathers and daughters from mothers). However, brown eyes are dominant over blue, so why do the three children not all have brown eyes? This is because we all carry two sets of genetic data for each characteristic, so we have two copies of each gene. The two copies might differ with one being a dominant allele and the other a recessive allele. If so, we display the one that is dominant, but still carry the other. We pass one of these on to our children, and it is just as likely to be the recessive one as the dominant one. The mother was carrying a 'blue eye allele' (from her mother) as well as a 'brown eye allele' and it was the first of those that she happened to pass on to the boys.

With other features it is not quite so obvious. Brown hair is dominant over blonde, so it is perhaps surprising that all three children are blonde, since their father has brown hair and their mother is fair. However, a look in the family photo album reveals that the father was fair when he was a child, and one of his grandfathers was fair, so he may be carrying a 'blonde hair gene'.

Task 1

Read the information to see how the children compare with their parents.

(a) Why is it that certain features of the children are similar to those of their parents?

(b) How does the information travel from parent to child?

(c) Why is the mother somewhat surprised to have three blonde children?

Task 2

Was it inevitable that at least some of the children would have blonde hair? Explain the reasoning behind your answer.

Task 3

Will is rather stockier in build than his brother and sister. He is like both his grandfathers in this respect. The other two are slimmer and slightly taller.

Suggest two reasons why Will is different from his brother and sister in this way.

Task 4

The inheritance of characteristics is a complex system. It means that humans vary, as do all organisms that reproduce sexually. This variation rarely amounts to anything significant – blondes are no better (or worse) at surviving than people with any other hair colour.

Explain why living things vary.

Maximise your grade

Answer includes showing that you...	
E	can describe how organisms that form as a result of sexual reproduction have a mix of the parents' characteristics.
	know that sexual reproduction gives rise to variation.
	can describe how different genes control different characteristics.
	know how genetic information is transferred from parent to child.
C	understand how characteristics are passed on from parent to child.
	interpret genetic diagrams.
	understand how genetic information is passed on through generations.
A	can explain clearly and in detail how genetic information is passed on through generations.

Fossils

You will find out:
> fossils provide evidence about early life-forms
> fossils of life-forms with soft bodies are not often found

When the Antarctic was hot

Some of the rocks in the Antarctic contain fossils of tropical plants. This shows that Antarctica was once a hot, humid land. Studying fossils increases knowledge about the history of the Earth.

Did you know?

Fossils of organisms that used to live in the sea have been found on Mount Everest. The rocks that make up Everest used to be deep under the oceans.

FIGURE 1: Fossils show that Antarctica must have had a tropical climate in the past.

 Fossils

Early life-forms

Fossils are the remains of plants and animals that lived millions of years ago. There are a number of ways that these remains can be preserved.

> Some hard parts of animals may not decay easily.

Examples: Bones and teeth of vertebrates remain for much longer than other parts.

> Some parts of organisms do not decay because conditions are not suitable for decay organisms.

Example: Mammoths that died more than 20 000 years ago have been found preserved in permanently frozen ground in northern Asia.

> Parts of an organism may be replaced by other materials as they decay.

Examples: Bones, shells and even soft body parts can be replaced by hard minerals to form a fossil.

> Traces of an organism may be preserved in rocks.

Examples: Burrows and footprints may be preserved as prints, called trace fossils, as the sediment in which they were made hardens and turns to rock.

Comparing fossil remains with present day plants and animals can show how much (or how little) organisms have changed over time.

An incomplete record

Most organisms do not form fossils when they die. Most of them just decay. This is especially true for things with soft bodies, such as jellyfish or worms. Sometimes, fossils of soft-bodied organisms are found. More often, though, scientists and archaeologists have to rely on traces from them, such as the burrow or the marks that the organism made when moving across the seabed.

This means that the fossil record is very incomplete. There are the remains of only a tiny proportion of all things that once lived on Earth.

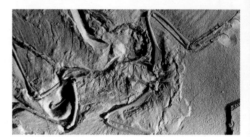

FIGURE 2: The animal that produced this fossil lived 150 million years ago. It has features of both reptiles and birds.

FIGURE 3: This may not look much like a fossil, but the white stripe is the remains of a worm burrow, millions of years old.

QUESTIONS

1 What is a fossil?

2 Suggest why fossils of organisms such as dinosaurs are likely to be found much more often than fossils of jellyfish.

Q fossil formation

The fossil record

Fossils provide evidence that suggests that all species of living things that exist today have evolved from simple life-forms. The fossil record for most species is incomplete. However, the fossil record for the horse does show how it has evolved over the last 50–60 million years. Table 1 shows the key stages in the evolution of the horse.

TABLE 1: Fossils provide clues about how modern horses evolved.

Time (years ago)	Height (m)	Appearance (an idea)	Bones of front leg	How it lived
1 million	1.6	modern horse		> on grassland > very fast runner
10 million	1.0			> in very dry places > fast runner
30 million	1.0			> in dry conditions > relied on speed to escape being eaten
40 million	0.6			> in dryer conditions > needed to be able to run away from predators
60 million	0.4			> on soft ground, near water > its feet could support its weight without sinking into mud

QUESTIONS

3 What has happened to the size of horses as they have evolved?

4 How has the horse's leg and foot changed?

5 Suggest how natural selection may have brought about the changes in horses' feet over time. Hint: use the last column in Table 1.

How did life evolve?

There is not enough valid or reliable evidence to be certain how life on Earth first evolved. At best, scientists can come up with suggestions, using the few pieces of evidence that there are.

Some scientists think that life first evolved in the seas, from the mix of chemicals that was present on the early Earth. Perhaps it happened in hydrothermal vents – deep cracks in the ocean floor from which hot mixes of chemicals pour out into the water. Others think that life, or the non-living substances from which life developed, may have been brought to Earth on meteorites. Some of these have been found to contain chemicals such as amino acids.

It is interesting that all living things – from the simplest archaea and bacteria to complex animals such as ourselves – have many things in common. For example, they all contain DNA, and the way that the code on the DNA works and is translated in all these organisms is almost identical. This is a very strong piece of evidence that all life on Earth has evolved from the same common ancestors.

QUESTIONS

6 If life was found on Mars, do you think it would have DNA? Explain your answer.

Extinction

Mass extinction

Dinosaurs lived between 225 and 65 million years ago. They ruled the Earth. Some were herbivores. Others, such as velociraptors, were fierce carnivores, killing their victims with their powerful jaws. Then, something happened to the world, and they all became extinct.

FIGURE 1: A giant meteorite may have changed the environment on Earth so that dinosaurs could no longer survive.

What is extinction?

Life on Earth began about 3.8 billion years ago. At first, everything lived in the sea. Living things first moved onto dry land about 550 million years ago.

Earth was very different then. As it slowly changed, many different plants and animal species died out. Then they were replaced by new types. When a species completely dies out, and there are no more individuals left, the species has become **extinct**.

Living things become extinct for several different reasons.

> The environment in which they live might change – either slowly, or suddenly as the result of a catastrophic event.

> All the animals in a species might be eaten by a new predator.

> A new disease might kill them off.

> A new, more successful, competitor might move in to the habitat.

The dodo lived on the island of Mauritius. Dodos became extinct about 300 years ago when people arrived on the island. The people brought dogs, pigs and rats. These predators, and humans themselves, killed and ate dodos and their eggs.

FIGURE 2: Dodos were not able to fly. They are now extinct.

QUESTIONS

1 (i) Give an example of an animal that is now extinct.

(ii) Explain what caused it to become extinct.

Causes of extinction

Environment change

This is when the place where a species lives changes – for example the temperature might increase.

It means that the individuals living there are not adapted to the new conditions and so, unless they can adapt or move away, they will die out.

Building cities and towns, cutting down rainforests or causing pollution can all lead to species becoming extinct because the natural habitat has been changed. A species will become extinct unless individuals can cope with the changes.

Did you know?

Since the year 1500, more than 190 species of birds are known to have become extinct.

Q extinction GCSE ... dodo

New predators

When a new predator is introduced into a habitat, it eats prey that may not be adapted to get away. This can lead to the extinction of the prey species. Often, this new predator has been humans.

New diseases

Plants and animals are naturally immune to lots of different diseases. However, if a new, fatal, disease is introduced to a habitat then they may not have immunity. The disease could kill all the individuals of a particular species.

A more successful competitor

If there is a limited amount of food in a habitat and two different species are competing for it:

> the better adapted species is more likely to survive

> the less well adapted one may become extinct.

FIGURE 3: The St Helena olive is now extinct, destroyed by a fungal disease.

QUESTIONS

2 Explain how human activities cause other species to become extinct.

3 Before humans arrived in New Zealand, there were no mammals there and no large predators. Many species of birds were flightless.

(i) Suggest why many birds in New Zealand had evolved to become flightless.

(ii) When humans arrived, they introduced many different animals, including cats, rats and possums. Cats are predators. Rats eat eggs. Possums eat leaves, fruits and flowers. Suggest why many species of birds in New Zealand became extinct.

Mass extinctions

The fossil record suggests that there have been several periods in Earth's history when huge numbers of species became extinct over a relatively short period of time. It is thought that these were probably caused when a sudden, catastrophic event produced massive changes on Earth.

FIGURE 4: The main mass extinctions thought to have occurred on Earth.

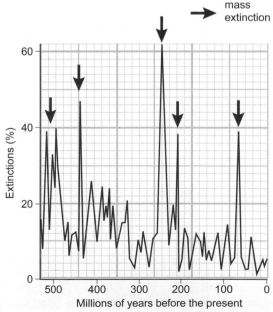

QUESTIONS

4 Research one of the mass extinctions shown in Figure 4.

(i) What is thought to have caused this extinction?

(ii) Explain why it is difficult to be absolutely sure about the exact cause.

New species

The St Kilda house mouse

The island of St Kilda lies 86 kilometres off the west coast of Scotland. People used to live on St Kilda. House mice also lived on the island. These mice were larger than mainland house mice. The mice became extinct when the last people left the island in 1930.

FIGURE 1: Animals on islands often evolve to become larger than those on the mainland. No one really knows why.

 ## Species

What is a species?

Many biologists define a species as a group of organisms that share similar characteristics, and that can breed together to produce fertile offspring.

So, to create a new species from an old one, there has to be something to stop individuals in the two groups from breeding together.

Isolation

Siberian tigers and Bengal tigers belong to the same species. Today, Siberian tigers live only along a strip of land on the far eastern coast of Russia. They never come into contact with Bengal tigers, which mostly live in India, Bangladesh and neighbouring countries. The two kinds of tigers are isolated from one another.

If this situation went on for long enough, the two kinds of tigers might gradually become more and more different from one another. Eventually, they might be so different that – if they came into contact with one another – they might not be able to breed together.

FIGURE 2: Since the beginning of the 20th century, tigers have disappeared from much of their range.

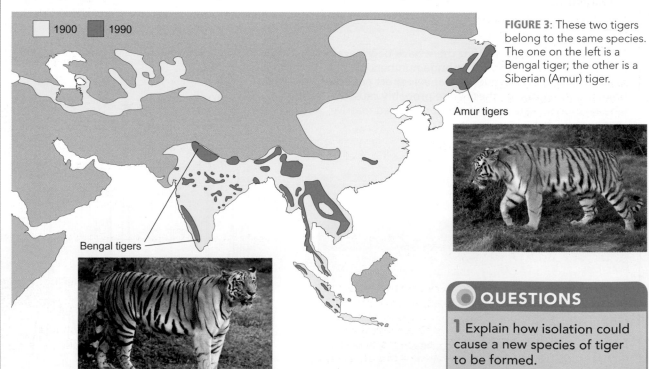

1900 1990

Amur tigers

Bengal tigers

FIGURE 3: These two tigers belong to the same species. The one on the left is a Bengal tiger; the other is a Siberian (Amur) tiger.

QUESTIONS

1 Explain how isolation could cause a new species of tiger to be formed.

Q evolution of species GCSE

How new species arise (Higher tier)

Many new species arise through a series of steps.

To illustrate this, here is an example of how two separate species of lizard evolved from one single species.

1 Geographical isolation

Different groups of organisms that belong to the same species become separated from one another.

A few lizards might drift away from the mainland on a floating log and end up on an island. They start to breed there and build up a population of lizards on the island.

2 Genetic variation

In both the mainland lizards and the island lizards, there are many different alleles of genes.

There might be genes that control the length and strength of the lizards' hind legs. Some alleles give them legs that help them to jump higher than lizards with other alleles.

3 Natural selection

In each population of lizards, natural selection takes place. Those lizards with alleles that give them a survival advantage are more likely to reproduce and pass on those alleles to their offspring.

For example, on the mainland, there are many predators of lizards. The individual lizards that can jump higher can escape by leaping into trees. On the island, however, there may not be any predators. There is no selection for lizards with strong legs.

4 Speciation

Over time, more and more differences build up between the two populations of lizards, because of different selection pressures. Eventually, even if they are reunited with each other, they do not breed successfully together.

The lizards may have become such different shapes that they do not recognise one another as potential breeding partners.

FIGURE 4: These two species of lizards have evolved on different islands in the Caribbean. They probably had a common ancestor.

Did you know?

New species are evolving today, but probably not as quickly as other species are becoming extinct.

QUESTIONS

2 Explain why islands often have species of animals and plants that are not found anywhere else.

Other kinds of isolation

Geographic isolation is not the only way that two populations of the same species can become isolated. For example, one kind of fruit fly can breed on either apples or hawthorn fruits. It seems that those that hatched on apples always go back to apples to breed, while those that hatched on hawthorn fruits always go back to hawthorn fruits.

QUESTIONS

3 Suggest how the 'apple flies' and 'hawthorn flies' might eventually develop into new species.

 species formation process ... factors influence evolution

Checklist B2.5–2.8

To achieve your forecast grade in the exam you will need to revise

Use this checklist to see what you can do now. Refer back to the relevant topics in this book if you are not sure. Look across the three columns to see how you can progress. **Bold** text means Higher tier only.

Remember that you will need to be able to use these ideas in various ways, such as:

> interpreting pictures, diagrams and graphs
> applying ideas to new situations
> explaining ethical implications
> suggesting some benefits and risks to society
> drawing conclusions from evidence you are given.

Look at pages 278–299 for more information about exams and how you will be assessed.

To aim for a grade E	To aim for a grade C	To aim for a grade A
Recall that protein molecules are long chains of amino acids.	List some examples of the different types of protein found in the human body.	Explain, with examples, how the shape of a protein can affect the way it behaves.
Recall that enzymes are proteins.	Explain that enzymes act as catalysts, increasing the rate of chemical reactions.	
Know that food is digested in the gut by enzymes.	List the different types of enzyme involved in digestion and where they are produced.	Explain how different parts of the digestive system create optimum conditions for enzyme activity.
Know that humans put enzymes to uses in the home and industry.	Describe at least one domestic and one industrial use of enzymes, including its advantages and disadvantages.	Describe in detail how the digestive function of enzymes has applications in the home and industry.
Know that aerobic respiration releases energy from glucose.	Describe the process of aerobic respiration using a word equation.	Explain the role of mitochondria in aerobic respiration.
Recall one way in which the human body uses energy.	Describe a range of ways in which the human body uses energy.	
Know that, if oxygen is not available, respiration is described as anaerobic.	Explain that anaerobic respiration produces a waste product called lactic acid.	**Explain how anaerobic respiration releases less energy than aerobic respiration and results in an 'oxygen debt'.**
Know that there are two types of cell division: mitosis and meiosis.	Describe the key difference between mitosis and meiosis.	**Explain what happens to the chromosomes during meiosis.**

To aim for a grade E

Recall that undifferentiated cells are called stem cells.

Recall that Gregor Mendel carried out experiments on the inheritance of colour in pea plants.

Know that certain characteristics are passed on from one generation to the next.

Know that many characteristics are controlled by single genes.

Recall that a gene is a small section of DNA.

Recall that sex is genetically determined.

Know that some disorders can be passed on from one generation to the next.

Recall that fossils are the 'remains' of organisms from many years ago.

Know what is meant by the term extinction.

Give a basic definition of the term species.

To aim for a grade C

Describe how the process of cell differentiation varies between plants and animals.

Describe how Mendel's work led him to propose the idea of 'factors' for flower colour being passed from parents to offspring.

Describe the results of simple genetic crosses by interpreting genetic diagrams.

Describe the relationship between genes and alleles.

Explain that the DNA of each individual is unique.

Describe the chromosome combinations that determine sex.

Describe at least one example of a genetic disorder.

Describe some of the ways in which fossils are formed.

Give some of the reasons that organisms become extinct.

Describe how a new species can be formed by isolation.

To aim for a grade A

Explain how stem cells can be obtained during IVF treatment.

Explain why the full significance of Mendel's work was not appreciated until after his death.

Explain the inheritance of single characteristics by constructing genetic diagrams.

Use the terms homozygous, heterozygous, phenotype and genotype.

Explain the difference between dominant and recessive alleles.

Explain how different genes result in the production of different proteins.

Use a genetic diagram to explain the inheritance of cystic fibrosis and polydactyly.

Explain the importance of the fossil record.

Explain in detail the possible reasons for extinction.

Explain the four key requirements in the formation of a new species.

1. As part of his training, a footballer runs on a treadmill. After an injury, he was unable to train for several months. He then began training again.

His oxygen consumption was measured as he ran on the treadmill at different speeds. The graph shows the footballer's oxygen consumption when he resumed training, and after training for 8 weeks.

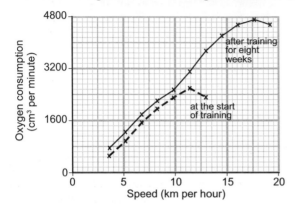

AO3 **(a)** Describe how the footballer's oxygen consumption changed as he ran faster, when he first began training. [2]

AO2 **(b)** Explain why the footballer's oxygen consumption changed when he ran faster. [2]

AO3 **(c)** Describe how the footballer's oxygen consumption, after training for 8 weeks, differed from his oxygen consumption when he first began training. [3]

AO3 **(d)** Use the information in the graph to explain how training can help a footballer to perform better. [2]

2. Alzheimer's disease is an illness in which a special type of neurone in the brain gradually dies. People with Alzheimer's disease lose the ability to remember.

In 2011, researchers took stem cells from a person's skin. They found a way of making the stem cells divide and differentiate into the special neurones that are lost in Alzheimer's disease. They hope that this could lead to a treatment for the disease.

AO1 **(a)** Explain what is meant by each of these terms:
(i) neurone **(ii)** stem cell **(iii)** differentiate [3]

AO1 **(b) (i)** Name the process by which a stem cell divides. [1]
(ii) Describewhat happens during this process. [2]

AO2 **(c)** The researchers have used the stem cells to make large numbers of the special neurones. They can grow the neurones in the laboratory, and test the effect of many new drugs on them.

Suggest why this is better than testing new drugs on people with Alzheimer's disease. [3]

3. Lipase breaks down oil to fatty acids and glycerol. A student learned that bile salts help lipase to do this, because they break up the oil into tiny droplets.

The student did an investigation to see if bile salts really do help lipase to digest oil. The diagram shows how she set up her experiment.

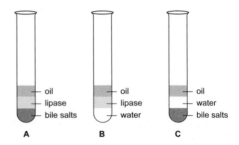

She tested the pH of the contents of the tubes at the start of the experiment and then at 5-minute intervals for 20 minutes. The table shows her results. Decreasing pH indicates more acid produced.

Tube	Time (minutes)				
	0	5	10	15	20
pH in A	7.0	6.8	6.6	6.3	6.1
pH in B	7.0	6.9	6.7	6.6	6.5
pH in C	7.0	7.0	7.0	7.0	7.0

AO2 **(a)** Suggest why the student added water to tubes B and C. [1]

AO2 **(b)** The student used a pH meter to measure the pH in the test tubes. Explain why this was better than using universal indicator paper. [1]

AO3 **(c)** Explain why the pH changed in tube A. [2]

AO3 **(d)** Compare the pH changes in tube B with the pH changes in tube A. [2]

AO2 **(e)** Use the information above, and your own knowledge of how enzymes work, to suggest an explanation for the differences between the results for tube A and tube B. [2]

✱ WORKED EXAMPLE – Foundation tier

In this question you will be assessed on using good English, organising information clearly and using specialist terms where appropriate.

Describe how fossils can be formed, and discuss the extent to which fossils provide information about the history of
life on Earth. [6]

Fossils are made when things die and their bones turn into rocks. They get buried deep underground.

Fossils tell us what kind of animals lived a long time ago. Fossils in deep-down rocks are older than ones near the surface, so we can see how long ago they lived. We can see how animals changed over time. Horses used to be very small and have lots of toes, but now they are bigger and have hooves.

Some fossils are animals that don't live on Earth any more, so they show us that some things have become extinct. The dinosaurs probably became extinct because a big asteroid crashed into the Earth. Scientists know about this by looking at the rocks that the dinosaur fossils are found in.

How to raise your grade!
Take note of these comments – they will help you to raise your grade.

This is true, but there are many other ways that fossils can be formed. The candidate could have mentioned that they are formed from the hard parts of animals that do not decay easily; from parts that have not decayed because the conditions were not right for decay organisms; or as traces of organisms such as footprints or burrows.

These are good points. However, the candidate wrote only about animals – there are also fossils of plants and other organisms. The candidate has included a specific example. This is a good idea.

Gaining information about extinction is an important point to include. Another good example is given here.

The candidate has included a range of ideas, but none are described in detail. For example, more descriptions of how fossils are formed would gain more marks.

The candidate does not mention that most organisms do not form fossils: there must have been many species that are not known about from fossils.

The answer is well structured - this candidate planned the answer before starting to write. Each paragraph deals with different ideas. Spelling and grammar are good, and the answer is easy to understand.

This answer will receive four marks.

1. An industrial enzyme, used all over the world in food manufacture, converts glucose into fructose. The enzyme is produced by a bacterium.

AO1 **(a)** What is the scientific term for an enzyme that converts glucose into fructose? [1]

AO2 **(b)** One particular enzyme has an optimum temperature of 55–60 °C. Suggest why this is more useful to the manufacturers than an enzyme with an optimum temperature of 37 °C. [2]

AO2 **(c)** Fructose tastes sweeter than glucose. Explain how using fructose instead of glucose, to sweeten food, could help a person to lose weight. [2]

(d) In the USA, fructose made from sweet corn, called high fructose corn syrup, is used in many food products. A study in 2002 investigated a possible link between the use of high fructose corn syrup, and the increase in the numbers of obese people in the USA. The graph shows some of their results.

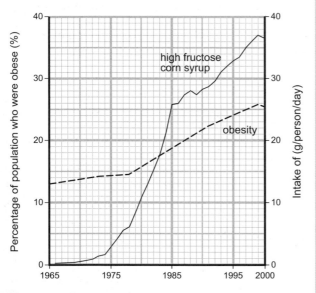

AO3 **(i)** Describe the change in the percentage of obese people between 1961 and 2000. [2]

AO3 **(ii)** Describe the change in consumption of high fructose corn syrup between 1961 and 2000. [3]

AO3 **(iii)** Evaluate whether these data support the idea that consumption of high fructose corn syrup may be a reason for the changes in the percentage of obese people in the USA. [3]

2. The leg muscles of a jogging person use energy.

AO1 **(a)** Describe how the cells in a jogger's muscles release energy from glucose. [2]

(b) When a person runs fast, their muscle cells may produce lactic acid. The graph shows how the concentration of lactic acid in a person's blood changes as they run at different speeds.

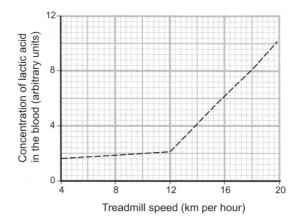

AO1 **(i)** Explain why lactic acid is produced in muscle cells. [3]

AO2 **(ii)** Suggest an explanation for the change in the gradient of the graph at a running speed of just below 12 kilometres per hour. [2]

AO3 **(iii)** When a person stops running, they may continue to breathe faster than normal, for several minutes. A person who has been running at 18 kilometres per hour for five minutes is likely to continue breathing faster for a longer period of time than a person who has been running at 8 kilometres per hour.

Use the information in the graph, and your own knowledge, to explain why. [4]

3. Chromosome 7 contains a gene that determines the number of fingers that grow on the hand as an embryo develops. A rare, dominant allele of this gene causes extra fingers to grow. This is called polydactyly.

AO1 **(a)** Explain the meaning of the term allele. [1]

AO1 **(b)** Explain the meaning of the term *dominant*. [1]

AO2 **(c)** Two parents both have polydactyly. They have two children, neither of whom have polydactyly. Explain how this could happen. [4]

AO1 recall the science   AO2 apply your knowledge   AO3 evaluate and analyse the evidence

✳ WORKED EXAMPLE – Higher tier

In this question you will be assessed on using good English, organising information clearly and using specialist terms where appropriate.

The Amazon rainforest in South America covers an enormous area. It contains hundreds of species of birds. Over time, new species have arisen from earlier species. The black-necked aracari and the green aracari are thought to have evolved from a single species in the rainforest a very long time ago.

The climate is wet and warm. Between one and two thousand million years ago, the climate in South America fluctuated. There were warm, wet periods when the rainforest covered the whole area, and dry periods when the rain forest grew only in small patches.

Use this information to explain how the black-necked aracari and the green aracari may have arisen from a single species. [6]

New species arise when one species splits into two groups and the two can't breed with one another any more.

This could have happened with the aracari. Long ago, when there was forest all over South America, one species of aracari might have lived in the forest. Then the climate got drier, so the forest was split into many small pieces. The aracaris living in one patch of forest were isolated from the aracaris living in other patches. This is called geographical isolation.

There would have been genetic variation in each group of aracaris. There may have been different selection pressures in different patches of forest. Perhaps in one forest patch the birds with black necks had a better chance of survival. In another patch it was the green ones that survived best. Over time, the aracaris in different patches may have evolved to become different from another.

When the climate changed, the forest expanded and the patches joined up into one big forest, so now all the aracaris could meet up with each other. But the green and black-necked ones had become so different that now they couldn't breed with one another any more. They had become two different species.

How to raise your grade!

Take note of these comments – they will help you to raise your grade.

This is a very good answer, and receives six marks. It is carefully planned, and follows a logical sequence. The candidate has used their knowledge of how new species arise, and applied it very clearly to this unfamiliar situation. Spelling and grammar are excellent, and the answer is easy to read and to understand.

This is a good way to begin the answer. It gives an overview of what the answer is will say.

This is also very good. The candidate uses knowledge of geographical isolation, and applies it to this particular example.

The fact that individual aracaris would have had slightly different genes is an important point. Again, the candidate is using their knowledge about natural selection and applying it clearly.

This is a good way to end the answer. The argument is clear, and it explains exactly how the two species came to be different.

Chemistry C2.1–2.3

What you should know

Matter

Everything around us is made up of particles. The simplest particles are atoms. Atoms can join together to make molecules.

Different elements provide different types of atoms. Together these atoms make all known substances.

Elements can be represented using symbols.

 Name any five elements and give their chemical symbols.

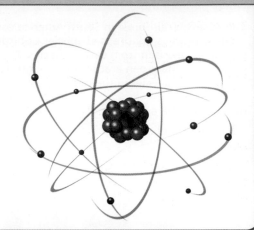

Using different materials

When deciding which substance to use for a particular job, you need to consider its properties. Different substances have different properties.

Metals are hard, strong and malleable. They conduct electricity and heat. This makes them useful for many different applications.

 Look around you and find five different uses of metals. Suggest which properties are important in each use.

Describing chemical reactions

When different substances react together, they form new substances.

Symbols, formulae and equations can be used to describe these reactions.

 When hydrogen and oxygen burn together, they make water. This reaction can be represented using: hydrogen + oxygen → water

Write the symbols for the two elements and the formula for the compound. Describe the three substances in as much detail as you can.

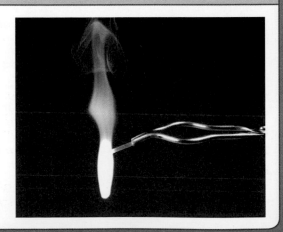

You will find out

Structure and bonding

> Ionic compounds can be described using formulae.

> Diagrams can represent the electronic structures of ions present in an ionic compound, covalent bonds in molecules and bonding in metals.

> The charge on simple ions relates to the group number of the element in the periodic table.

> Non-metal elements join together by sharing electrons.

> Metal elements join with non-metal elements by transferring their electrons to achieve a noble gas electronic structure, forming particles called ions.

Structure, properties and uses of substances

> The uses of substances depend on their properties.

> A substance's properties depend on the bonding between atoms and how they are arranged.

> Substances consisting of simple molecules have low melting points. Those with covalent structures have very high melting points. Ionic substances have giant lattices and high melting points.

> Metals conduct electricity by the transfer of electrons.

> A polymer's properties depend on what it is made from and how it is made.

> Understanding structure has led to the development of new materials.

Atomic structure, analysis and quantitative chemistry

> An atom's atomic number is the number of protons in its nucleus and this equals its number of electrons.

> An atom's mass number is the number of protons and neutrons in its nucleus.

> Atoms of different elements have different masses.

> Forensic scientists use a wide variety of instruments to identify substances present and their amounts.

> The mole is a means of measuring the number of particles in substances.

> Simple calculations about numbers of particles are useful in medicine and manufacturing industries.

Investigating atoms

You will find out:

> atoms have a small central nucleus surrounded by electrons
> more about protons, neutrons and electrons
> about the discoveries that led to theories of atomic structure

An atom is mainly empty space

Until 1909, people believed that atoms were the smallest particles of matter, but in 1909 scientists performed an experiment that produced surprising results. It showed that an atom is mostly empty space. Even more astonishing, they found that nearly all of an atom's mass is squashed into a tiny 'core' in the centre. They had discovered the nucleus.

Did you know?

The Ancient Greeks believed that atoms were joined together with tiny hooks.

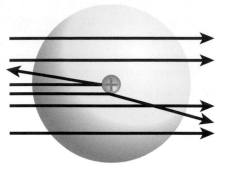

FIGURE 1: Most alpha particles pass through an atom, but some bounce back. They hit an obstacle – the nucleus.

What's inside an atom?

About a century ago, scientists discovered that atoms contain even tinier **sub-atomic particles** – protons, neutrons and electrons.

At the centre of an atom is its nucleus – a group of protons and neutrons. The nucleus is much smaller than the atom. Electrons orbit the nucleus in 'shells'. The shells are at various distances from the nucleus.

Relative mass and relative charge

'Relative' means the mass or charge of sub-atomic particles compared with each other.

> Protons and neutrons have similar mass, but the mass of an electron is much less.

> Protons and electrons have equal but opposite charges.

> Neutrons are neutral.

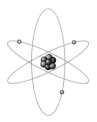

● neutron
● proton
○ electron

FIGURE 2: How can you tell that this atom is not drawn to scale?

Particle	Relative mass	Relative charge
proton	1	+1
neutron	1	0
electron	very small	−1

Atomic number and mass number

In the periodic table, each element is represented by a symbol and two numbers.

> The **atomic number** is the number of protons in the atom. This is the same as the number of electrons. In the periodic table the elements are arranged in order of increasing atomic number.

> The **mass number** is the total number of protons and neutrons in an atom.

mass number ⟶ 23
atomic number ⟶ 11 **Na**

QUESTIONS

1 List the differences between a proton, a neutron and an electron.

2 How many neutrons are there in the nucleus of a sodium atom?

3 Give the atomic number and mass number of the atom shown in Figure 2.

4 The atomic number and mass number of hydrogen are equal. Both are 1. What does this tell you?

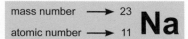

atom structure ... proton neutron electron

How ideas about atoms have changed

The idea of atoms as tiny, indivisible and indestructible particles was suggested by Greeks in the 5th century BC. The theory was revived and developed by John Dalton in the early 19th century. Dalton believed that:

> all matter consists of atoms, which cannot be divided up

> all atoms of the same element are identical, but different from atoms of every other element

> atoms cannot be created nor destroyed

> atoms can join together to form molecules.

In the early 20th century, knowledge of atoms increased rapidly.

QUESTIONS

5 Find out why the Greeks named the smallest particles of matter 'atoms'.

6 Explain why scientific ideas change over time.

7 Suggest how a scientist would have described an atom in 1915.

1900

1920

1940

1897 Discovery of electrons
~1900 Scientists realised that atoms contained smaller particles.
1909 Discovery of the nucleus.
1911 Rutherford's theory that atoms have a central nucleus orbited by electrons
1913 Theory that electrons orbit in a series of 'shells'
1919 Discovery of protons
1932 Discovery of neutrons

Scientists still study atoms to find out more about them. It is now known that there are particles even smaller than protons, neutrons and electrons.

Discovering the nucleus

Rutherford was investigating **alpha (α) particles** given off by radioactive elements. He knew that they passed through very thin metal. In 1909, his research team fired α particles at gold foil. Flashes on a fluorescent screen showed where the α particles hit (Figure 3).

Most α particles went straight through, showing that gold atoms are mostly empty space. However, some were deflected and came out at an angle. To the team's surprise, a few came backwards.

Rutherford realised that the α particles were bouncing off something inside the atoms. He called it the 'nucleus' (see Figure 1). Although the foil was a few hundred atoms thick, most α particles did not hit any nuclei. So nuclei must be very small compared with the size of atoms. Rutherford later calculated that a nucleus is about one ten-thousandth the diameter of a whole atom.

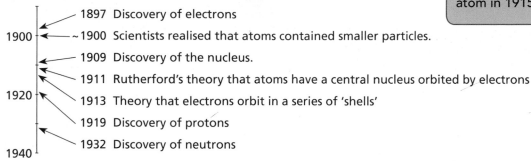

radioactive sample emits beam of alpha particles

some alpha particles bounce back

fluorescent screen

most alpha particles hit here

some alpha particles are deflected

FIGURE 3: Only α particles that come close to a nucleus get deflected off course or bounced back.

QUESTIONS

8 How does the diagram in Figure 1 explain what is happening in Figure 4?

Mass number and isotopes

You will find out:
> isotopes are atoms of the same element, but with different numbers of neutrons
> why isotopes are important

Dating the round table

According to legend, King Arthur ruled Camelot in the 6th century, with his Knights of the Round Table. A round wooden table hangs in Winchester Castle. Could it be King Arthur's? Scientists examined the ratio of carbon isotopes in the wood. This showed that the table was made around 1270, many centuries after King Arthur.

Did you know?

Some nuclear reactors use 'heavy water'. It is 10% heavier than normal water. The difference is caused by a heavier isotope of hydrogen.

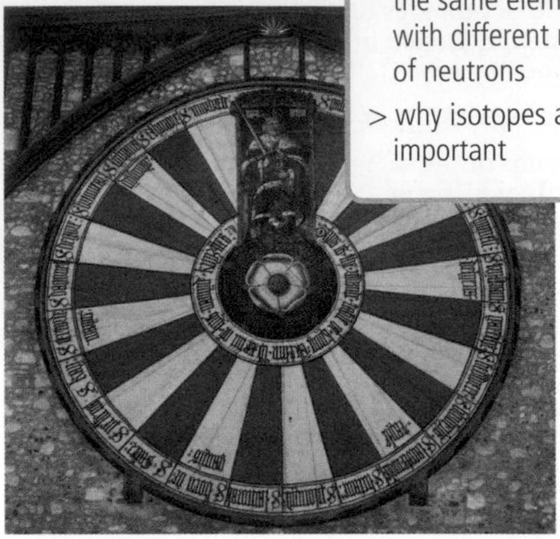

FIGURE 1: Carbon isotopes showed that this is not old enough to be King Arthur's round table.

Mass number and isotopes

Mass number

Protons and neutrons are much heavier than electrons. They make up almost all the mass of an atom. Both have a relative mass of 1. So, a carbon atom with six protons and six neutrons has a total mass of 12. In general:

Mass number = number of protons + number of neutrons

Atomic number = number of protons (different for each element)

mass number ⟶ 12
(6p + 6n)
C
atomic number ⟶ 6
(6p)

$^{1}_{1}H$
Hydrogen-1

$^{2}_{1}H$
Hydrogen-2

Hydrogen-3

● neutron
● proton
○ electron

FIGURE 2: The three isotopes of hydrogen.

Isotopes

All atoms of the same element have the same number of protons. However, they can have different numbers of neutrons. Atoms of the same element with different numbers of neutrons are called **isotopes**.

Because isotopes are the same element, they occupy the same place in the periodic table. Isotopes are named by giving the mass number after the element's name or symbol, for example, carbon-14. Carbon has three isotopes:

Isotope	Symbol	Number of protons	Number of neutrons
carbon-12	$^{12}_{6}C$	6	6
carbon-13	$^{13}_{6}C$	6	7
carbon-14	$^{14}_{6}C$	6	?

Scientists used the ratio of carbon-14 to carbon-12 to determine the age of Winchester's round table.

● QUESTIONS

1 Why is the mass number of carbon-12 double its atomic number?

2 State how many neutrons there are in an atom of carbon-14.

3 Write the symbol for hydrogen-3.

Properties of isotopes

Most elements have at least two isotopes. Some have up to ten.

> The proportion of each isotope is the same in the element and in its compounds.

> An element's isotopes have the same atomic number and the same number of electrons in the same electronic structure. So, the different isotopes react chemically in the same way.

> Atoms of isotopes of the same element have different masses. The average mass of all the atoms in a sample of the element is called its **relative atomic mass (A_r)**.

Calculating A_r (Higher tier)

Scientists measure relative atomic mass on a scale where the mass of one carbon-12 atom is 12.0000 exactly. A_r takes into account the masses and proportions of each isotope.

75% of chlorine atoms are isotope Cl-35 (mass=35). The other 25% are isotope Cl-37 (mass=37). Chlorine's relative atomic mass (A_r) is the average mass of all the atoms, which is 35.5.

In every 100 chlorine atoms:

75 have mass 35: $75 \times 35 = 2625$

25 have mass 37: $25 \times 37 = 925$

Total mass of 100 atoms = 3550

Average mass = $3550 \div 100 = 35.5$

$A_r = 35.5$

B	C	N
A_r 10.8	A_r 12.0	A_r 14.0
boron	carbon	nitrogen
Al	Si	P
A_r 27.0	A_r 28.1	A_r 31.0
aluminium	silicon	phosphorus

FIGURE 3: A section of the periodic table. Which two of these elements can you be sure have more than one isotope? How can you tell?

QUESTIONS

4 Why do all isotopes of the same element react in the same way?

5 Explain why mass numbers are always whole numbers, but relative atomic masses are not.

6 Silicon has three isotopes: Si-28 (92%), Si-29 (5%) and Si-30 (3%). Calculate A_r for silicon.

Dating with isotopes

Some elements have radioactive isotopes, which decay over time, changing into atoms of other elements. Scientists can measure the amount of radioactive isotope remaining and the amount of decay products formed. Knowing these and how quickly the isotopes decay, they can work out the age of a specimen.

> The decay of carbon-14 shows the age of archaeological specimens that were once alive, such as the Winchester round table and Egyptian mummies.

> The much slower decay of uranium isotopes into lead was used to calculate Earth's age – about 4.5 billion years.

> Similar methods can date rocks and therefore the fossils that they contain.

QUESTIONS

7 Why is carbon-14 dating suitable for dating items that were once alive, such as wood or mummified bodies?

8 Suggest other types of specimen that could be dated by this method.

FIGURE 4: Taking a tiny sample of bone to determine its age. Which isotope will be measured?

Compounds and mixtures

You will find out:

> compounds are made of two or more elements chemically combined

> what chemical formulae show about compounds

> the meaning of relative formula mass

Spot the difference

Put a lighted splint into a mixture of hydrogen and oxygen gases – the mixture burns with a loud 'pop'. Put a lighted splint into water – a compound of hydrogen and oxygen – the compound extinguishes the flame. Is a compound the same as a mixture?

Did you know?

Fluorine reacts with almost anything. Yet PTFE, a compound of fluorine and carbon, does not react with anything. It is used for non-stick pans and replacement parts inside the body.

FIGURE 1: A mixture of sodium and chlorine is dangerous, but sodium chloride is essential to life.

Compounds and their formulae

A **mixture** can be separated into the substances from which it is made by physical methods such as filtering, evaporation or distillation. Seawater is a mixture of water, salt and other dissolved solids. Distilling it separates out the water.

A chemical **compound**:

> consists of two or more different elements chemically combined

> can be broken down into simpler substances only by chemical reaction

> has different properties from the original elements.

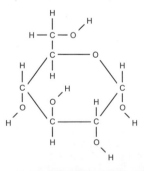

FIGURE 2: How can you tell that this cup contains a mixture?

Chemical formulae

A chemical formula shows the type and number of atoms in each molecule.

A glucose molecule is made from 6 carbon, 12 hydrogen and 6 oxygen atoms, so its formula is $C_6H_{12}O_6$.

Some compounds have giant structures, containing vast numbers of atoms or ions. Their formulae show the ratio of the elements involved. Sodium chloride contains equal numbers of sodium ions and chloride ions – a ratio of 1:1. Its formula is NaCl.

FIGURE 3: The arrangement of atoms in a glucose molecule.

QUESTIONS

1 Salt is obtained by evaporating brine. Does this show that brine is a mixture or a compound?

2 How does distillation show that crude oil is a mixture, but each hydrocarbon is a compound?

3 A molecule contains two hydrogen atoms, one sulfur atom and four oxygen atoms. Give its chemical formula.

FIGURE 4: Why is the formula of salt, NaCl, not as shown here?

Q element vs compound vs mixture

Relative formula mass

An element has a relative atomic mass (A_r). A compound has a **relative formula mass (M_r)**. It is the total mass of all the atoms in the formula.

To work out the M_r of a compound, multiply A_r for each element by the number of its atoms shown in the formula, then add up the results.

Examples

Water, H₂O

atoms	H	O
number in formula	2	1
A_r	1	16
M_r of water	(2 × 1) + (1 × 16) = 18	

Calcium carbonate, CaCO₃

atoms	Ca	C	O
number in formula	1	1	3
A_r	40	12	16
M_r of calcium carbonate	(1 × 40) + (1 × 12) + (3 × 16) = 100		

Chemists use relative formula masses to work out how much of each reactant is needed to make the required amount of a product.

Measuring in moles

Atoms are far too small to count. Even the smallest speck of powder contains billions of billions of atoms. This is why chemists measure in **moles**.

A 'mole' is like a 'dozen'. A dozen eggs and a dozen bricks are very different, but have one thing in common – there are twelve of each. Similarly, a mole of any substance contains the *same number* of particles (atoms, molecules or ions) – but the number is enormous.

Although you cannot count atoms directly, measuring in moles is quite simple:

1 mole of a substance is its M_r in grams.

> 9 g of water contains 0.5 mole of H₂O molecules (9 ÷ 18)

> 1 kg of calcium carbonate contains 10 mole CaCO₃ particles (1000 ÷ 100).

FIGURE 5: Moles vary in mass and volume. What do these samples all have in common?

◉ QUESTIONS

4 Explain the difference between relative atomic mass and relative formula mass.

5 Calculate M_r for
(a) calcium oxide, CaO
(b) ethene, C₂H₄
(c) glucose, C₆H₁₂O₆.
(A_r values are given above).

6 Calculate the mass of
(a) 3 moles of ethene
(b) 0.25 moles of calcium carbonate.

More complicated formulae

Some formulae have brackets enclosing a group of atoms. The number after the bracket shows how many of those groups are present. For example, calcium nitrate:

Ca(NO₃)₂ means 1 calcium atom + 2 nitrate groups, each with 1 nitrogen and 3 oxygen atoms.

So it could be written CaN₂O₆

A_r Ca = 40, N = 14, O = 16

So M_r = (1 × 40) + (2 × 14) + (6 × 16) = 40 + 28 + 96 = 164

◉ QUESTIONS

7 Work out the relative formula mass of
(a) calcium hydroxide, Ca(OH)₂
(b) copper nitrate, Cu(NO₃)₂
(c) ammonium sulfate, (NH₄)₂SO₄
(A_r not given before: N = 14, S = 32, Cu = 64).

Remember

One mole of any substance is its relative formula mass in grams.

Electronic structure

You will find out:

> electrons occupy shells or energy levels

> how to represent the electronic structures of the first 20 elements

Glowing with energy

Light sticks do not have batteries, so how do they produce light? It is all to do with electrons in atoms. Bending a light stick mixes chemicals that react, releasing energy. This gives the electrons extra energy, raising them to higher shells or energy levels. As the electrons drop back down, they release the energy as light.

Did you know?

'Neon' signs work by using electrical energy to boost electrons to higher energy levels. They emit light as the electrons drop down again. Neon glows red – other gases emit different colours.

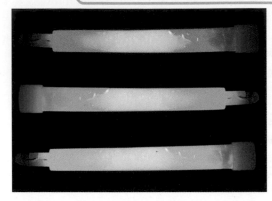

FIGURE 1: Some chemical reactions give out light.

Sub-atomic particles

Inside atoms there are sub-atomic particles:

> protons and neutrons in a central nucleus

> electrons orbiting in shells around the nucleus.

How an atom reacts depends on its number of electrons (atomic number) and the way they are arranged in the shells or **energy levels**.

How are the electrons arranged?

The electrons lie in shells at various distances from the nucleus – the farther out, the higher the energy level. These shells are complete layers surrounding the nucleus, but are normally drawn in cross-section as circles.

Each shell is larger than the one before, so has room for more electrons. The first shell can hold one or two electrons, and the second up to eight. The others can hold more.

Electrons fill up each shell in turn, starting nearest the nucleus. Their arrangement is called the **electronic structure**, or electron configuration, of the atom. Figure 3 shows the electronic structure of sodium (atomic number = 11).

Structures by number

Instead of drawing electronic structures, you can just write the number of electrons in each energy level. Sodium's structure (Figure 3) is 2.8.1. Calcium (atomic number = 20) is 2.8.8.2 – it has eight electrons in its third shell and two in its fourth shell.

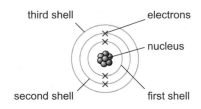

FIGURE 2: The structure of an atom. What is the atomic number of this element?

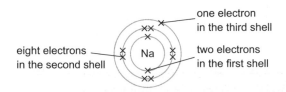

FIGURE 3: The electronic structure of a sodium atom. Why are sodium's electrons arranged in this way?

QUESTIONS

1 Explain what 'electron shells' means.

2 Draw the electronic structure of sulfur (atomic number = 16) and write it as numbers.

Q electron shells GCSE ... draw electronic structures

Electronic structures

Elements 1 to 20

The electronic structures of the first 20 elements are straightforward if you remember the rules:

> Electrons occupy the lowest available energy levels. They fill up each shell in turn.

> 2 electrons in the first (except H which has only 1); up to 8 in the second.

> The third shell then accepts up to 8 electrons before starting the fourth.

> The process stops when the total = atomic number.

This gives:

H = 1   He = 2
Li = 2.1   Be = 2.2   B = 2.3 and so on up to Ne = 2.8
Na = 2.8.1   Mg = 2.8.2   Al = 2.8.3 . . . . Ar = 2.8.8
K = 2.8.8.1   Ca = 2.8.8.2

Beyond calcium, electronic structures become more complicated. The third shell can hold up to 18 electrons, not just 8. This gives an extra 10 elements (mainly transition metals) in each period in the periodic table.

Group number

An atom's highest occupied energy level is called its **outer shell**. The number of outer electrons shows to which periodic table group that element belongs. It also determines how the element reacts, since it is the outer electrons that take part in chemical reactions.

Lithium, sodium and potassium atoms have one electron in their outer shell and so belong to Group 1. They have similar properties, including violent reactions with water.

FIGURE 4: Dolomite (top) is similar to limestone (bottom), but has magnesium in place of some of the calcium atoms. Why are magnesium and calcium so similar?

> ⦿ **QUESTIONS**

3 An atom has 8 electrons. Describe how they are arranged.

4 An atom has 19 electrons. Name the group to which the element belongs.

5 Which number shell is potassium's outer shell?

Noble gas structures

Noble gases are unreactive. Their electronic structures are stable. When elements combine, their atoms gain, lose or share electrons in order to have a noble gas electronic structure. For example:

> Metal atoms with 1, 2 or 3 outer electrons give these electrons to non-metal atoms, forming ions.

> Non-metals atoms with 6 or 7 outer electrons get 8 by gaining electrons from metal atoms, forming ions, or by sharing with another non-metal atom, forming covalent bonds.

> Non-metals with 4 or 5 outer electrons share with other non-metals, forming covalent bonds.

⦿ **QUESTIONS**

6 When the atoms that are described in questions 3 and 4 react together, what happens to their electrons?

Ionic bonding

You will find out:

> ionic bonding involves the transfer of outer electrons

> ions have the same electronic structure as a noble gas

Ions are 'go' particles

Copper is purified by electrolysis. It collects on the cathode, so copper particles must move through the solution. Non-metal particles move in the opposite direction, towards the anode. Michael Faraday named these moving particles 'ions' from the Greek word for 'go'.

FIGURE 1: Electrolysis of tin bromide solution – tin ions go to the cathode; bromide ions go to the anode.

What is ionic bonding?

Any atom has an equal number of protons and electrons, so it has no overall charge. An **ion** forms when an atom gains or loses electrons and becomes charged.

When a metal reacts with a non-metal:

> metal atoms lose electrons and become positive ions

> non-metal atoms gain electrons and become negative ions.

The strong attraction between these oppositely charged ions holds the compound together. It is called **ionic bonding**.

Sodium and chlorine

Sodium metal reacts with chlorine gas forming the ionic compound sodium chloride. Figure 3 shows how the electrons in sodium and chlorine atoms are rearranged. Both elements form ions with a stable noble gas structure of eight outer electrons.

A sodium atom (Na) loses an electron to form a sodium ion (Na$^+$) with a 1+ charge because it has 11 protons (as before), but only 10 electrons. It is now a positive ion.

A chlorine atom (Cl) gains an electron to form a chloride ion (Cl$^-$) with a 1– charge because it has 17 protons (as before), but 18 electrons. It is now a negative ion.

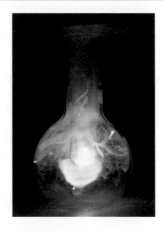

FIGURE 2: Sodium burning in chlorine. What is the white smoke?

Remember

All electrons are the same, so dots and crosses are used to show in which atoms the electrons were, originally.

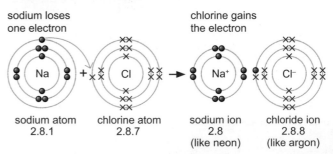

sodium loses one electron

chlorine gains the electron

Na + Cl → Na$^+$ Cl$^-$

sodium atom
2.8.1

chlorine atom
2.8.7

sodium ion
2.8
(like neon)

chloride ion
2.8.8
(like argon)

FIGURE 3: Rearranging electrons to form ions. Why does the electron go from the metal to the non-metal, not the other way?

Did you know?

Metal ions are called cations because they move towards the cathode during electrolysis. Non-metal ions move towards the anode, so are called anions.

QUESTIONS

1 Why do atoms gain or lose electrons?

2 Explain why metal ions are positive, but non-metal ions are negative.

3 Describe an ionic bond.

Q ionic bonding GCSE

More ionic compounds

An ionic compound forms when electrons transfer from metal atoms to non-metal atoms. This gives both ions the same electronic structure as noble gas atoms.

> For metals: electrons lost = Group number

> For non-metals: electrons lost = 8 – Group number

In periodic table	Electrons lost	Charge on ion
Group 1 metal	1	+1
Group 2 metal	2	+2
Group 3 metal	3	+3
	Electrons gained	Charge on ion
Group 6 non-metal	2	-2
Group 7 non-metal	1	-1

Calcium chloride $CaCl_2$

Calcium is in Group 2, so its atoms each lose two electrons to form calcium ions, Ca^{2+}.

Chlorine is in Group 7, so its atoms each gain one electron to form chloride ions, Cl^-.

Calcium ions and chloride ions have the same electronic structure as the noble gas argon: 2.8.8

A calcium atom gives one electron to each of two chlorine atoms. Therefore, the formula of calcium chloride is $CaCl_2$

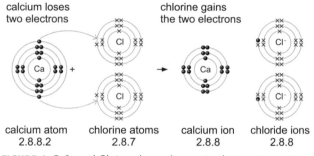

calcium loses two electrons   chlorine gains the two electrons

| calcium atom 2.8.8.2 | chlorine atoms 2.8.7 | calcium ion 2.8.8 | chloride ions 2.8.8 |

FIGURE 4: Ca^{2+} and Cl^- ions have the same electronic structure.

Magnesium oxide MgO

Magnesium is in Group 2. Magnesium atoms each lose two electrons to form magnesium ions, Mg^{2+}.

Oxygen is in Group 6. Oxygen atoms each gain two electrons to form oxide ions, O^{2-}.

A magnesium atom gives two electrons to one oxygen atom. Therefore, the formula of magnesium oxide is MgO.

The force of attraction between double charged Mg^{2+} and O^{2-} ions is stronger than between 1+ and 1– ions. Thus the ionic bonding in magnesium oxide is stronger than in sodium chloride.

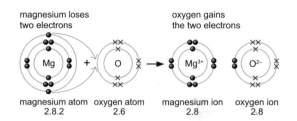

magnesium loses two electrons   oxygen gains the two electrons

| magnesium atom 2.8.2 | oxygen atom 2.6 | magnesium ion 2.8 | oxygen ion 2.8 |

FIGURE 5: Mg^{2+} and O^{2-} ions have the same electronic structure. Which noble gas also has this electronic structure?

QUESTIONS

4 Explain why magnesium oxide has the formula MgO, but magnesium chloride is $MgCl_2$.

5 Explain why the formula of aluminium fluoride is AlF_3.

6 Predict the formula of the ionic compound aluminium oxide.

Conducting electricity

The bulb in Figure 6 is not lit because pure water does not conduct electricity – no current can flow. However, if some sodium chloride is added to the water, the bulb will light because the solution does conduct.

It is the same when any ionic compound dissolves in water. When it dissolves, ions are free to move. Because they are charged, the ions carry the current through the water.

Solid ionic compounds do not conduct electricity because the ions are not free to move.

distilled water

FIGURE 6: Distilled water does not conduct electricity.

QUESTIONS

7 Suggest why molten sodium chloride does conduct electricity.

Alkali metals

Street lights

Orange street lights contain sodium metal and a little neon and argon. When first switched on, they glow red owing to the neon. The orange glow gradually appears as the sodium warms up and vaporises. Electricity excites sodium atoms, raising electrons to higher energy levels. They drop back, emitting energy as orange light.

Did you know?

Before 1807, potash and soda, two compounds, were listed as elements – substances that cannot be decomposed. Then Sir Humphry Davy found that they could, and discovered potassium and sodium.

FIGURE 1: A sodium lamp – an everyday use of an alkali metal.

Properties of Group 1 elements

The **alkali metals** are a family of very reactive metal elements. They make up Group 1 of the periodic table.

They are all:

> silver-coloured metals, but soft enough to cut with a knife

> good electrical and thermal conductors

> stored under oil to prevent them reacting with air or moisture

> shiny when first cut, but rapidly **tarnish** in air.

Alkali metals react vigorously with water. These reactions produce hydrogen gas and form metal hydroxide solutions, which are alkaline. This is one reason for calling them 'alkali metals'.

Lithium, sodium and potassium are less dense than water – they float while reacting. Rubidium and caesium are denser than water – they sink. They react so violently that they can only be used safely by experts.

The alkali metals react readily with non-metal elements to form ionic compounds. Like Na^+, all alkali metal ions have a 1+ charge.

FIGURE 2: Part of the periodic table. Can you see a pattern in the atomic numbers and atomic masses of Li, Na and K?

QUESTIONS

1 List properties of Group 1 elements that are (a) typical for metals and (b) unusual for metals.

2 Explain why sodium floats on water.

3 Write down the symbol (including its charge) for potassium ion.

FIGURE 3: Alkali metals in oil. Which of these metals has the lowest density?

Q alkali metal properties

Similarities and differences in reactions

Similarities

Group 1 metals react in similar ways because their atoms all have just one electron in their outer shell. This electron can be transferred easily, leaving a 1+ ion with a noble gas electronic structure (Figure 4). Because the atoms of Group 1 metals achieve noble gas electronic structures more easily than atoms of other metals, Group 1 metals are more reactive.

Alkali metals form ionic compounds with similar formulae. For example, they:

> burn to form the metal oxides, Li_2O, Na_2O and K_2O, which react with water to form hydroxides, LiOH, NaOH and KOH

> react with chlorine to form the metal chlorides, LiCl, NaCl and KCl.

Differences

Although alkali metals react in similar ways, there are differences between them. They show trends (gradual changes) in their properties from one metal to the next. They become more reactive down the group. For example, when reacting with water:

> lithium just floats and fizzes gently

> potassium bursts into flame, zooms across the water and spits

> caesium explodes on contact with water.

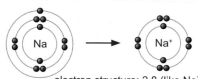
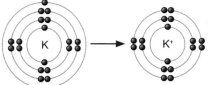

loses one electron to non-metal

Li → Li⁺

electron structure: 2 (like He)

Na → Na⁺

electron structure: 2.8 (like Ne)

K → K⁺

electron structure: 2.8.8 (like Ar)

FIGURE 4: Ion formation. Why does losing one electron give ions a 1+ charge?

FIGURE 5: When sodium reacts with water it fizzes vigorously and skates across the surface. Why is this less violent than the reaction of potassium with water?

> ### QUESTIONS
>
> **4** Which is the least reactive alkali metal?
>
> **5** Bromine is a non-metal in Group 7. Predict the formula of potassium bromide.
>
> **6** How would you expect rubidium to react with water?

Discovering alkali metals

Some metals, like copper, have been used since prehistoric times. However, alkali metals were not discovered until the 19th century, even though sodium and potassium are three hundred times more abundant than copper in Earth's crust.

Unlike copper, alkali metals are much too reactive to be found 'native' (uncombined in nature). Their high reactivity means that they form very stable compounds, making it difficult to extract them from these compounds.

When electrolysis was developed in the early 19th century, Sir Humphry Davy found that it would decompose alkalis such as potash and soda. This led to his discovery of potassium and sodium. The other alkali metals followed later.

> ### QUESTIONS
>
> **7** Suggest two reasons why Group 1 metals are called 'alkali metals'.
>
> **8** John Dalton's 1806 list of elements included soda and potash, which are actually compounds. What compounds are they and why were they classified originally as elements?

Q reactivity alkali metals

Halogens

You will find out:

> halogens are reactive non-metals with similar properties
> about trends in these properties

Iodine deficiency

Iodine deficiency can lead to serious health problems that affect more than 10% of the world's population. Adding tiny amounts of iodine compounds, such as potassium iodide, to the salt that people eat can prevent these diseases. It costs just a few pence a year, per person, to produce enough iodised salt for this.

FIGURE 1: This person has a goitre (swollen thyroid gland in the neck), caused by a lack of iodine in their diet.

Properties of Group 7 elements

The **halogens** are a family of reactive non-metal elements. They make up Group 7, at the right-hand side of the periodic table.

The family members are:

> fluorine, F – pale yellow, poisonous gas

> chlorine, Cl – green, poisonous gas; the most common halogen

> bromine, Br – dark red poisonous liquid with orange–brown vapour

> iodine, I – dark grey crystalline solid, which **sublimes** to form a purple vapour when heated.

All the halogens exist as molecules, made from atoms joined together by a single covalent bond. Their formulae are F_2, Cl_2, Br_2 and I_2.

FIGURE 2: The halogens. Fluorine is the most reactive non-metal – which is why it is not in the line-up.

QUESTIONS

1 Name the halogen with the darkest colour.

2 Which halogen is liquid at room temperature?

3 All the halogen elements exist as diatomic molecules rather than atoms. Describe how the atoms are joined together.

FIGURE 3: Diatomic halogen molecules.

The chemical reactivity of halogens

Halogens react readily with metals to form salts. In fact, the word 'halogen' comes from the Greek for 'salt-makers'. These salts are all white crystalline solids like table salt (sodium chloride). They are ionic compounds. Halide ions have a 1- charge: F^-, Cl^-, Br^- and I^-.

Group 7 elements react in similar ways because they all have seven electrons in their outer shell. Gaining one extra electron from a metal gives them a stable electronic structure like a noble gas. This gaining of an electron changes each atom into an ion with a 1- charge.

🔍 halogen group properties

Halogen atom	Electronic structure	Number of electrons in outer shell	Halide ion	Electronic structure	Number of electrons in outer shell
fluorine	2.7	7	fluoride, F⁻	2.8 (like Ne)	8
chlorine	2.8.7	7	chloride, Cl⁻	2.8.8 (like Ar)	8
bromine	2.8.18.7	7	bromide, Br⁻	2.8.18.8 (like Kr)	8
iodine	2.8.18.18.7	7	iodide, I⁻	2.8.18.18.8 (like Xe)	8

Note: You do not need to remember the electronic structures of bromine or iodine.

Sodium metal reacts with chlorine gas to form the ionic compound sodium chloride.

$$2Na(s) + Cl_2(g) \rightarrow 2NaCl(s)$$

Other halogens react in a similar way to form similar ionic compounds with the same type of formula, such as lithium bromide, LiBr, or potassium iodide, KI. You can show the changes to electronic structure by writing as in the following examples.

sodium atom, Na	chlorine atom, Cl		sodium ion, Na⁺	chloride ion, Cl⁻
2.8.1	2.8.7	→	2.8 (like Ne)	2.8.8 (like Ar)

lithium atom, Li	bromine atom, Br		lithium ion, Li⁺	bromide ion, Br⁻
2.1	2.8.18.7	→	2 (like He)	2.8.18.8 (like Kr)

potassium atom, K	iodine atom, I		potassium ion, K⁺	iodide ion, I⁻
2.8.8.1	2.8.18.18.7	→	2.8.8 (like Ar)	2.8.18.18.8 (like Xe)

Relative reactivity

Although halogens react in similar ways, their reactivity decreases down the group.

> Fluorine, at the top, is more reactive than any other element. It reacts violently with almost anything.

> Chlorine is not far behind. It even attacks unreactive metals such as gold.

> Iodine is much less dangerous. Doctors put iodine solution on cuts and grazes.

QUESTIONS

4 Explain why halide ions have a charge, while noble gases with the same electronic structure do not.

5 Write the word equation and balanced symbol equation for the reaction between potassium and bromine.

6 How reactive is bromine, compared with the other halogens?

Remember

When halogen atoms form salts, they become halide ions – fluoride, chloride, bromide and iodide.

Did you know?

Chlorine kills. Yet you drink some every day. A tiny dose of chlorine does you no harm, but is enough to kill harmful bacteria. This tiny dose, added to tap water, makes the water safe to drink.

Reactive elements, unreactive compounds

Many people think that, because chlorine is reactive and poisonous, anything containing chlorine must be dangerous.

It is not true. Sodium chloride is not poisonous. In fact, it is essential for life. Polyvinyl chloride (PVC plastic) is not reactive. It is used because neither air nor water affects it.

Similarly, fluorine is the most reactive element, but sodium fluoride in toothpaste is safe to put in your mouth.

QUESTIONS

7 Use your knowledge of alkali metals and halogens to explain why sodium chloride is unreactive and safe, although it is made from two dangerously reactive elements.

halogen reactivity

Ionic lattices

Rock solid

Earth's crust is made of rocks – hard, solid materials with melting points well above 1000 °C. The mantle below the crust is hotter than this, so consists of molten rocks. When these come to the surface, they cool and solidify. Rocks that were once molten are called igneous rocks.

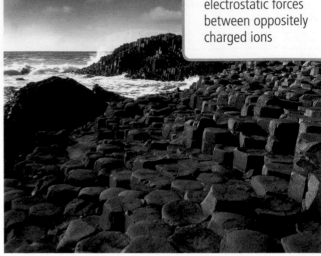

FIGURE 1: When molten rock cools suddenly, it forms a glass-like solid, not crystals. If it cools very slowly, it can grow enormous crystals, such as those in the Giant's Causeway.

 Ionic compounds

Most rocks and minerals consist of ionic compounds of metals combined with non-metals. Many common substances are the same: sodium chloride (table salt), potassium nitrate (in fertilisers) and calcium sulfate (plaster of Paris, used to set broken bones).

Properties of ionic compounds

Ionic compounds:

> are solid at room temperature

> have high melting and boiling points

> conduct electricity when molten, or dissolved in water, but not when solid.

Structure of ionic compounds

Ionic compounds contain positively charged metal ions and negatively charged non-metal ions. Opposite charges attract each other.

The action of these electrostatic forces of attraction is called **ionic bonding**.

The forces act in all directions.

> Each positive ion attracts negative ions around itself.

> Each negative ion attracts more positive ions, and so on.

They build up into rows and layers, holding billions of ions together in a **giant structure**.

FIGURE 2: The structure of sodium chloride. Suggest why salt crystals are cubic.

QUESTIONS

1 State whether ionic compounds are solid, liquid or gaseous at room temperature.

2 What are ionic bonds?

3 Give the term that chemists use for the 3-D arrangement of ions in a compound.

🔍 ionic bonding properties ... ionic lattice

Ionic lattices

Within giant structures, ions are arranged in rows, with positive and negative ions alternating in each row. This regular arrangement is a **lattice**.

Sodium chloride is an ionic compound. There are no molecules of sodium chloride, just sodium ions and chloride ions. In a sodium chloride lattice, each ion is surrounded by six ions of the opposite charge, as shown by the red and green lines in Figure 3.

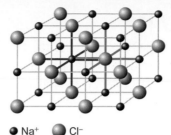

FIGURE 3: The lattice for sodium chloride. How many Cl⁻ ions surround each Na⁺ ion?

● Na⁺   ● Cl⁻

Melting and boiling points

Ionic bonds are strong. They hold the ions in fixed positions in the lattice. This is why ionic compounds are hard solids.

To melt the solid, the lattice must be broken down so that the ions become free to move and flow. It takes a lot of energy to break the bonds (to pull oppositely charged ions away from one another). Therefore, ionic compounds melt and boil only at high temperatures. Most ionic compounds have melting points above 500 °C.

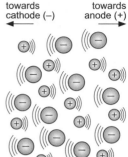

towards cathode (−) ◀   towards anode (+) ▶

FIGURE 4: Ions moving in a liquid. In which direction will each type of ion travel through this liquid?

Conduction

When an ionic compound is molten (melted by heating) or dissolved in water, its ions are free to move. Because ions are charged, moving carries their charge through the liquid. In other words, they can conduct electricity.

However, they do not simply conduct like metals do. Energy from an electrical current causes ionic compounds to decompose. This is **electrolysis**. The metal from the compound collects at the cathode, and the non-metal is released at the anode.

QUESTIONS

4 Describe how sodium and chloride ions are arranged in (a) a salt crystal (b) salt solution.

5 Magnesium chloride melts at 987 °C. Explain why its melting point is so high.

6 Explain why molten magnesium chloride conducts electricity but the solid does not.

Multi-atom ions

Many non-metal ions are formed not from single atoms, but from two or more atoms (often including oxygen). The atoms join together by covalent bonding, gain one or more electrons and, therefore, have a negative charge. These clusters are also known as ions. Examples include:

> nitrate ion, NO_3^-

> sulfate ion, SO_4^{2-}

> carbonate ion, CO_3^{2-}

There is even a non-metal ion with a positive charge. The ammonium ion, NH_4^+ forms compounds such as ammonium chloride and ammonium sulfate, just as metal ions do.

QUESTIONS

7 Here are the chemical formulae of some compounds: sodium hydrogencarbonate is $NaHCO_3$; potassium cyanide is KCN; ammonium phosphate is $(NH_4)_3PO_4$. Sodium ions, potassium ions and ammonium ions all have a +1 charge. Write formulae and charges for hydrogencarbonate, cyanide and phosphate ions.

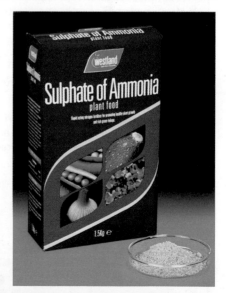

FIGURE 5: How many elements in ammonium sulfate (sulphate)? How many different ions?

Covalent bonding

Sublime dry ice

In films and on stage, an artificial mist sometimes swirls around. This special effect is produced by 'dry ice', which is frozen carbon dioxide, a covalent compound. It is dry because, unlike normal ice, it does not melt into a liquid. It changes straight from solid to gas. This is called sublimation.

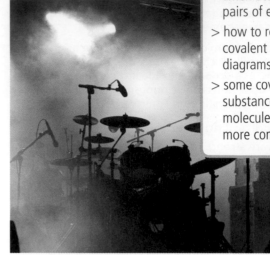

FIGURE 1: When dry ice is dropped into water it sublimes, producing very cold carbon dioxide gas that condenses water vapour in the air – the mysterious mist you see.

You will find out:

> covalent bonds form when atoms share pairs of electrons

> how to represent covalent bonds in diagrams and models

> some covalent substances are simple molecules, others are more complex

Covalent bonds

Forming a covalent bond

When atoms of two non-metals combine, they share electrons. They do this to achieve noble gas electronic structures, which are stable.

The outer electron shells of the atoms overlap. This allows the atoms to share some of their outer electrons with each other. This sharing forms a **covalent bond** which holds the two atoms together. Covalent bonds are very strong.

The simplest covalent molecule is hydrogen (H_2). Each hydrogen atom has only one electron. By sharing, they both have two electrons in their outer shells. This is the same as the noble gas helium.

Representing covalent bonds

To represent a covalent bond: draw the overlapping outer shells and put two electrons in the overlap.

Covalent bonds are often represented simply as a line between the two symbols. In ball and stick molecular models, the stick represents the bond.

FIGURE 2: Three ways to represent the covalent bond in a hydrogen molecule.

FIGURE 3: A model of a sucrose molecule. What do the sticks represent?

Molecules

A molecule is formed when a group of atoms are held together by covalent bonds. Some elements and many compounds consist of molecules.

Element	Molecule
hydrogen	H_2
chlorine	Cl_2
sulfur	S_8
Compound	**Molecule**
water	H_2O
carbon dioxide	CO_2
ammonia	NH_3
sucrose (sugar)	$C_{12}H_{22}O_{11}$
DNA	very large and complex

TABLE 1: Examples of molecules.

QUESTIONS

1 Explain why non-metal atoms form covalent bonds, not ionic, when they react together.

2 What is a covalent bond?

3 Describe the bonding in a hydrogen molecule.

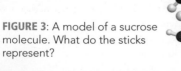

sharing electrons ... sucrose bonding GCSE

Sharing by numbers

Each atom in a molecule shares electrons to have eight in its outer shell (two for hydrogen), like a noble gas. The number shared depends on how many outer electrons it already has.

A chlorine atom has seven outer electrons, so it shares one more from another atom, as shown in Figures 4 and 5. Note that, in these drawings, only the outer shells are shown.

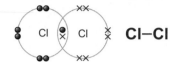 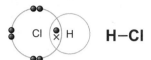

FIGURE 4: A chlorine molecule. How many covalent bonds are there?

FIGURE 5: A molecule of hydrogen chloride. Why does the hydrogen not have eight electrons in its outer shell?

An oxygen atom has six outer electrons, so needs two more to make eight. It can do this by sharing one electron each with two other atoms. Water is an example. An oxygen atom forms *two* covalent bonds, while each hydrogen atom forms one bond each. This is why water is H_2O, not HO or HO_2.

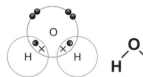

FIGURE 6: A molecule of water. Why does oxygen bond onto two hydrogen atoms?

Alternatively, an oxygen atom can share two of its electrons, forming two covalent bonds to the same atom. This is a **double covalent bond**, which is represented by a double line. Oxygen gas molecules have a double bond.

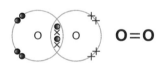

FIGURE 7: An oxygen molecule. Why are there two lines between the atoms?

Diagrams of covalent molecules are like Venn diagrams in maths. Each circle represents a set of electrons in the outer shell of an atom. Electrons in the overlap area belong to both sets – that is, to both atoms.

> **Remember**
> Electrons are all the same. Dots and crosses only show that they come from different atoms.

Did you know?
The 'co' in covalent bonding means 'shared', as in 'co-operation'. Non-metals cannot form ionic bonds with each other, because their atoms cannot lose electrons. Instead, they co-operate with each other and share their electrons.

⬤ QUESTIONS

4 Nitrogen is in Group 5. How many electrons does a nitrogen atom share?

5 Draw the bonding in NH_3 (ammonia).

6 What is the difference between a single covalent bond and a double covalent bond?

How do covalent bonds work?

Nuclei are positively charged, but electrons are negative. Opposite charges attract. The positive nuclei of both atoms attract the negative electrons in the overlap area between them.

It is like two people (the nuclei) pulling on a rope (the shared electrons). Pulling in opposite directions holds them together a fixed distance apart. This is called the bond length.

⬤ QUESTIONS

7 Suggest why the atoms end up a fixed distance apart. What stops them moving closer still?

Covalent molecules

You will find out:

> substances with simple molecules have low melting and boiling points

> some other characteristic properties of covalent molecules

Sand, sea and air

Sand, water and air are all covalent substances containing oxygen. So why is one a solid, one a liquid and one a gas? Why doesn't sand dissolve in the water? After the tide has gone out, the wet sand dries out. Why does the water evaporate, but not the sand? The answers to these questions lie in their different structures.

FIGURE 1: A beach poses many chemical questions.

 ## Molecules

Molecules of elements

Some non-metal elements exist as simple molecules – small groups of atoms joined together by covalent bonds.

Element	Molecule
hydrogen	H_2
oxygen	O_2
nitrogen	N_2
chlorine	Cl_2
iodine	I_2
phosphorus	P_4
sulfur	S_8

These molecules consist of single covalent bonds, except for:

> oxygen, which is a double covalent bond, and

> nitrogen, which is a triple covalent bond.

Molecules made from just two atoms are diatomic.

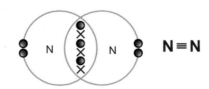

FIGURE 2: A nitrogen molecule. How many of its electrons does nitrogen share, to achieve a noble gas structure?

Molecules of compounds

Non-metal compounds come in various shapes and sizes.

Many common compounds have small, simple molecules. Examples are: HCl (hydrogen chloride), H_2O (water), NH_3 (ammonia) and CH_4 (methane).

Others, such as those in detergents, are more complex. Their molecules contain dozens of atoms.

Polymers, such as plastics and proteins, have thousands of atoms covalently bonded in long chains.

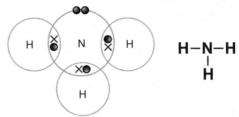

FIGURE 3: An ammonia molecule. Why does nitrogen form three covalent bonds?

Properties of covalent substances

Covalent elements and compounds usually:

> have low melting and boiling points

> do not conduct electricity

> are insoluble in water, but may dissolve in other liquids.

Substances with simple molecules (made from only a few atoms) are often liquids or gases at room temperature. Those with larger molecules are solid, but melt easily.

Some substances, such as diamond (carbon) and sand (silicon dioxide), have **giant covalent structures**. Each atom has covalent bonds to several others in various directions. They are solids with high melting points.

QUESTIONS

1 Which of the compounds named above has diatomic molecules?

2 Draw a diagram to show the bonding in CH_4 (methane).

3 CO_2 (carbon dioxide) has double bonds, O=C=O. Draw a diagram showing the bonding.

Q diatomic molecule GCSE

Properties

Conducting electricity

Unlike ionic compounds, substances consisting of covalent molecules do not **conduct** electricity, even when molten or dissolved. This is because there are no charged particles able to move and carry the charge around the circuit.

Solubility of substances

As a general rule:

> ionic compounds are soluble in water

> covalent substances are soluble in organic solvents.

Melting and boiling points (Higher tier)

Covalent bonds are very strong. They hold atoms firmly together within each molecule. However, bonds *between* molecules are very much weaker. They are called **intermolecular bonds**. Because they are much weaker than covalent bonds, they can be broken easily and molecules separated from one another. That is why molecular substances have low melting and boiling points.

Intermolecular bonds between larger or heavier molecules are stronger, resulting in higher melting and boiling points, but even solid molecular substances melt easily. Here are some examples.

> Oxygen and methane are gases. The intermolecular bonds are very weak.

> Bromine and hydrocarbons in petrol are liquid. The intermolecular bonds are weak.

> Sulfur and sucrose are solids. The intermolecular bonds are stronger.

Sulfur

A sulfur molecule is made from eight sulfur atoms covalently bonded in a ring-like structure. These bonds are strong, but the attractions between molecules are much weaker. Little energy is required to overcome them: sulfur melts easily, at 113 °C.

FIGURE 4: Sulfur crystals in rock.

FIGURE 5: Arrangement of S_8 molecules in sulfur crystals. What happens when sulfur melts?

Remember

Melting or boiling separates molecules from each other. It does not break covalent bonds within them.

QUESTIONS

4 Explain why salt solution conducts electricity, but sugar solution does not.

5 Why is chlorine a gas, bromine a liquid and iodine a solid at room temperature?

6 Describe the difference between covalent bonds and intermolecular bonds.

Water

Water is the only substance that exists naturally on Earth as a solid, a liquid and a gas.

Water molecules are small and light. The relative molecular mass of H_2O is only 18, compared with 32 for O_2. Therefore, water should be a gas, with a boiling point about −200 °C. It is not, because the intermolecular attractions between water molecules are stronger than normal for small molecules.

QUESTIONS

7 Use your understanding of covalent molecules and bonding to answer the questions in *Sand, sea and air* at the top of the page opposite.

Covalent lattices

You will find out:

> some covalent substances form giant structures

> why diamond and graphite are so different

> why graphite conducts electricity

Diamond hard

Diamond is the hardest natural substance. It has the highest melting point and is the best thermal conductor. It is also the most 'sparkly', making diamond the best jewellery stone. Diamonds are crystals of carbon formed at high pressure and temperature in Earth's mantle. They are brought to the surface by volcanoes.

FIGURE 1: The only way to cut and polish a diamond is to use diamond paste.

Giant covalent structures

Not all covalent substances exist as molecules. Some have giant structures or **macromolecules**. Each atom shares electrons to form covalent bonds with several other atoms around it in regular patterns. This three-dimensional regular arrangement is called a **lattice**.

Examples of giant covalent structures include diamond and graphite (different forms of carbon) and sand (silicon dioxide).

Properties

The whole diamond crystal or grain of sand is one enormous macromolecule, containing billions of atoms. Every atom is joined to all the others through a network of strong covalent bonds.

Materials with giant covalent lattices:

> have very high melting points (over 3500 °C for diamond and graphite)

> are very hard (silicon dioxide is used in sandpaper)

> do not conduct electricity

> are insoluble in water or other liquids.

Graphite

Graphite is an odd one out. Although it is a giant covalent lattice, it is softer and does conduct electricity. These properties make it useful in pencils and electric motors.

FIGURE 2: Part of a model of diamond. How many other atoms are bonded to each carbon?

QUESTIONS

1 Give three examples of giant covalent structures.

2 What is a lattice?

3 Which is the only giant covalent structure that conducts electricity?

Bonding, structure and properties

Silicon dioxide

Sand is impure silicon dioxide. Every sand grain is a macromolecule of silicon dioxide. Its structure is similar to diamond, with silicon and oxygen atoms held together by covalent bonds. The strong lattice makes it hard and unaffected by air and water, so sandy beaches are long lasting.

Silicon dioxide, SiO_2, has four bonds per silicon atom, but only two per oxygen. Its melting point, at 1610 °C, is lower than diamond at 1610 °C, but still high.

FIGURE 3: In silicon dioxide, each silicon atom is bonded to four oxygen atoms. How many bonds does each oxygen atom form?

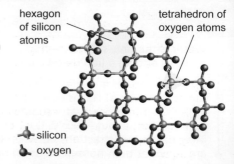

hexagon of silicon atoms

tetrahedron of oxygen atoms

silicon
oxygen

Q giant covalent structures

Diamond and graphite

Diamond and graphite are covalent lattices of carbon atoms arranged in different patterns. This gives them different properties.

Property	Diamond	Graphite
Appearance	transparent, colourless crystal	black, opaque irregular lump
Hardness	hardest natural substance	soft enough to rub off on paper
Electrical conductor	no	yes
Thermal conductor	best thermal conductor conducts heat in all directions	conducts heat along the layers, but not between layers

In diamond, each carbon bonds to four other atoms in a **tetrahedral** arrangement. This gives a rigid structure, making diamond extremely hard. Diamond-tipped drills can cut through glass, stone or concrete.

To melt diamond, you have to break all the bonds in the lattice – four per atom. That needs a lot of energy and, therefore, a very high temperature (over 3500 °C).

Graphite has layers. Each carbon bonds to only three others in flat hexagons. There are no covalent bonds between the layers, which are only loosely held together by weak bonds. Layers can slide over each other, making graphite soft and slippery. When dragged across paper, the layers rub off, leaving a mark. Graphite is used as a lubricant. It doesn't flow away like oil.

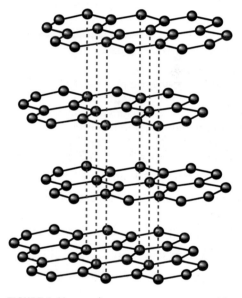

FIGURE 4: How carbon atoms are arranged in graphite. What do the dotted lines represent?

QUESTIONS

4 Explain why diamond has a very high melting point.

5 Explain why graphite can mark on paper.

6 Explain why SiO_2 is a hard solid, but CO_2 a gas.

Explaining graphite's properties (Higher tier)

In graphite, each carbon atom uses three of its four outer electrons to form covalent bonds to three other atoms within a layer. The fourth electron from each atom is free to move. These **delocalised electrons** (like those in metals) can carry charge through the structure. This means that graphite conducts electricity.

Materials conduct heat by atoms vibrating, passing energy from atom to atom through the structure. Graphite conducts heat along the layers, but not between them. As with metals, graphite also conducts heat by the movement of delocalised electrons.

QUESTIONS

7 Describe the differences between graphite and diamond, and explain how their structures account for these differences.

diamond graphite comparison

Polymer chains

You will find out:
> the properties of polymers depend on how they are made
> the difference between thermosoftening and thermosetting polymers

Save a flush

Save-a-flush® is a packet of a super-absorbent polymer called poly(acrylate). It absorbs up to 400 times its own mass of water. When placed in a toilet cistern, the polymer swells, taking up space. As a result, less water is needed to refill the cistern after each flush. This could save twelve litres of water a day.

Did you know?

Fibres made from certain plastic polymers, such as *Kevlar*®, are stronger than steel.

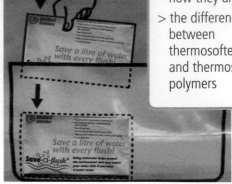

FIGURE 1: Save water with a packet of polymer.

Chain molecules

Polymers are very large molecules. They contain thousands of atoms joined with covalent bonds. However, they are not giant covalent structures. Plastics are made up of billions of separate polymer molecules.

Each polymer molecule is a long chain of small molecules (**monomers**) joined end to end. The monomers repeat along the chain, like a string of beads.

> Some polymers are made from a single starting material, often an alkene. They contain only one type of monomer.

    A   →   – – –A–A–A–A–A–A–A–A–A–A–A–A– – –

  monomer   polymer

Examples include poly(ethene), poly(propene) and poly(chloroethene).

> Others, such as nylon and polyester, are made from two different monomers. These alternate along the chain.

   A + B   →   – – –A–B–A–B–A–B–A–B–A–B–A–B– – –

two different   polymer
  monomers

The effects of heat

Plastic polymers are divided into two groups.

> Many plastics gradually soften before melting. While soft, they can be moulded into shape. These are **thermosoftening** polymers.

> Others are soft only when first made. When heated during moulding, instead of softening they set hard. These are thermosetting polymers. They do not melt and so they are useful for items that get hot.

Thermosoftening polymers can be recycled: they are heated to soften them and then moulded into new objects. Thermosetting polymers cannot be recycled.

FIGURE 2: Do these beads represent poly(ethene) or polyester?

FIGURE 3: Which type of polymer is used to make this pan handle?

QUESTIONS

1 Describe the difference between a polymer and a monomer.

2 Why are polymer molecules like chains?

3 Suggest why thermosetting polymers cannot be recycled.

Q polymers GCSE ... polymer AND monomer

Polymers vary

Plastic bags need different properties from plastic kettles or DVD cases. A polymer's properties, and therefore uses, depend on which monomer(s) it is made from.

Even the same polymer can have different properties, depending on the process used to make it.

> High density poly(ethene) (HDPE) has a waxy appearance. It is used for plastic bottles, water pipes and toys.

> Low density poly(ethene) (LDPE) has polymer chains packed less tightly. This decreases the density and tensile strength. Because this allows LDPE to stretch more easily, it can be made into thin film for plastic bags.

Both are made by polymerising ethene, but using different catalysts, temperatures and pressures.

Thermosoftening and thermosetting polymers

Polymer chains form a tangled mass, like a bowl of cooked spaghetti. If the chains can slide around easily, the plastic is flexible. If not, it is more rigid.

> Some thermosoftening plastics are flexible at room temperature. Others need warming. Stretching softened plastic helps to untangle the molecules. This allows the polymer to be drawn out into a fibre.

> When a thermosetting plastic is heated during moulding, covalent bonds form between the chains. These cross-links hold the chains together, making the polymer hard and rigid. Because its molecules cannot slide past each other, the polymer cannot melt.

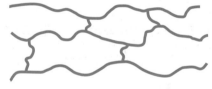

FIGURE 4: Why are the bottle and bag made from different types of poly(ethene)?

QUESTIONS

4 Suggest why LDPE is less dense than HDPE.

5 Give two ways that a polymer is like spaghetti.

6 Describe the effects cross-linking has on a polymer.

FIGURE 5: How can you tell that this polymer is thermosetting?

Intermolecular forces (Higher tier)

Polymer molecules are long chains made from carbon atoms, with various side groups attached. All the atoms are held together by strong covalent bonds.

The intermolecular forces (intermolecular bonds) between the polymer molecules are weak. However, because the molecules are very large, the effect of these forces is stronger than between small molecules. So, polymers are solids, with higher melting points than compounds containing smaller molecules.

A plastic contains polymer chains of various lengths. The intermolecular forces between smaller chains are easier to overcome than forces between longer chains, which have a greater area of contact. That is why a plastic has no definite melting point. It softens and melts over a range of temperatures.

QUESTIONS

7 Melamine is an example of a thermosetting polymer. Use the idea of intermolecular forces to explain why nylon can be made into fibres, but melamine cannot.

 thermosoftening thermosetting plastic

Metallic properties

Marvellous metals

Metals are strong and bend without breaking, allowing an almost limitless range of uses. Without electrical conduction through aluminium power lines, copper wiring and tungsten light bulb filaments, we would be in the dark. Metals also conduct heat, so we can keep warm and cook food. Without metals, we would still be in the Stone Age.

Did you know?

Computer chips are connected with gold wires – because gold is one of the best electrical conductors and does not corrode.

You will find out:

> metals consist of giant structures of atoms
> how this structure gives metals their special properties
> why metals conduct heat and electricity

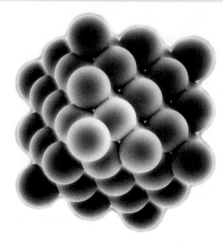

FIGURE 1: Tungsten glowing in a filament light bulb.

Metals: properties, structure and bonding

Properties

Metals have a range of very useful properties. They are:

> usually hard and dense, with high melting and boiling points

> good conductors of heat and electricity

> **flexible** and **malleable** (can be bent or hammered into shape without breaking)

> **ductile** (can be drawn out into thin wires).

All these properties are due to their structure, and the **metallic bonding** that holds the structure together.

Structure

Like ionic compounds and some covalent compounds, metals have giant structures. Metal atoms are arranged in a regular lattice of rows and layers, which can slide over each other. This lets metals bend without breaking. The atoms pack closely together so metals have high densities.

Bonding

Atoms in metals are held together by metallic bonding. Unlike ionic and covalent bonds where electrons are in fixed positions, some electrons in a metal are delocalised and free to move. This allows conduction.

FIGURE 2: Model of a metal lattice. How many different sets of layers can you see?

QUESTIONS

1 Why are most metals denser than non-metals?

2 Explain why metals are malleable and flexible.

3 Describe the difference between 'malleable' and 'flexible'.

FIGURE 3: In what way is a metal like this book?

🔍 metal GCSE

Alloys

In a pure metal, all the atoms are the same size. They fit into perfectly regular layers, which can slide easily. Alloys are mixtures of two or more metal elements. Larger or smaller atoms in alloys disrupt the regular lattice. The layers slide less easily, so alloys are harder and less malleable than pure metals.

Most metals in common use are alloys, because they are stronger than pure metals.

Because disrupting the lattice affects bonding, alloys have lower melting points than pure metals. For example,

Metal	tin	lead	solder (an alloy of tin and lead)
Melting point °C	232	328	about 185

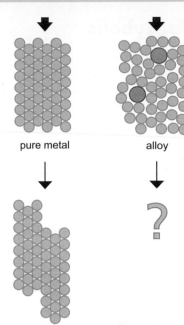

pure metal   alloy

layers slide

FIGURE 4: The layers in an alloy lattice cannot slide as easily as in pure metal.

Shape memory alloys

Special alloys, such as nitinol (a nickel/titanium alloy), have an amazing property. They 'remember' their shape. The arrangement of atoms in the lattice depends on temperature.

To make a dental brace, for example, nitinol is cooled, stretched and attached to the teeth. At body temperature, the atoms return to their original unstretched arrangement, pulling the brace tight.

What is metallic bonding? (Higher tier)

In metal lattices, the outer electrons are delocalised. Instead of being attached to particular atoms they form a 'sea of electrons' between the positive ions (atoms without their outer electrons). The ions repel each other, so become arranged at equal distances, forming the regular lattice.

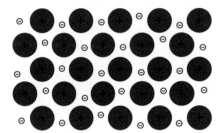

FIGURE 5: A diagram of metallic bonding. What do the + and − represent?

Electrostatic attraction between electrons and ions holds the lattice together. Because there are no fixed bonds, both the ions and electrons can move – giving metals their characteristic properties.

> A physical force (bending, hitting or pulling) moves the layers of ions over each other.

> A potential difference (voltage) moves electrons through the lattice.

QUESTIONS

4 Explain why bronze is less malleable than pure copper.

5 Describe the 'sea of electrons' in a metallic lattice.

6 Why does a sheet of steel not break, when bent into the shape of a car body?

Conduction (Higher tier)

Electrical

When a potential difference is applied to a piece of metal, the delocalised electrons flow from the negative connection towards the positive connection. The metal conducts electricity. The metal ions do not move.

Thermal

Heating a metal makes its particles vibrate more. These vibrations are passed on to nearby particles, conducting energy through the material. Both the ions and electrons vibrate, helping conduction. Since the delocalised electrons are 'free', they move faster and transmit the energy more quickly. Thus, metals conduct heat better than non-metals.

QUESTIONS

7 Explain the differences between conducting heat and electricity.

 alloy GCSE ... metallic bonding animation

Modern materials

You will find out:

> smart materials respond to changes around them

> nanoparticles show different properties from normal particles of the same materials

> about some uses of smart materials and nanoparticles

Buckyballs

In the 1980s, scientists accidentally discovered a new form of carbon. It is made up of 60 carbon atoms arranged in hexagons and pentagons like a football. They named this C_{60} molecule *buckminsterfullerene* after architect Buckminster Fuller, who designed buildings of this shape. The molecules are often called buckyballs.

FIGURE 1: C_{60} has 60 carbon atoms joined into hexagons and pentagons, like a football.

Did you know?

Making a lucky discovery by chance is called serendipity. There have been many such discoveries in science, including electric currents and phosphorus.

Smart materials and nanoscience

Smart materials respond to changes in conditions around them, such as light, temperature, electricity or magnetism. For example:

> **Photochromic** sunglasses darken in sunshine, but lighten again in the shade.

> **Thermochromic** materials change colour according to the temperature.

> **Shape-memory alloys** can be stretched, but return to their original shape when warmed.

Nanoscience is the study of new materials called **nanoparticles**. These are extremely small – between 1 and 100 **nanometres** (nm) in size.

A nanometre is 0.000 000 001 metre. That is about 100 000 times smaller than the width of a human hair.

Nanoparticles consist of just a few tens or hundreds of atoms.

Nanoparticles can be made of various substances. It is their size and not their composition that makes them nanoparticles. **Buckminsterfullerene** (C_{60}) is a form of carbon. Nanotubes can made from carbon, metal oxides or even DNA.

Nanoparticles behave differently from normal materials. They have some unusual properties that could improve everyday lives in the future. Potential applications include catalysts, sensors, coatings, lighter construction materials and medicines.

FIGURE 2: Which type of smart material is in use here?

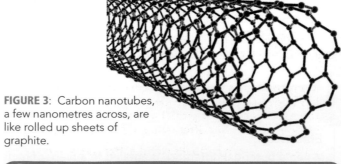

FIGURE 3: Carbon nanotubes, a few nanometres across, are like rolled up sheets of graphite.

QUESTIONS

1 Explain what is 'smart' about photochromic sunglasses.

2 How big are nanoparticles?

3 What are buckyballs made of?

Q nanoparticles uses ... buckminsterfullerene

Applications of modern materials

Thermochromic materials

Materials that reversibly change colour on heating or cooling have some fun applications.

> Dark mugs reveal a picture when a hot drink is poured in.

> T-shirts change colour where the body is in close contact.

> Mood rings respond to changes in body temperature.

Surface area and catalysis

Nanoparticles have special properties because they have very large surface areas compared with their volume. Table 1 shows that 1 cm³ of nano-sized cubes have a surface area a million times larger than a single 1 cm cube.

Most industrial catalysts are solids. Reactions occur on their surface, so nanoparticles should make excellent catalysts.

Length of one side	Surface area (cm²)	Volume (cm³)	Ratio surface area : volume
1 cm	6	1	6 : 1
1 mm (= 10^{-1} cm)	6×10^{-2}	10^{-3}	60 : 1
10 nm (= 10^{-6} cm)	6×10^{-12}	10^{-18}	6 000 000 : 1

TABLE 1: Properties of cubes.

Improved biosensors

Some pregnancy test kits use gold nanoparticles to detect a hormone that is present in a pregnant woman's urine. Military scientists hope to use similar materials as **biosensors** to detect very low levels of chemical or biological agents.

Protective coatings

Nanoparticle coatings are extremely thin and transparent. Uses being developed include:

> anti-reflective, scratch-resistant surfaces for spectacles and windscreens

> colourless sunscreen lotions instead of white

> anti-bacterial coatings on plastic food-wrappings.

Lighter, stronger materials

Weight for weight, carbon fibres are stronger than steel. Carbon nanotubes are even stronger. Although ten thousand times thinner than a hair, they can be made several centimetres long. Such fibres could be used in ultra lightweight **composites** for cars and aeroplanes.

FIGURE 4: Titanium dioxide particles in sunscreens are white. If they were nano-sized, they would be colourless.

QUESTIONS

4 Describe how 'mood rings' work.

5 Explain why nanoparticles make excellent catalysts.

6 Suggest how nanoparticles could help to keep food fresher, for longer.

Fullerenes (Higher tier)

Many molecules similar to buckminsterfullerene (C_{60}) have been made. Collectively known as **fullerenes**, they too are based on hexagonal and pentagonal rings of carbon atoms. They offer many interesting possibilities.

> Various sized buckyballs could act as cages to carry medicines to specific points in the body. For example, atoms of radioactive radon gas trapped in C_{60} could find and kill cancer cells.

> Flat sheets, like single layers of graphite, would make good lubricants.

> Nanotubes can reinforce items such as tennis racquets and wind turbine blades.

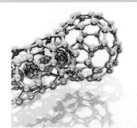

FIGURE 5: Using a nanotube to load pharmaceutical molecules into fullerenes.

QUESTIONS

7 Use your imagination to suggest new ways in which nanoscience might be used in future.

Identifying food additives

You will find out:

> some uses of food additives

> why additives need to be identified

> about chromatography – a technique used to analyse chemicals

Looking closely at food

People are more likely to buy food that looks attractive. However, its appearance is often due to additives, particularly colourings. People with allergies want to know whether a food contains additives that trigger that allergy. Scientists need to know how to identify additives in foods, to ensure that the additive is permitted.

Did you know?

The oldest types of food additives are preservatives. The Romans and Greeks used vinegar for pickling. From the Middle Ages, saltpetre (potassium nitrate) has been used to preserve meat.

FIGURE 1: Colour matters. Does this meal look appetising?

Additives and analysis

Food additives improve colour, texture and flavour. Some additives are **synthetic** (manufactured). Many are natural substances. Only substances known to be safe are allowed in food. In Europe, approved additives are given E numbers. For example, E120 is cochineal, a pink colouring, and E330 is citric acid.

Some people are allergic to certain additives and suffer unpleasant side effects from them, even though they are judged to be safe. There are other problems. For instance, a yellow dye, tartrazine (E102), is linked to hyperactivity in children. This is why it is important that additives are shown on food labels.

Mmm... *Gravy*

ingredients:
Potato starch, Maltodextrin, Vegetable Oils, Salt, Colour (E150c), Flavourings (contain Celery, Soya, Wheat), Wheatflour, Flavour Enhancers (E621, E635), Emulsifier (E322)(Soya), Spice & Herb Extracts, Onion Extract

ALLERGY ADVICE: Celery, Soya, Wheat

FIGURE 2: How many additives are there in this gravy powder?

Analysing food

Occasionally, illegal and/or **toxic** additives are found in foods. Scientists regularly perform **chemical analysis** on food samples to check what they contain.

One important method of analysis is **chromatography**. It separates mixtures, so that the components can be identified.

QUESTIONS

1 What is an additive?

2 Explain what an E number shows.

3 Explain why it is important for scientists to be able to analyse food.

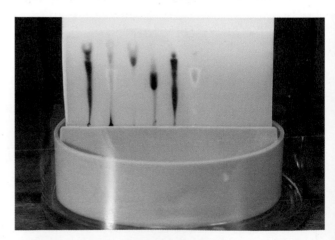

FIGURE 3: Chromatography in progress. How can you tell whether each sample contained a mixture?

Chromatography

Paper chromatography

Paper chromatography can show whether the colourings in a food are those listed on the label. This involves:

> extracting a sample of colour from the food

> drawing a start line, in pencil, across the chromatography paper near the bottom

> putting dots of the extracted colour and listed colourings along the line and labelling them

> standing the paper in a little **solvent** that dissolves the colourings.

As the solvent rises up the paper, it carries the colours with it, and separates them. The pattern of spots produced is called a **chromatogram**.

Interpreting the chromatogram

Chromatography relies on a simple fact: each chemical in a mixture travels up the paper at the same speed as it does when alone. Different chemicals travel at different speeds.

Look at Figure 4. Compare the colours of the spots and their distances from the start line. Three spots in the food sample means there were three dyes in the food. Two spots correspond with E102 and E133. The food contained both these dyes.

The third spot corresponds with none of the known dyes: it is unidentified. It may be a colour naturally present in the food, an approved dye not listed on the label, or it could be an illegal dye. Further analysis is needed.

Other uses

Chromatography means 'colour writing', but it is not limited to coloured materials. It is also used to separate and identify many types of substance, such as medicines and drugs.

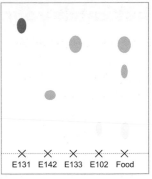

FIGURE 4: A chromatogram of food dyes. Which colouring in this food needs further investigation?

QUESTIONS

4 Suggest why the chromatogram start line should be drawn in pencil.

5 Suggest why the solvent level must be below the start line.

6 A chromatogram has two spots the same shade of red, but at different heights. What does this tell you?

 # More about chromatography

Retention factors

The solvent rises up chromatography paper faster than the spots, because substances are absorbed and retained by the paper. The **retention factor** (R_f) compares the movement of the substance and the solvent. It is always the same for a given substance under the same conditions.

$$R_f = \frac{\text{distance moved by the substance}}{\text{distance moved by the solvent}}$$

The distance is measured to the centre of the spot.

The R_f value for E102 in Figure 5 is $1.7 \div 10 = 0.17$

Thin layer chromatography

Thin layer chromatography (TLC) is similar to paper chromatography. Instead of paper, it uses 'plates' coated with a thin layer of absorbent, such as silica. It is quicker, but more expensive.

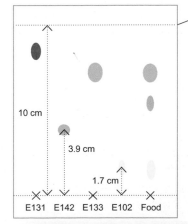

solvent front: height reached by solvent

10 cm

3.9 cm

1.7 cm

E131 E142 E133 E102 Food

FIGURE 5: What is the R_f value for E142 in this chromatogram?

QUESTIONS

7 Suggest some reasons why different substances have different R_f values.

Q chromatography GCSE ... retention factor definition

Instrumental methods

You will find out:

> how gas chromatography and mass spectrometry work

> these methods are quick and accurate and often used together

First catch your sample

Racehorses and greyhounds are anti-doping tested. Just like athletes, it is their urine that is tested. However, you cannot just ask a horse or dog to provide a sample. The vet has to persuade the animal to urinate and then catch the sample in a cup. A sealed sample is taken for analysis. The laboratory will use instrumental methods such as gas chromatography.

Did you know?

The Horseracing Forensic Laboratory performs anti-doping tests on racehorses and greyhounds. It also analyses animal feeds, to check for accidental contamination with illegal substances (such as poppy seeds containing heroin-like chemicals).

FIGURE 1: Greyhounds have to provide a urine sample, just like any other athlete.

Types of chromatography

Column chromatography uses a solid substance packed in a tube instead of paper or a **thin layer chromatography** (TLC) plate. A solvent carries the mixture down the column. Each compound moves at a different speed. The components separate and emerge one at a time. They are then collected and analysed.

Separating a complicated mixture needs a longer column. Analytical chemists use a method called **gas chromatography**. It uses a solid substance packed in a long, thin tube coiled into a spiral.

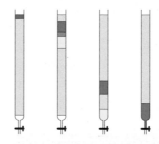

FIGURE 2: Column chromatography of green food colouring. What made the dyes move down the column?

1. The spiral is heated in an oven to vaporise the mixture.

2. Instead of a liquid solvent, a gas carries the vaporised mixture through the column.

3. Components travel through the material at a different speeds. Each reaches the detector (at the end of the column) after a different time.

4. The detector produces peaks on a graph, one peak for each component.

5. The number of peaks shows the number of substances in the mixture.

6. The height of a peak shows the amount of that substance present.

Chromatography is just one of many instrumental methods that analysts use. Analysing substances by hand takes a lot of time and effort. Instruments are faster, more accurate and more reliable. They can detect tiny amounts, so are especially useful when only a small sample is available.

FIGURE 3: The spiral column in a gas chromatograph is often 2 to 3 metres long.

QUESTIONS

1 Give the two main differences between column chromatography and gas chromatography.

2 Explain why the spiral column is enclosed in an oven.

3 What do the peaks on a gas chromatogram tell the analyst?

Q gas chromatography GCSE

Retention times and mass spectrometry

The time taken for a substance to pass through the column is called its **retention time** (R_t). Like **retention factor** (R_f) in paper chromatography, retention time for a particular substance is always the same under the same conditions. This helps to identify the substance.

Retention time for a particular peak can be read off the gas chromatogram. It can then be compared with known retention times for a list of substances, to find a match.

Some substances have similar retention times, so are difficult to distinguish. A more reliable identification method is needed. Most often it is mass spectrometry.

As each component emerges from the gas chromatograph, it passes by the detector and then into a **mass spectrometer**. This instrument identifies each component quickly and accurately by measuring its relative molecular mass.

The combined technique is called **GC–MS** (gas chromatography–mass spectrometry). It can detect very small quantities, so all the components in a mixture are detected and identified.

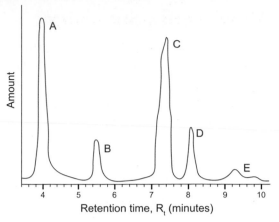

FIGURE 4: An enlarged section of a gas chromatogram. What is the retention time of component C?

QUESTIONS

4 Explain the term 'retention time'.

5 How do retention times help to identify components?

How does mass spectrometry identify substances? (Higher tier)

The mass spectrometer knocks an electron off the molecule, leaving a positively charged **molecular ion**. The instrument measures its relative mass (M_r), which is the same as the M_r for the whole molecule.

The mass spectrometer also breaks molecules into smaller fragments, records their pattern and measures their relative masses. This helps distinguish between compounds with the same M_r: for instance, propanone, CH_3COCH_3, and its isomer propanal, C_2H_5CHO. Both have $M_r = 58$. Propanone gives a large peak at mass 43 (fragment CH_3CO-) but not 29. Propanal gives one at 29 (C_2H_5- and -CHO) but not 43.

In GC–MS:

> the gas chromatograph separates the mixture

> the mass spectrometer identifies the components by their M_r and fragmentation patterns.

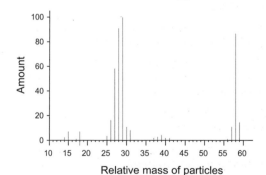

FIGURE 5: A mass spectrum.

QUESTIONS

6 For the mass spectrum in Figure 5, explain why M_r for the molecular ion is the same as for the whole molecule.

Making chemicals

You will find out:
> making chemicals needs lots of planning
> some of the aspects that must be considered
> about various types of reaction that can be used

Who needs chemistry?

'You can't live without us' is the motto of the UK Chemical Industries Association. Chemical companies make the materials for almost everything we use. Even the use of natural materials relies on manufactured products. Crops need fertilisers. Animals need veterinary products. Water must be sterilised. Wood is painted. Wool and cotton are dyed.

Did you know?

Soliris® is one of the most expensive chemicals available. It is a pharmaceutical product for treating a rare blood disease and costs about £11 000 per gram – four hundred times more expensive than gold.

FIGURE 1: A chemical plant needs lots of different types of equipment.

Making chemicals is complicated

Making chemicals is essential – but it is not as simple as mixing a few reactants together. It takes a lot of thought and care to get it right. Chemists must think about:

> the best way to make the product – which reactions to use

> what starting materials and reagents are needed for these reactions

> how much of each reactant is needed, to produce the required amount of product

> reaction conditions, such as temperature and pressure.

They must also consider:

> the economics – the cost of making the product, and how much they can sell it for

> health and safety – including hazards and risks for the chemicals used, and for the process

> environmental problems – avoiding pollution and how to dispose of waste materials.

Unless correct procedures are followed carefully, the process will not work properly. At best, the product will be low quality. At worst, the process may run out of control and cause a disaster.

Mistakes can be fatal. Investigators think that the accident shown in Figure 2 was due to incorrect labelling of a chemical. It was stored with the fertiliser, reacted and caused the explosion.

FIGURE 2: This fertiliser factory exploded in 2001, killing 30 people. The accident was probably the result of someone's carelessness.

QUESTIONS

1 Suggest three reasons why manufacturing a chemical is more complicated than preparing some in a laboratory.

2 Name two types of reaction conditions.

3 Apart from the risk of an accident, why is it essential to follow procedures carefully?

Routes and conditions

Often there are several ways to make a chemical. Different raw materials or different reaction routes may be used. Reaction types include:

> **oxidation** and **reduction** – one reactant gains oxygen atoms from another reactant that loses them

> **neutralisation** – reacting acids with bases to produce salts

> **precipitation** – mixing two solutions to produce an insoluble product

> **electrolysis** – passing electricity through a liquid, to form products at the electrodes.

For example, Table 1 shows three ways to make sodium hydroxide solution:

Equation for the reaction	Type of reaction
$2Na(s) + 2H_2O(\ell) \rightarrow 2NaOH(aq) + H_2(g)$	oxidation / reduction
$Na_2CO_3(aq) + Ca(OH)_2(aq) \rightarrow 2NaOH(aq) + CaCO_3(s)$	precipitation
$2NaCl(aq) + 2H_2O(\ell) \rightarrow 2NaOH(aq) + Cl_2(g) + H_2(g)$	electrolysis

TABLE 1: Making sodium hydroxide.

The way a reaction is carried out affects its speed, costs and safety. It may also alter the final product.

Reaction conditions include:

> temperature
Higher temperatures make reactions faster, but may cause a product to decompose.

> concentration
More concentrated solutions react faster, but may also give different products (Figure 3).

> pressure
Higher pressure squeezes gas molecules closer together, so increases the concentration.

> particle size
Small particles of solid react faster than large lumps.

> catalysis
A catalyst may speed up a reaction.

FIGURE 3: Copper reacting with concentrated nitric acid. How might the reaction be different with dilute nitric acid?

 QUESTIONS

4 Why is the reaction of sodium carbonate with calcium hydroxide to make sodium hydroxide described as a precipitation?

5 Does pressure affect reactions between liquids? Explain your answer.

6 Suggest factors that chemists consider when deciding which reaction routes and conditions to use for a particular product.

Chemical calculations

Chemists need to calculate the correct quantities of reactants needed to make the required amount of product. Using the wrong proportions is uneconomic and wasteful. Suppose 12 tonnes of X needs to react with 10 tonnes of Y to make the product.

> If 11 t of Y is used, only 10 t can react, so the other 1 t is wasted.

> If only 9 t of Y is used, it can make only 90% ($\frac{9}{10}$) of the correct amount of product.

Reacting the wrong proportions can sometimes even give a different product.

 QUESTIONS

7 Look at the example on the left. If only 9 t of Y is used, how much X is left unreacted, and therefore wasted?

🔍 oxidation AND reduction ... neutralisation ... precipitation reaction

Preparing for assessment: Applying your knowledge

To achieve a good grade in science, you not only have to know and understand scientific ideas, but you need to be able to apply them to other situations and investigations. These tasks will support you in developing these skills.

✳ Hit and run

Michael is a police forensic investigation officer. He has been asked to examine a car that has just been brought into the vehicle compound after a serious accident where a pedestrian has been killed by a hit and run driver. The hit and run driver has also hit a parked car as it left the scene. Michael is examining the parked car.

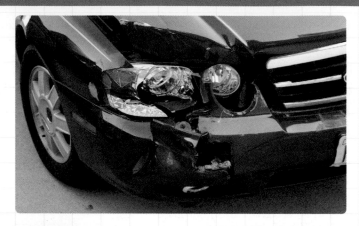

Michael finds several pieces of blue paint on the red car, and also a small piece of plastic, which is probably from the hit and run driver's car. He places the paint and the plastic in sample tubes, ready to analyse them in the laboratory.

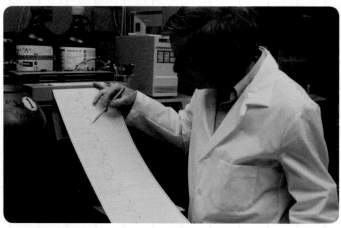

Back at the lab, Michael dissolves each sample in propanone, which is a good solvent. In turn, he injects each dissolved sample into a gas chromatograph attached to a mass spectrometer.

The gas chromatograph has a tube packed with an unreactive solid powder. The injected sample vaporises and is blown through the column by an inert gas. Chemicals in the sample are separated and each one passes through the mass spectrometer where it can be identified.

✳ Task 1

(a) Describe how Michael could use information from the gas chromatograph to find out how many chemicals are in the blue paint.

(b) Use Graph 1 to decide how many chemicals are present in the paint. Remember that Michael dissolved the blue paint sample in propanone.

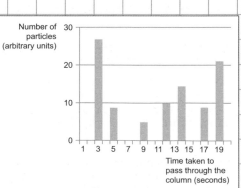

GRAPH 1:The gas chromatograph of one of the blue paint samples that Michael analysed.

✸ Task 2

(a) Explain how gas chromatography separates a mixture into each of the different chemicals present.

(b) Explain how the time taken for a chemical to pass through the column (its retention time) can be used to help identify the chemical.

(c) Explain why mass spectrometry is a better way to identify a chemical than using just the chemical's retention time.

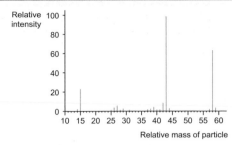

GRAPH 2: Mass spectrum of the first chemical through the gas chromatograph shown in Graph 1.

✸ Task 3

(a) Graph 2 shows the mass spectrum for the first chemical through the column.

What was the retention time of this chemical on the gas chromatography column?

(b) The mass spectrum shows a number of peaks due to particles of different masses. The peak at 58 is the 'molecular ion' and corresponds to the chemical. The others are fragments of the chemical that also form in the mass spectrometer.

Use the relative atomic masses below to show that the chemical is propanone, the solvent used to dissolve the paint sample.

Element	A_r
C	12
H	1
O	16

✸ Task 4

Car manufacturers have to supply the police with the chemical composition of all the paints that they use.

Describe how Michael could use this information and his results to identify the make of the hit and run car.

✸ Task 5

Suggest why using gas chromatography with mass spectroscopy is better than using paper chromatography, when investigating crimes such as the one that you have looked at here.

✸ Maximise your grade

Answer includes showing that you...
know how to use a gas chromatograph to identify chemicals in a mixture.
can work out the number of chemicals in paint from its chromatogram.
can briefly describe how the car manufacturers' data and gas chromatography/mass spectroscopy can identify the make of car.
can describe how a particle's mass affects its speed through a gas chromatography column.
can calculate the relative formula mass of propanone.
can explain the advantages of gas chromatography over paper chromatography.
recognise that a solvent used in the gas chromatography may also be present in the sample.
can explain in detail how the car manufacturers' data and both gas chromatography and mass spectroscopy can identify the make of car.

Chemical composition

You will find out:
> how to calculate the percentage of each element in a compound
> the meaning of 'empirical formula' and how to calculate it

An elementary mistake

John Dalton introduced the idea of atomic weights (or masses) in 1806. He worked them out by measuring the masses of elements that combined with each other. Several of Dalton's results were way out because he did not know the chemical composition of compounds – he just guessed. He assumed that water was HO, which gave the atomic weight of oxygen as 7. The correct composition of H_2O would have given him 14 – much closer to the true value, 16.

FIGURE 1: Relative atomic masses can be determined accurately to several decimal places using a high resolution mass spectrometer.

Percentage composition

To grow properly, plants need:

> nitrogen (N) for healthy leaves

> phosphorus (P) for good roots

> potassium (K) for flowers and fruit.

Plants take these elements from the soil. Farmers and gardeners use fertilisers to replace them. Carrots (roots), cabbages (leaves) and strawberries (fruit) each need different proportions of each element. Fertiliser bags show the proportions of N, P and K.

To calculate the composition by mass of a compound you need to know its formula and each element's **relative atomic mass** (A_r).

Step 1
Calculate the relative formula mass of the compound.

For example, for potassium nitrate, KNO_3, the A_r values are K = 39, N = 14, O = 16

So the **relative formula mass** (M_r) = 39 + 14 + (3 × 16) = 101

Step 2
Use the formula:

$$\text{Percentage (by mass)} = \frac{\text{(element's relative atomic number)} \times \text{(number of atoms in the formula)}}{\text{relative formula mass of compound}} \times 100$$

For KNO_3:

Element	Percentage by mass	
potassium	$\frac{(39 \times 1)}{101} \times 100$	38.6%
nitrogen	$\frac{(14 \times 1)}{101} \times 100$	13.9%
oxygen	$\frac{(16 \times 3)}{101} \times 100$	47.5%

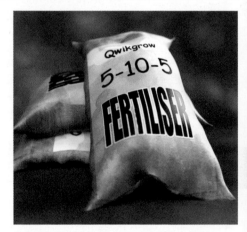

FIGURE 2: 5–10–5 on this label refers to the proportion of nitrogen, phosphorus and potassium in the fertiliser. Would the fertiliser be suitable for carrots?

QUESTIONS

1 Explain why farmers need to know the composition of fertilisers.

2 Calculate the percentage of nitrogen in ammonium nitrate, NH_4NO_3. (A_r: H = 1, N = 14, O = 16)

3 Calculate the percentage of copper in copper sulfate, $CuSO_4$. (A_r: O = 16, S = 32, Cu = 64)

Q fertiliser GCSE

Empirical formula (Higher tier)

Before instrumental methods were available, chemists analysed a sample of a compound and measured the mass of each element. From these data, they calculated the ratio of atoms present. This gave the compound's **empirical formula**. Empirical means based on experiments.

Example

46 g of a mineral contained 16 g copper, 14 g iron and 16 g sulfur. (A_r Cu = 64, Fe = 56, S = 32)

1. Divide each mass by A_r for that element.
   Cu, 16 ÷ 64 = 0.25; Fe, 14 ÷ 56 = 0.25; S, 16 ÷ 32 = 0.5

2. Divide each result by the smallest one (0.25 in this case).
   Cu, 0.25 ÷ 0.25 = 1; Fe, 0.25 ÷ 0.25 = 1; S, 0.5 ÷ 0.25 = 2

3. Thus the ratio of atoms Cu:Fe:S is 1:1:2
   This gives the empirical formula $CuFeS_2$. The mineral is the copper ore, chalcopyrite.

The same calculation also works using percentages instead of masses.

For a molecular compound, the actual formula of the molecule may be a multiple of the calculated empirical formula. For examples, see Table 1.

Compound	Empirical formula	Molecular formula
hydrogen peroxide	HO	H_2O_2
hydrazine	NH_2	N_2H_4
ethanoic acid	CH_2O	CH_3COOH
glucose	CH_2O	$C_6H_{12}O_6$

TABLE 1: Empirical and molecular formulae of compounds.

QUESTIONS

4 What is the empirical formula of butane, C_4H_{10}?

5 Recalculate the formula for chalcopyrite from its composition: 34.8% copper, 30.4% iron and 34.8% sulfur.

6 Analysis shows that a compound contains 12.8 g oxygen, 6.4 g sulfur and 0.4 g hydrogen. Calculate its empirical formula.

Analysing hydrocarbons

The proportions of carbon and hydrogen in a **hydrocarbon** can be measured by burning the compound and measuring the masses of carbon dioxide and water produced.

Example

Burning 1.12 g of a hydrocarbon produced 3.52 g of carbon dioxide and 1.44 g of water.

A_r H = 1, C = 12, O = 16

M_r CO_2 = 44, H_2O = 18

1. Mass of carbon present = $\frac{12}{44} \times 3.52 = 0.96$ g

2. Mass of hydrogen present = $\frac{2}{18} \times 1.44 = 0.16$ g

3. Divide each mass by A_r for that element:
   C, 0.96 ÷ 12 = 0.08; H, 0.16 ÷ 1 = 0.16

4. Ratio of atoms is C:H = 1:2, so the empirical formula is CH_2

5. The compound could be any alkene (C_2H_4, C_3H_6, and so on) or cyclohexane, C_6H_{12}.

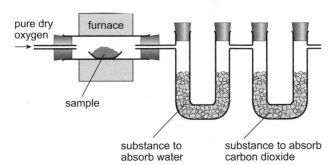

pure dry oxygen → furnace

sample

substance to absorb water

substance to absorb carbon dioxide

FIGURE 3: Analysing carbon and hydrogen in a compound. What measurements are needed before and after burning the sample?

QUESTIONS

7 Calculate the empirical formula of a hydrocarbon that gives 1.32 g carbon dioxide and 0.72 g water when burned.

Quantities

The *Discworld* ironmaker

When Sir Terry Pratchett was awarded his knighthood, he thought that a knight should have a sword. So, he decided to make his own. He dug the iron ore from a local field, and smelted it in a homemade furnace in his garden. He hammered the iron flat and, with a blacksmith's help, made it into a sword. Pratchett needed to know how much iron ore to use, and how much charcoal and limestone to smelt it with. He needed to learn some chemistry.

FIGURE 1: Sir Terry Pratchett made his own sword from iron ore.

Quantities from equations (Higher tier)

Balanced equations show the proportions of substances involved in a reaction. Look at the neutralisation reaction:

$$2NaOH(aq) + H_2SO_4(aq) \rightarrow Na_2SO_4(aq) + 2H_2O(\ell)$$

This balanced equation shows that:

2 sodium hydroxide formula units react with 1 sulfuric acid formula unit to produce 1 sodium sulfate formula unit and 2 water formula units (molecules).

To convert these quantities into masses, chemists use relative atomic and formula masses (A_r and M_r).

A_r H = 1, O = 16, Na = 23, S = 32

	NaOH	H_2SO_4	Na_2SO_4	H_2O
M_r	23 + 16 + 1 = **40**	$(2 \times 1) + 32 + (4 \times 16) =$ **98**	$(2 \times 23) + 32 + (4 \times 16) =$ **142**	$(2 \times 1) + 16 =$ **18**

Combining these masses with the equation shows that:

$(2 \times 40) = 80$ mass units of sodium hydroxide react with $(1 \times 98) = 98$ mass units of sulfuric acid

to form

$(1 \times 142) = 142$ mass units of sodium sulfate and $(2 \times 18) = 36$ mass units of water.

Mass units could be, for example, grams, kilograms or tonnes. However, they must be the same for all substances.

In a laboratory:

80 g of sodium hydroxide react with 98 g of sulfuric acid to form 142 g of sodium sulfate and 36 g of water.

In an industrial process:

80 t of sodium hydroxide react with 98 t of sulfuric acid to form 142 t of sodium sulfate and 36 t of water.

Crosscheck

In any chemical reaction, the mass of products and reactants must be equal. 80 + 98 = 142 + 36, so this is OK.

Remember

A formula unit is a molecule or the empirical formula of a compound with a giant structure.

Did you know?

Research chemists make a few grams in the laboratory. Manufacturers make thousands of tonnes a year. Both calculate their quantities in the same way, using equations and relative atomic masses.

Q word symbol equations GCSE

How much is needed? (Higher tier)

Knowing the proportions of reactants and products allows any other quantities to be calculated.

Example 1

How much sulfuric acid is needed to neutralise 20 g of sodium hydroxide?

80 g of NaOH needs 98 g of H_2SO_4

1 g of NaOH needs $\frac{98}{80}$ g of H_2SO_4

20 g of NaOH needs $\frac{20 \times 98}{80}$ g of H_2SO_4

= 24.5 g

Example 2

How much sodium hydroxide is needed to manufacture 710 tonnes of sodium sulfate?

142 t of Na_2SO_4 needs 80 t of NaOH

1 t of Na_2SO_4 needs $\frac{80}{142}$ t of NaOH

710 t of Na_2SO_4 needs $\frac{710 \times 80}{142}$ t of NaOH

= 400 t

In a reaction, the **theoretical yield** is the quantity of a reactant that is obtained if all of the reactants react and all of the product is collected.

In the neutralisation above, 142 g of sodium sulfate is the theoretical yield from 80 g of sodium hydroxide and 98 g of sulfuric acid.

For various reasons, the actual yield is always lower.

 QUESTIONS

1 Explain why balanced equations are essential.

2 Calculate the relative formula mass for (a) calcium carbonate, $CaCO_3$, (b) calcium oxide, CaO.

3 What mass of calcium carbonate, $CaCO_3$, must be decomposed to give 14 g of calcium oxide, CaO?

Sir Terry Pratchett's sword (Higher tier)

Sir Terry Pratchett dug up 80 kg of iron ore. Assuming the ore was 50% iron oxide (Fe_2O_3) how much iron could he make?

The balanced equation below summarises a series of reactions that lead to the reduction of iron ore to iron by carbon:

$2Fe_2O_3(s) + 3C(s) \rightarrow 4Fe(\ell) + 3CO_2(g)$

A_r Fe = 56, O = 16, C = 12

	Fe_2O_3	CO_2
M_r	$(2 \times 56) + (3 \times 16) = \mathbf{160}$	$12 + (2 \times 16) = \mathbf{44}$

Using the balanced equation:

$(2 \times 160) = 320$ mass units of iron oxide react with $(3 \times 12) = 36$ mass units of carbon

to form

$(4 \times 56) = 224$ mass units of iron and $(3 \times 44) = 132$ mass units of carbon dioxide

Working in kilograms:

320 kg of iron oxide react with 36 kg of carbon to form 224 kg of iron and 132 kg of carbon dioxide

Assuming 80 kg of ore contained 40 kg of Fe_2O_3, the theoretical yield is $\frac{40 \times 224}{320} = 28$ kg

 QUESTIONS

4 Calculate how much charcoal (carbon) is needed to reduce 40 kg of Fe_2O_3.

5 Suppose the ore contained 58 kg of magnetite, Fe_3O_4, instead of 40 kg of haematite, Fe_2O_3. Repeat the calculations to find out the theoretical yield of iron and the mass of charcoal needed.

Q chemical reaction equation quantities

How much product?

You will find out:

> that the amount of product formed is called the yield

> why the actual yield is always less than the theoretical yield

> how to calculate percentage yield

Missing magnesium

Sam weighed some magnesium ribbon, burned it, and weighed the magnesium oxide formed. The product weighed only about half of what Sam calculated should form. Possible reasons for this included: the dirty ribbon was not pure magnesium, some magnesium oxide escaped as smoke, or the product was not pure, because some magnesium had reacted with nitrogen instead.

FIGURE 1: When magnesium burns, some magnesium oxide escapes as smoke.

 How much product

Theory and reality

How much product a reaction gives depends on the amounts of reactants used.

> The **theoretical yield** is calculated from the equation. It is the maximum possible yield.

> The **actual yield** is found by weighing the product obtained.

The actual yield is always less than the theoretical yield:

> Reactants are rarely 100% pure. The mass of actual reactant is less than the mass weighed out, so forms less product.

> The same reactants form different products (Figure 2).

> It is impossible to separate all of the product from the reaction mixture.

> If the reaction is **reversible**, some of the products turn back into reactants, reducing the yield.

Percentage yield

To judge the effectiveness of making a product in a particular way, you compare the quantity of product obtained (actual yield) with what the equation predicts (theoretical yield). This is its **percentage yield**.

$$\text{percentage yield} = \frac{\text{actual yield}}{\text{theoretical yield}} \times 100$$

Example

If a reaction has a theoretical yield of 4.0 g but only 3.6 g is actually obtained:

$$\text{percentage yield} = \frac{3.6}{4.0} \times 100 = \textbf{90\%}$$

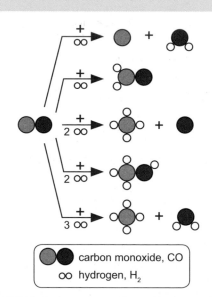

carbon monoxide, CO

hydrogen, H_2

FIGURE 2: Carbon monoxide and hydrogen form several different products.

QUESTIONS

1 Explain the terms theoretical yield and actual yield.

2 Which yield is always higher? Explain why.

3 Calculate the approximate percentage yield of Sam's magnesium oxide reaction.

theoretical actual yield

Calculating percentage yield (Higher tier)

Actual yield is obtained by collecting, purifying and weighing the product formed. Actual yields cannot be predicted from theory – they must be measured. Even for the same reaction, actual yield will vary. It depends on purity of the reactants, reaction conditions and method of separating the product from the reaction mixture.

Theoretical yield is calculated from the balanced equation (see previous page).

Example 1 Preparing sodium sulfate

$$2NaOH(aq) + H_2SO_4(aq) \rightarrow Na_2SO_4(aq) + 2H_2O(\ell)$$

Actual yield Neutralising 10.0 g of sodium hydroxide produced **14.2 g** of sodium sulfate.

Theoretical yield

M_r NaOH = 40; Na_2SO_4 = 142

10.0 g of NaOH gives $10.0 \times \frac{142}{(2 \times 40)}$ g of Na_2SO_4 = **17.75 g**

Percentage yield $= \dfrac{\text{actual yield}}{\text{theoretical yield}} \times 100 = \frac{14.2}{17.75}$ g × 100 = **80%**

Example 2 Manufacturing methanol

$$CO(g) + 2H_2(g) \rightarrow CH_3OH(\ell)$$

Actual yield Reacting 7.0 t of carbon monoxide with hydrogen produced 3.6 t of methanol.

Theoretical yield

M_r CO = 28; CH_3OH = 32

7.0 t of CO gives $7.0 \times \frac{32}{28}$ t of CH_3OH = **8.0 t**

Percentage yield $= \dfrac{\text{actual yield}}{\text{theoretical yield}} \times 100 = \frac{3.6}{8.0}$ t × 100 = **45%**

It is low because the reactants also form other products (see Figure 2).

Remember

Percentage yield cannot be predicted. It can be measured only by performing the reaction.

QUESTIONS

4 A reaction has a theoretical yield of 5.5 g but produces only 3.96 g. Calculate the percentage yield.

5 A process has a theoretical yield of 25 t and a percentage yield of 76%. How much product is obtained?

6 Reducing 160 t of haematite, Fe_2O_3, gives a theoretical yield of 112 t of iron. Because the ore is impure, the percentage yield is only 75%. What mass of ore is needed to make 10 t of iron?

Percentage yields, economics and the environment

Percentage yield indicates how effectively reactants are converted into the required product. If there are several ways to make a chemical, it is more economic to use a reaction with a high percentage yield. It gives more product from a given mass of reactant.

Finding ways to maximise percentage yields makes economic sense. It also helps the environment. Less raw materials are needed and less are wasted by ending up in unwanted by-products, which must be disposed of.

Inefficient combustion of fuels is a particular problem. It occurs on a huge scale worldwide and pollutes the environment with soot and carbon monoxide.

QUESTIONS

7 Explain why 'Finding ways to maximise percentage yields' makes economic sense.

FIGURE 3: Comment on the percentage yield of carbon dioxide when diesel is burned in this truck's engine.

calculation percentage yield

Reactions that go both ways

Red blood cells

Red blood cells contain haemoglobin, an important compound that carries oxygen around your body. In the lungs, each haemoglobin molecule combines with four oxygen molecules to form oxyhaemoglobin. This reaction is reversible. The oxyhaemoglobin is carried in the bloodstream to other cells throughout the body. There the reaction reverses. Oxyhaemoglobin molecules release the oxygen molecules, re-forming haemoglobin. Cells use the oxygen for respiration. The haemoglobin is carried back to the lungs to start again.

Did you know?

Rechargeable batteries rely on reversible reactions. Chemicals inside react, producing electricity. Recharging reverses this reaction, re-forming the original chemicals.

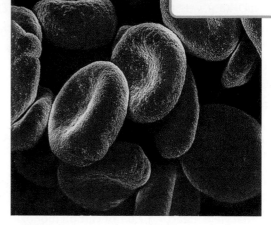

FIGURE 1: Haemoglobin in red blood cells combines with oxygen in a reversible reaction.

Reversible reactions

Lemonade and other fizzy drinks contain carbon dioxide. You cannot see bubbles in unopened bottles, because the gas is dissolved in the liquid. Carbon dioxide dissolves in water under pressure, forming a weak acid. That is why sparkling mineral water tastes sharper than still water.

carbon dioxide + water → carbonic acid

$$CO_2(g) + H_2O(\ell) \rightarrow H_2CO_3(aq)$$

Opening the bottle releases the pressure, producing a stream of bubbles. The reaction goes backwards, forming the original reactants again.

carbonic acid → carbon dioxide + water

$$H_2CO_3(aq) \rightarrow CO_2(g) + H_2O(\ell)$$

This is therefore a **reversible reaction**. It can go either way.

The forward and reverse reactions can be combined in one equation using a double-ended arrow:

$$CO_2(g) + H_2O(\ell) \rightleftharpoons H_2CO_3(aq)$$

The double ended arrow means that the reaction is reversible. The **forward reaction** goes from left to right. The **reverse reaction** goes from right to left.

FIGURE 2: Why do bubbles appear when a fizzy drink is opened?

QUESTIONS

1 Explain why fizzy drinks are slightly acidic.

2 Dissolving carbon dioxide in water is called a reversible reaction. Why?

3 Draw the sign that shows that a reaction is reversible. Explain why this sign is used.

🔍 carbon dioxide water reversible reaction

Examples in the laboratory

These examples of reversible reactions can be shown easily, in the laboratory. You may have seen some in class.

> ### pH indicators

Indicators change colour in acidic solutions, but change back again if the acid is neutralised and the solution is made alkaline. The reaction can go forward and back again as many times as you like.

FIGURE 3: These are both copper sulfate. Why do they look different?

> ### Heating ammonium chloride

Solid ammonium chloride 'disappears' when warmed, because it decomposes into two colourless gases. When the gases meet a cold surface, the solid reappears, as the gases recombine.

$NH_4Cl(s) \rightleftharpoons NH_3(g) + HCl(g)$

> ### Copper sulfate

Although blue copper sulfate crystals appear dry, they contain **water of crystallisation**, as their formula, $CuSO_4.5H_2O$, shows. The water is held within the crystal structure. Heating drives off this water, leaving a white powder, anhydrous copper sulfate. ('anhydrous' means 'without water'.) Adding water turns the white powder blue again, because the copper sulfate becomes rehydrated.

FIGURE 4: A drop of liquid on cobalt chloride paper. What does this tell you about the liquid?

$CuSO_4.5H_2O(s) \rightleftharpoons CuSO_4(s) + 5H_2O(\ell)$

Anhydrous copper sulfate is used to test for the presence of water. If it turns blue, the test liquid contains water, but is not necessarily just water.

> ### Cobalt chloride paper

This is another test for water. Filter paper is soaked in pink cobalt chloride solution and then dried. The anhydrous (dry) compound is blue, but turns pink again when wet.

$CoCl_2.6H_2O(s) \rightleftharpoons CoCl_2(s) + 6H_2O(\ell)$

◉ QUESTIONS

4 Describe how you could show that indicators undergo reversible reactions.

5 What would you see if ammonium chloride was warmed in the bottom of a test tube?

6 Some cobalt chloride paper has turned pink by absorbing moisture from the air. Suggest how you could turn it blue again.

Reversible reactions used in industry

Ammonia is manufactured by reacting nitrogen with hydrogen at about 400 °C. Unfortunately, hot ammonia decomposes again.

$N_2(g) + 3H_2(g) \rightleftharpoons 2NH_3(g)$

Sulfuric acid is one of the most important industrial chemicals, since it is needed to make a wide range of other chemicals. To manufacture it, sulfur dioxide is oxidised to sulfur trioxide:

$2SO_2(g) + O_2(g) \rightleftharpoons 2SO_3(g)$

followed by:

$SO_3(g) + H_2O(l) \rightarrow H_2SO_4(\ell)$

When any reversible reaction is used, the reverse reaction will always occur. Manufacturers must control reaction conditions, so that it occurs as little as possible, to maximise the percentage yield.

Remember

If a reaction is reversible, it will go both ways. Some products will turn back into reactants.

◉ QUESTIONS

7 Suggest why it is not possible to prevent the reverse reaction of a reversible reaction.

Preparing for assessment: Planning an investigation

To achieve a good grade in science, you not only have to know and understand scientific ideas, but you need to be able to apply them to other situations and investigations. These tasks will support you in developing these skills.

✳ Investigating the yield of a reaction

The amount of a product obtained from a chemical reaction is known as the yield. The theoretical yield is the maximum amount of product that can be obtained from the mixture. When compared with the actual yield obtained, a percentage yield can be worked out.

Kim knew that lots of factors could affect the yield of a reaction and decided to investigate one of them.

Kim mixed solutions of calcium nitrate and sodium sulfate. They reacted to produce a precipitate of calcium sulfate. She filtered this precipitate, dried it in a hot oven and then weighed it to find out how much was produced.

Kim carried out the investigation three times at each of five different temperatures. She made sure that she filtered the solution whilst it was at the temperature she chose, and weighed the dried calcium sulfate.

Before carrying out the practical work, Kim calculated how much calcium sulfate should be produced by the reaction, if it all reacted. She calculated that she should obtain from the reaction 1.20 g of calcium sulfate.

The table, below, shows the results that Kim obtained.

Temperature (°C)	Mass of calcium sulfate obtained (g)		
	experiment 1	experiment 2	experiment 3
20	0.22	0.21	0.23
30	0.31	0.36	0.35
40	0.49	0.52	0.53
45	0.63	0.62	0.51
50	0.76	0.77	0.75

✳ Planning

1. Suggest the hypothesis that Kim was testing.

The hypothesis should be a statement that states a possible relationship between the independent variable and the dependent variable.

2. Suggest two control variables that Kim needs to consider, and how they should be controlled.

Control variable are factors that could affect the results if they are not kept the same.

3. Suggest two likely sources of error present in Kim's method of investigation.

Consider all the steps that Kim is carrying out. Where could things go wrong?

✹ Assessing risks

4. When carrying out work involving chemicals, there are hazards.

Suggest two possible hazards that should be considered in this investigation.

5. Suggest how you could control the hazards that you suggested in question 4.

> Hazards are possible harmful things. You should consider the chemicals used, and also the processes being carried out.

> Control measures are methods to reduce the risk. You should research the hazard and think of ways to avoid the hazard actually happening.

✹ Processing data

6. (a) Construct a results table for Kim's results. Add an extra column for the calculation of the mean.

(b) Calculate the mean value for each temperature and complete the table.

7. Draw a graph to show the relationship between the temperature and the mean mass of calcium sulfate produced.

9. Construct a table to show the percentage yield at each temperature.

10. Draw a graph to show the relationship between the temperature and the percentage yield of calcium sulfate produced.

11. Think back to the hypothesis that you proposed in question 1. Explain whether or not it is supported by the results.

> Make sure that your table has headings and units for each column, and also a column to allow for the mean to be calculated.

> Exclude anomalous results from the calculation.

> Make sure that you label the axes clearly and include the units.

> Remember to divide each mean mass by the expected mass (1.2 g) and multiply by 100.

> If the relationship is linear, use a ruler to construct a line of best fit.

> Refer to the data that Kim obtained that either supports or challenges the hypothesis. You can use either of the graphs or tables here.

✹ Connections

How Science Works

• Plan practical ways to develop and test scientific ideas

• Assess and manage risks when carrying out practical work

• Select and process primary and secondary data

• Analyse and interpret primary and secondary data

Science ideas

C2.3.3 Quantitative chemistry

Checklist C2.1–2.3

To achieve your forecast grade in the exam you will need to revise

Use this checklist to see what you can do now. Refer back to the relevant topics in this book if you are not sure. Look across the three columns to see how you can progress. **Bold** text means Higher tier only.

Remember that you will need to be able to use these ideas in various ways, such as:

> interpreting pictures, diagrams and graphs

> suggesting some benefits and risks to society

> applying ideas to new situations

> drawing conclusions from evidence you are given.

> explaining ethical implications

Look at pages 278–299 for more information about exams and how you will be assessed.

To aim for a grade E	To aim for a grade C	To aim for a grade A
Represent an atom by a symbol. Recall the three sub-atomic particles.	Calculate the atomic mass of an element. Explain what an isotope is.	**Recall the relative masses of protons, neutrons and electrons.** **Explain that the relative atomic mass (A_r) compares the mass of atoms with the ^{12}C isotope.**
Recall that compounds are made from atoms of two or more elements.	Know that chemical bonds form, when electrons are transferred or shared in the outer energy levels of atoms, to achieve the electronic structure of a noble gas.	
The number of outer electrons of an element is its group number.	When atoms form chemical bonds by transferring electrons, they form ions.	Ions have the electronic structure of a noble gas atom (Group 0).
Recall that Group 1 elements produce 1^+ ions and Group 7 elements produce 1^- ions.	Know that ionic compounds have giant ionic lattices and high melting and boiling points because large amounts of energy are needed to break the chemical bonds.	
Recall that non-metal atoms share pairs of electrons, forming strong covalent bonds.	Know that some covalently bonded substances consist of simple molecules.	Know that diamond and silicon dioxide have giant covalent structures.
Recall that substances consisting of simple molecules are gases, liquids or solids.	**Explain that intermolecular forces are weak attractive forces between molecules.** **Explain that intermolecular forces are overcome when substances boil and melt.**	
Describe metals as giant structures of regularly arranged atoms.	**Describe electrons in the highest occupied energy levels (outer shell) of metal atoms as delocalised and free to move.** **Explain that, in metals, positive ions and delocalised electrons are held together by strong electrostatic attractions.**	

To aim for a grade E   To aim for a grade C   To aim for a grade A

To aim for a grade E	To aim for a grade C	To aim for a grade A
Recall that simple molecules have no electric charge and cannot conduct electricity.	Know that charged ions conduct electricity when they are free to move.	**Explain that metals are good conductors because of their delocalised electrons.**
Know that layers of atoms in metals can slide over each other. Recall that alloys are usually made from two or more metals.	Know that the different atoms of the metals in an alloy distort the layers, making it harder to slide over each other and making the alloy harder than the metal. Recall that shape memory alloys can return to their original shape.	
Recognise diamond and graphite macromolecules from diagrams.	Know that diamond has a very strong giant covalent structure. Know that, in graphite, the carbon atoms form layers.	**Explain the properties of graphite in terms of weak intermolecular forces between the layers.** **Explain why graphite conducts heat and electricity.**
	Know that fullerenes are based on hexagonal rings of carbon atoms. **Give uses for fullerenes.**	
Know that the properties of polymers depend on what they are made from and how they are made.	Know that thermosoftening polymers have individual, tangled polymer chains. Know that thermosetting polymers have cross-linked polymer chains.	**Explain the properties of thermosoftening polymers in terms of intermolecular forces.**
Know what is meant by nanoscience and nanoparticles.	Recall the size of nanoparticles and their different properties.	Explain how nanoparticles are useful in new materials.
Calculate the relative formula mass (M_r) of a compound.	Recall that the relative formula mass, in grams, is one mole. Calculate the percentage of an element present in a compound.	**Calculate the empirical formula of a compound from the masses or percentages of the elements in a compound.**
Recall that the amount of product obtained is known as the yield.	Know that, in reversible chemical reactions, the products can react to produce the original reactants.	**Calculate the masses of reactants and products from balanced equations and calculate yields.**
Know that chemical analysis can identify additives in foods.	Know that small quantities of elements and compounds can be identified using instrumental methods.	
Recall that the output from a gas chromatography column is often linked to a mass spectrometer. Know that this instrumental method is used to analyse a mixture of compounds.		**Know that a mass spectrometer gives the relative molecular mass of each substance in a mixture.**

*In the examination, data and the periodic table will be given on a separate sheet.
You will be expected to select appropriate data from the sheet.*

1. Graphite is a form of carbon. Here is its structure.

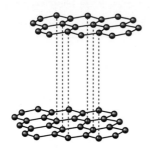

AO2 **(a)** How many other carbon atoms is each carbon atom joined to?   [1]

AO1 **(b)** Name and describe the type of bonding that joins carbon atoms in graphite.   [2]

AO2 **(c)** Graphite is soft and slippery.
Use information from the diagram to explain why.   [2]

AO2 **(d)** Explain why this structure has a high melting point.   [2]

2. An atom of magnesium can be represented like this:

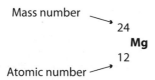

Mass number → 24
Mg
12
Atomic number →

AO2 **(a) (i)** How many electrons does an atom of magnesium have?
(ii) How many neutrons does an atom of magnesium contain?   [1]

AO1 **(b)** Magnesium is used in the construction of aircraft. It is usually mixed with aluminium to make an alloy.
Explain why the alloy is used rather than the pure metal. Your answer should refer to the alloy's structure.   [3]

AO3 **(c)** If an aircraft crash is caused by metal fatigue, air accident investigators examine the damaged metal. They send samples for testing to at least two different laboratories.
Suggest why.   [2]

3. Nano particles are small structures of a few hundred atoms. They have different properties from the same material as a larger structure.

Fake suntan creams are being made that use nano particles. The smaller size of the colour particles means that the colouring effect looks more natural. Tests have shown that the compounds in the creams remain on the surface of the skin. They are not absorbed by the skin and into the blood. However, the compounds have not been tested in this way as nano particles.

AO3 **(a)** Suggest why cosmetic manufacturers are looking for more natural looking colour effects.   [2]

AO2 **(b)** Explain why it is important that the colour particles do not pass through the skin.   [2]

AO3 **(c)** The compounds used are not harmful. Suggest why consumers, using fake suntan creams that contain nano particles, may be putting their health at risk.   [2]

4. Methane has the formula CH_4. It is made from carbon and hydrogen.

AO2 **(a)** How many atoms is a molecule of methane made from?   [1]

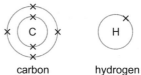

carbon   hydrogen

The diagrams show the electronic arrangement in a carbon atom and in a hydrogen atom.

AO2 **(b) (i)** Sketch a diagram of a molecule of methane showing how the atoms bond together.   [2]

AO1 **(ii)** What type of bonding is this?   [1]

AO1 recall the science   AO2 apply your knowledge   AO3 evaluate and analyse the evidence

✱ WORKED EXAMPLE – Foundation tier

Sodium fluoride has the formula NaF. It is present in some toothpastes.

(a) Use the periodic table to help you to answer these questions.

(i) Sodium has 11 electrons. Draw the electronic structure of sodium. [2]

(ii) Draw the electronic structure of fluorine [2]

(iii) Calculate the relative formula mass of sodium fluoride (Na = 23, F =19) [1]

23 + 19 = 42

(b) Sodium and fluorine both form ions when they make the compound sodium fluoride.

What happens to the electronic structures of each atom as they form ions? [2]

The sodium and fluorine each share an electron to form a bond.

(c) Explain why sodium chloride can conduct electricity when dissolved in water, but does not when solid. [3]

To conduct electricity the ions have to be free to move. In the solid they cannot as they are joined to four other atoms. When dissolved in water the atoms become free to move and carry the electricity.

How to raise your grade!
Take note of these comments – they will help you to raise your grade.

The good, clear diagrams show the electrons as crosses on each shell. In this type of question it is a good idea to label the centre of each atom, so that there is no confusion as to which atom is which. This also applies to ions. Two marks would be awarded to each diagram.

The clear calculation shows the working out. If you have a more complex calculation and make a simple error, you will still get credit for showing that you know how to calculate the answer. A simple number is either full marks or no marks. This answer receives one mark.

The candidate has failed to realise that this is ionic bonding, transferring electrons, rather than sharing electrons. The answer should state that an electron is transferred from a sodium atom to a fluorine atom, forming a sodium ion and a fluoride ion. As a rule, if the compound contains a metal, then it is always ionic bonding. No mark would be given.

The candidate has the right idea, but is confused about using atoms and ions. In an ionic lattice, the particles are always ions, never atoms. This answer receives two marks.

*In the examination, data and the periodic table will be given on a separate sheet.
You will be expected to select appropriate data from the sheet.*

AO1 **1.** The diagrams show how electrons are arranged in a lithium atom and a fluorine atom.

(a) Explain in terms of electrons what happens when lithium reacts with fluorine [3]

(b) Explain as fully as you can why lithium fluoride has a high melting point. [3]

2. Gas chromatography linked to mass spectrometry (GC-MS) is a useful analytical method. The method separates a mixture of compounds and helps with rapid identification of the compounds present.

It is often used to identify substances found at the scene of a crime, to help with the investigation.

AO3 **(a)** In investigating crimes, why is gas chromatography-mass spectroscopy a better method than other types of analysis? [2]

AO2 **(b)** A driver has caused a serious road accident and has supplied a urine sample. The urine contains urea and a drug residue THC-COOH.

Atomic masses are given in the table below.

(i) THC-COOH has the formula $C_{21}H_{28}O_4$

What relative formula mass would the GC-MS record for THC-COOH from the sample?

Show clearly how you worked out your answer. [2]

(ii) Urea has the formula $(NH_2)_2CO$. Its relative formula mass is 60.

Calculate the percentage by mass of oxygen in urea. [2]

Mass numbers are given in the table below.

Element	H	C	N	O
Mass number	1	12	14	16

AO3 **(c)** The graph shows the GC-MS results.

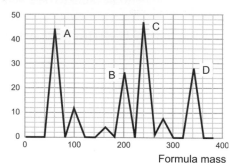

(i) Which peak corresponds to urea? [1]

(ii) Which peak corresponds to THC-COOH? [1]

(iii) Explain why there are other peaks in the graph. [1]

AO3 **3.** Nano particles are small structures of a few hundred atoms. Their properties are different from those of the same material as a larger structure.

Catalysts using nano particles are being developed to use in car exhaust systems. They will improve the emissions from the catalytic converters.

The table, below, shows costs and emissions per kilometre for different catalytic converters.

Catalyst	Cost (£/g)	Mass of nitrogen oxides emitted (g/km)		Mass of carbon monoxide emitted (g/km)	
		original converter	nano particle converter	original converter	nano particle converter
platinum	50	0.05	0.03	0.055	0.050
palladium	20	0.04	0.03	0.065	0.025

(a) (i) Which nano converter is best at lowering nitrogen oxide emissions? [1]

(ii) Which nano converter is the poorest at lowering carbon monoxide emissions? [1]

(b) Some scientists think nitrogen oxides are more damaging to the environment than carbon monoxide. These scientists are suggesting we should use platinum nano particle converters in the future. Give two advantages and two disadvantages of using platinum nano particle converters compared with using palladium nano particle converters. [4]

✳ WORKED EXAMPLE – Higher tier

A group of students carried out an investigation into two different polymers.

One polymer was thermosoftening and the other was thermosetting.

The students:

- poured boiling water into cups made of each polymer
- attempted to stretch identical shaped strips of each polymer by adding a mass of one kilogram.
- heated each one in a Bunsen flame, in a fume cupboard.

The table shows their results.

Test	Result of test	
	Polymer A	Polymer B
boiling water into cup	cup collapses	cup remains rigid
stretch test extension (mm)	15	2
effect of Bunsen heat	melted very quickly	remained hard and charred slightly

Explain each of the test results in terms of the structure and intermolecular forces of the two polymers. [4]

Polymer A is thermosoftening and when hot water is poured in the tangled polymer chains untangle, and the structure weakens and so collapses. Polymer B is thermosetting and has polymer chains that are linked together. The bonds do not break or weaken much on heating.

On stretching the tangled chains in polymer A are more able to stretch by straightening out, but the bonds between chains in Polymer B do not allow the same flexibility.

When heated polymer A's tangled chains become free to move. In polymer B the cross linking bonds do not break.

How to raise your grade!
Take note of these comments – they will help you to raise your grade.

The candidate has clearly identified both types of polymer correctly from the results given. This is worth one mark.

The answer for each test clearly relates the structure of each type of polymer to the property. The explanation for the boiling water in polymer B is rather limited, as it does not relate back to the result.

This is worth one mark. The answer for stretching of both polymers is in enough detail to convey the right ideas. Again, one mark would be given.

The effect of heat response does not give enough explanation. There are two statements about the structure which should then be related the test results, but are not. The candidate may be trying to finish the answer quickly.

Make sure that you give as much detail as you can, even at the end of the answer.

No marks would be given for this part of the answer.

Chemistry C2.4–2.7

What you should know

Chemical reactions

New substances are made by chemical reactions.

Chemical reactions often need help, in the form of heat or light, to start to happen.

Some reactions take a long time to happen and others are very quick.

 What is meant by a 'chemical reaction'?

Energy from chemical reactions

Burning magnesium ribbon is a very fast chemical reaction. It produces lots of heat and light.

Natural gas is burned, in homes, to provide heat energy for cooking and for warmth.

Fireworks are controlled explosions that transfer energy in a spectacular way.

 Describe another chemical reaction that results in useful energy being released.

Acids, bases and salts

Acids have a pH of less than 7. Alkalis have a pH of between 7 and 14.

Alkalis are bases that can dissolve in water.

Acids can react with bases or alkalis to form new compounds called salts.

Salts are neutral. They have a pH of 7.

The type of acid used decides the type of salt made.

 Name three common laboratory acids and the salts that each will make.

You will find out

Rates of reaction

> Concentration, particle size and temperature can be used to control the rate of a chemical reaction.

> Catalysts can alter the rate of a chemical reaction.

> Chemical reactions occur when reacting particles collide with one another and with sufficient energy.

Exothermic and endothermic reactions

> When chemical reactions take place, there is a change in the energy held by the reactants as they change into the products. This can be measured in the temperatures of the reactants when compared to the products.

> Reactions making salts from acids and bases produce heat through the transfer of energy.

> Sometimes, the salts made by neutralisation are insoluble in water.

> Exothermic reactions are used, for example, in self-heating cans. Endothermic reactions are used, for example, in sport injury packs.

Acids, bases and salts

> Soluble salts can be made from the reaction of acids with bases. Insoluble salts can be made from solutions of ions.

> A neutralisation reaction happens when an acid reacts with an alkali (soluble bases).

> Hydrogen ions, $H^+(aq)$, make solutions acidic and hydroxide ions, $OH^-(aq)$, make solutions alkaline.

Electrolysis

> Passing electricity through molten substances or solutions can separate out the elements in the substances.

> Many useful elements are obtained in this way. They are then used to make other substances.

> Aluminium is made from its ore by electrolysis. Salt (sodium chloride) is broken down to produce chlorine gas and hydrogen gas.

> Many items, such as mobile phones, have an electroplated layer of metal over a much cheaper material.

Rates of reaction

Fast and slow reactions

How quickly a reaction happens depends on temperature and other conditions. Industrial chemists know how to control the conditions in a reaction to make it happen just fast enough. If a reaction happens too quickly, it could run out of control and become dangerous. If it happens too slowly, the company will never make any money.

You will find out:

> chemical reactions occur at different rates
> the factors that affect reaction rates
> how to measure reaction rates

FIGURE 1: Iron filings in a flame. Iron turns to iron oxide very slowly at normal temperatures, but instantly in the flame.

Did you know?

Many industrial reactions are done by passing the reactants through tubes. The reaction must be fast enough to be complete before the mixture reaches the end of the tube.

Rates of reaction can vary

What does 'rate' mean?

When chemists talk about **rate of reaction** they mean how much chemical reacts, or is formed, in a given time.

$$\text{rate of reaction} = \frac{\text{amount of reactant used up}}{\text{time taken}} = \frac{\text{amount of product formed}}{\text{time taken}}$$

> Fast reactions have high reaction rates – a lot of reactants turn into products in a very short time.

Examples include explosions and neutralising acids with alkalis.

> Slow reactions have low reaction rates – only a little reaction takes place over a long time.

Examples include iron rusting and fruit ripening.

Altering reaction rates

Zinc reacts with sulfuric acid, forming hydrogen and zinc sulfate solution. How long the zinc takes to dissolve or how quickly hydrogen bubbles form indicates the reaction rate.

The rate can be increased (the reaction made faster) by:

> increasing the **temperature**

> increasing the **concentration** of the acid

> using **smaller pieces** of zinc – powder for instance

> adding a **catalyst** – copper sulfate for this reaction.

Note: If the reactants include a gas, increasing the **pressure** increases reaction rate.

FIGURE 2: Zinc and dilute sulfuric acid reacting at room temperature. What would you see if the mixture was warmed?

QUESTIONS

1 When gunpowder burns, is the rate of reaction high or low?

2 Which has the higher reaction rate, zinc reacting with sulfuric acid, or cement setting?

3 Give four ways of making a reaction between a solid and a liquid go faster.

Remember
The units for reaction rate depend on the units used to measure *amount* and *time*.

Measuring reaction rates

Measuring the rate of a reaction needs two measurements:

1. quantity (mass or volume) of either a reactant or a product – a reaction rate must say which substance was measured

2. time.

Here are some methods for reactions that produce a gas.

A: Time a solid reacting to give a solution

Weigh the solid and time how long it takes to react completely.

Measuring mass and time only once gives an **average reaction rate**.

B: Weigh the gas given off

Stand the apparatus on a top-pan balance. Take readings at regular intervals. As gas is given off, the mass of the mixture decreases. The decrease equals the mass of gas given off.

This allows comparison of the **initial reaction rate** with rates at other stages of the reaction.

C: Measure the volume of gas given off

Connect the apparatus to a gas syringe or upturned measuring cylinder. Read the volume of gas evolved at known times.

Like method B, this gives reaction rates at various times.

QUESTIONS

4 In Figure 3, method B, how could you make the balance read the mass of gas given off directly?

5 In Figure 3, method C, using an upturned burette would make volume measurements more accurate, but more difficult. Explain why.

6 Explain what *initial reaction rate* means.

FIGURE 3: Ways of measuring reaction rate. What other item is needed?

A

sulfuric acid
zinc

B

181.05 g

C
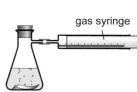
gas syringe
or

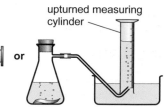
upturned measuring cylinder

Calculating reaction rates

Units

Units of quantity may be, for example, grams (g) or centimetre cubed (cm^3). Units for time may be minutes or seconds. Rates of reaction can have various units.

Examples

> 0.9 g of limestone reacts completely with acid in $7\frac{1}{2}$ minutes (450 s).

Average reaction rate = $\frac{0.9}{7.5}$ = 0.12 g/min of $CaCO_3$

or $\frac{0.9}{450}$ = 0.002 g/s of $CaCO_3$

> Magnesium reacting with acid gives off 6.3 cm^3 of hydrogen in the first 30 seconds.

Initial reaction rate = $\frac{6.3}{30}$ = 0.21 cm^3/s of H_2 or 12.6 cm^3/min

Changing rates

As a reaction proceeds, it gradually slows down. Its rate of a reaction decreases – eventually down to zero. Figure 4 shows this. The volume of gas per half minute gets less. When the reaction is complete, the volume will stay constant and the graph will become horizontal.

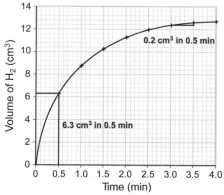

0.2 cm^3 in 0.5 min

6.3 cm^3 in 0.5 min

FIGURE 4: Measurements from a magnesium–acid reaction. What is the rate between 3 and 3.5 minutes?

QUESTIONS

7 Suggest why reaction rates decrease as the reaction proceeds.

Collision theory

Chemical omelettes

You cannot make an omelette without breaking eggs. Imagine eggs banging into each other. A gentle collision does not break them. The eggs must be moving faster, with more energy. After breaking, the whites and yolks can be mixed to make the omelette. In a chemical reaction, particles must collide with enough energy to break their chemical bonds. The broken bits can then mix and make something different.

You will find out:

> particles must collide in order to react

> to cause a reaction, particles must collide with enough energy

> how to get more successful collisions and therefore faster rates of reaction

FIGURE 1: A chemical reaction is a process similar to making an omelette.

Chemical collisions

Collisions and reaction rate

In order to react, reactants must collide. The more collisions, the higher the reaction rate will be. Making reactant particles collide more frequently increases the reaction rate.

The **collision frequency** (number of collisions per second) can be increased by:

> increasing the temperature – reactant particles move faster, so collide more often

> increasing the concentration – in each centimetre cubed of solution there are more particles to collide

> increasing the pressure – squeezes gas molecules closer together, increasing their concentration

> using smaller particles of a solid – increases the surface area of solid that gases or liquids can react with.

Catalysts

Adding a catalyst does *not* affect the collision frequency. It increases the reaction rate by making the reaction easier.

FIGURE 2: Both beakers contain 25 cm^3 hydrochloric acid and 1 g calcium carbonate. The beaker on the left contains powdered limestone. In the beaker on the right, the limestone is in two lumps. Why is one reaction much faster?

QUESTIONS

1 Does magnesium react faster with hot or cold hydrochloric acid? Why?

2 Give three ways to speed up the reaction between nitrogen and hydrogen.

3 Explain why iron filings burn in a Bunsen flame, but an iron nail does not.

Explaining collision theory

It is not just collision frequency that matters. Reactant particles colliding gently just bounce off, without causing any damage. Collisions cause reactions only if there is enough energy to break the reactants' bonds. The fragments can then recombine to form new substances.

Therefore, increasing the rate of reaction depends on increasing the collision frequency, and/or the proportion of successful collisions.

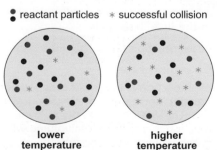

• reactant particles   * successful collision

lower temperature   higher temperature

FIGURE 3: Effect of temperature on the number of successful collisions per second in a liquid or gas.

Q factors rate of reaction ... collision theory GCSE

> Increasing the temperature increases the particles' kinetic energy. They collide more often and with more energy, so the chance of collisions being successful is greater.

> Increasing the concentration of one or more reactants increases the number of particles in a given volume of solution. This makes collisions more likely, so increases the number of successful collisions, but not the proportion.

> Increasing the pressure of a gas is the same as increasing the concentration of a solution and has the same effect. Pressure changes affect all the gases present.

> Increasing the surface area gives more area where collisions can occur. When a solid reacts, collisions can only occur at its surface. Using smaller pieces increases the surface area. Again, this increases the number, but not the proportion, of successful collisions.

> Using a catalyst changes how the reaction happens. A catalyst takes part in the reaction, but can be recovered chemically unchanged afterwards. With a catalyst present, reactants need less energy for a collision that gives products. So, the proportion of successful collisions is higher.

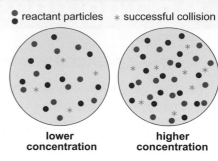

• reactant particles   * successful collision

lower concentration   higher concentration

FIGURE 4: Effect of concentration on the number of successful collisions per second in a liquid or gas.

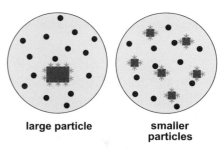

large particle   smaller particles

FIGURE 5: Effect of surface area of a solid on the number of successful collisions per second.

QUESTIONS

4 Explain why the proportion of successful collisions increases with temperature.

5 How would diluting a solution affect the reaction rate?

6 Explain why 1 g of powdered zinc reacts with acid much faster than a 1 g lump.

How much change?

> Doubling the concentration of one reactant often doubles the rate.

> Cutting a cube of solid reactant into eight, doubles its surface area, and thus doubles the rate.

> Increasing temperature always increases the rate. In many reactions a 10 °C rise in temperature roughly doubles the rate.

For a new reaction, the effect of a catalyst cannot be predicted. Some are more effective than others. Different reactions need different catalysts. Not all reactions can be catalysed.

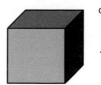 cut into eight equal-sized cubes

2 cm × 2 cm × 2cm cube   1 cm x 1 cm × 1 cm each cube

FIGURE 6: Calculate the surface area of the large cube and the total surface area of the eight smaller ones.

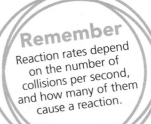

Remember
Reaction rates depend on the number of collisions per second, and how many of them cause a reaction.

QUESTIONS

7 Suggest three ways to slow down a reaction between a solid and a solution.

8 Why are metal catalysts often used in the form of wire gauze?

Q catalyst types

Adding energy

You will find out:

> how and why temperature affects reaction rates

> how temperature changes are used in everyday reactions

> what is meant by activation energy

Cookery is chemistry

Why do baked potatoes, chips and boiled potatoes taste different? Why do they take different times to cook? Heating potatoes to different temperatures in contact with air, fat or water causes different chemical reactions. Baking dehydrates the potato's skin. It takes longer because the potato is not cut into smaller pieces. Frying causes reactions with fats, forming products with different flavours. It is faster because of the higher temperature. Cookery is chemistry, and temperature is an important factor.

FIGURE 1: Raw potato cooks in different ways at different temperatures.

Did you know?

In the 1850s, Robert Bunsen was studying emission spectra – the light given out by hot elements. He needed a better method of heating them. This is why he developed the Bunsen burner.

Why temperature affects reaction rates

The Bunsen burner is used worldwide for one purpose – to make chemical reactions go faster. Transferring energy to reactants by heating increases reaction rates. It makes reactant particles move around faster. This has a double effect:

> more collisions per second

> more successful collisions causing reaction.

At home, cooker hobs and ovens serve the same purpose – transferring energy.

Kitchen chemistry

Cooking causes reactions in foods. It alters their chemical content, which changes their flavours. Cooks use temperature to control these reactions. At higher temperatures, reaction rates are higher, so food cooks more quickly. For example:

> in a pressure cooker, water boils at about 120 °C instead of 100 °C, so food cooks faster than at normal air pressure.

> frying or grilling meat is quicker than boiling it in a liquid.

FIGURE 2: Why is 0 to +5 °C the best temperature for a fridge?

Method	Temperature (°C)	Cooking time (minutes)
boil	100	15–20
pressure cooker	120	4–5
stir-fry	150	2–3

TABLE 1: Cooking carrots.

Fridges and freezers slow down the chemical reactions that cause food to 'go off'.

> Keeping food below 5 °C in a fridge keeps it 'fresh' about four times longer.

> At –18 °C in a freezer, the reaction rate is almost zero.

QUESTIONS

1 Explain why temperature has such an effect on reaction rates.

2 Approximately how many times higher is the reaction rate in a pressure cooker?

3 Explain why a freezer can preserve food for months.

Q temperature reaction rate relationship

Investigating the effects of temperature

Magnesium reacts with sulfuric acid, giving off hydrogen gas.

$$Mg(s) + H_2SO_4(aq) \rightarrow MgSO_4(aq) + H_2(g)$$

The progress of this reaction can be followed by measuring the volume of gas produced.

Figure 3 shows results using the same amounts of magnesium and acid at three temperatures.

> The reaction is complete after about 4 min at 20 °C, 2 min at 30 °C and 1 min at 40 °C.

> The first 2 cm³ of hydrogen was given off in about 18 s, 9 s and 4 s, respectively. (Each grid line = 6 s.)

In both cases, each 10 °C rise halves the time taken, so the reaction is going twice as fast. The rate doubles. This is true for many reactions. For example, performing a reaction at 90 °C, instead of room temperature, makes it over a hundred times faster.

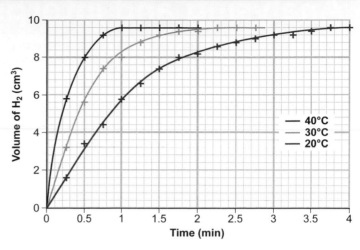

FIGURE 3: Production of hydrogen in magnesium–sulfuric acid reaction. Why do these graphs end up at the same height?

QUESTIONS

4 From the times given, calculate
(a) the average reaction rate (in cm³ H₂/min) for the complete reaction at each temperature
(b) the initial reaction rate (in cm³ H₂/s) for producing the first 2 cm³ at each temperature.

5 Predict the times needed for all the magnesium to dissolve at (a) 50 °C and (b) 10 °C.

4 Allow the adhesive to set for at least 16 hours. Full strength is reached after 3 days.

Warming the joint shortens the setting time. For example, at 40–60°C (hot-water radiator) the setting time to reach full strength is 4–6 hours; whereas at 80–100°C (open oven) it is 30–60 minutes.

FIGURE 4: Instructions for an adhesive. How can you tell that setting involves a chemical reaction?

Activation energy

To cause a reaction, reactant particles must collide with sufficient energy to break bonds. The minimum energy needed for this to happen is called the **activation energy**. Below a certain temperature, the reactants have insufficient energy to react. For instance, vegetables will not cook at all in cold water.

Reactions that give out energy, such as combustion, need only a 'kick' from a spark or match to set them off.

The 'kick' breaks a few bonds, allowing some reaction to occur. The energy given out by that reaction provides more reactants with activation energy, so the reaction continues by itself.

QUESTIONS

6 Explain why methane from a Bunsen burner will not burn until ignited, but then keeps burning.

Q activation energy GCSE

Preparing for assessment: Analysing and interpreting data

To achieve a good grade in science, you not only have to know and understand scientific ideas, but you need to be able to apply them to other situations and investigations. These tasks will support you in developing these skills.

✳ Investigating rates of chemical reactions

Some students were carrying out an investigation into how temperature affects the rate of reaction. They were testing the hypothesis that the higher the temperature of the hydrochloric acid, the faster the reaction would take place.

The students reacted calcium carbonate pieces with dilute hydrochloric acid and measured the volume of carbon dioxide gas produced. The equation for the reaction is:

$$CaCO_3(s) + HCl(aq) \rightarrow CaCl_2(aq) + H_2O(\ell) + CO_2(g)$$

In each experiment, the students used an excess (far more than needed for the reaction to complete) of calcium carbonate.

✳ Analysing data

Temperature of the hydrochloric acid (°C)	15	21	29	40	51
Time to produce 200 cm³ of carbon dioxide (s)	210	152	77	42	19

TABLE 1: Results from the first group of students.

1. What conclusion can the students draw from their investigation?

2. Does this conclusion support the hypothesis? Explain your answer using data from the chart or graph.

A second group of students repeated the investigation. Here are their results.

Temperature of the hydrochloric acid (°C)	10	15	19	24	31
Time to produce 200 cm³ of carbon dioxide (s)	206	152	109	78	52

TABLE 2: The second group's results.

3. Plot a graph of the second group's results.

4. Do the results from the second group of students support the hypothesis? Explain your answer.

5. How would you change the second group's investigation to check your answer to question 4?

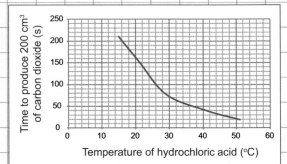

FIGURE 1: Graph of the first group's results.

Look at both the table and the graph to reach your conclusion.

Consider whether your answer to question 1 is the same as the hypothesis. Look at the graph to judge the trend.

The shape of your graph in question 3 is different from Figure 1. You should state how it is different and suggest a reason why the two investigations have produced different conclusions.

Look at the range of the data in both investigations and think why the pattern might be slightly different.

✳ Interpreting data

The teacher had explained to the class that sometimes results are repeatable and at other times they are reproducible.

6. The teacher told each group that its results were neither repeatable nor reproducible. Explain why.

A third group investigated the reaction in a slightly different way. They used hydrochloric acid at 30 °C and measured the loss in the reaction mixture's mass over time. Here are their results.

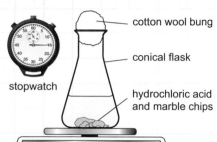

cotton wool bung

conical flask

stopwatch

hydrochloric acid and marble chips

− 2.05 g

FIGURE 2: Monitoring the reaction between calcium carbonate and hydrochloric acid by mass loss.

Time (s)		0	30	60	90	120
Loss in mass (g)	trial 1	0.00	0.45	0.86	0.97	1.08
	trial 2	0.00	0.62	0.87	0.99	1.06
	trial 3	0.00	0.58	0.85	1.01	1.05
	mean of trials	0.00	0.55	0.86	0.99	1.06

TABLE 3: Results from the third group's investigation.

7. Explain why the third group was able to check the repeatability of its results.

8. One of the results is anomalous. Identify this result by the time, and the trial number.

9. Look at the results from all three groups. How far does the data support or contradict the hypothesis?

Finally, a fourth group chose to investigate the hypothesis that raising the temperature of the reaction by 10 °C will make the reaction rate double. Their results are shown below.

Temperature of the hydrochloric acid (°C)	20	30	40	50	60	70
Rate of reaction (cm³/s of carbon dioxide produced)	1.8	3.1	4.9	8.9	16.3	32.4

TABLE 4: Results from the fourth group's investigation.

10. Do the results support the hypothesis over the temperature range 35–45 °C?

11. For which range of temperatures is their hypothesis correct?

✳ Connections

How Science Works

- Select and process primary and secondary data

- Analyse and interpret primary and secondary data

- Use scientific models and evidence to develop hypotheses, arguments and explanations

Science ideas

C2.4.1 Rates of Reaction

Look for a result that does not match the pattern in each trial.

Compare all three sets of data against the hypothesis. Is the hypothesis right or could it be rephrased to make a better hypothesis? Refer to all three sets of data in your answer.

Read from the graph the rate at these two temperatures. Has the rate doubled?

Look carefully at the graph, and decide where adding 10 °C to the temperature doubles the rate of reaction. Use data to support your answer.

Concentration

You will find out:
> how and why concentration affects reaction rates
> why reaction rates decrease during a reaction
> for gases, pressure is equivalent to concentration

Acid rain

This limestone statue's face and other features have been corroded away by acid rain. It has survived for centuries, but the corrosion has become much worse over the last fifty years. That is because pollution has increased the concentration of acid in rainwater. Increased concentration has increased the rate of corrosion.

Did you know?

Writing 'use more concentrated acid' is ambiguous. It could mean 'use a larger volume of concentrated acid', or 'use acid with a higher concentration'. Be careful to write what you mean.

FIGURE 1: Corrosion due to increased acid concentration in rain.

Investigating the effects of concentration

Concentration is the amount of solute in a given volume of solution. Chemists measure concentration in moles per decimetre cubed (mol/dm^3).

1 mole = 1 relative formula mass (in grams) of the substance

1 decimetre cubed (dm^3) = 1000 cm^3 or 1 litre

For example, 0.5 mol/dm^3 NaOH contains 0.5×40 g in each 1 dm^3 of the solution.

Higher concentration means that there are more reactant particles in the same volume. This results in more collisions per second, increasing the reaction rate. However, increasing concentration does not affect the energy of collisions. The proportion of collisions causing reactions is unchanged.

Magnesium and acid

Figure 2 shows results for magnesium reacting with different concentrations of sulfuric acid.

> Doubling the concentration halves the time taken to complete the reaction – 8 minutes for 0.5 mol/dm^3, 4 minutes for 1.0 mol/dm^3 and 2 minutes for 2.0 mol/dm^3.

> The time taken to produce the same volume of gas also halves – for example, 1.8 minutes, 0.9 minutes and 0.45 minutes for 4 cm^3.

If the reaction takes only half the time, it must be going twice as fast. The rate has doubled.

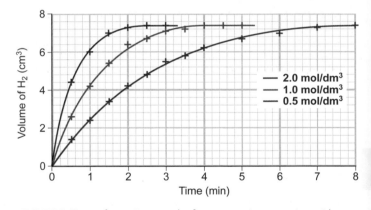

FIGURE 2: Rate of reaction graphs for magnesium reacting with sulfuric acid. Suggest how the measurements were made.

QUESTIONS

1 What was the average reaction rate (for the complete reaction) using 2.0 mol/dm^3 acid?

2 What average rate would you expect for 4 mol/dm^3 acid?

3 Doubling the concentration usually doubles the reaction rate. Explain why.

concentration reaction rate

Further concentration

What else do the graphs show?

The rates of reaction described on the previous pages are **average reaction** rates over the whole reaction. They are worked out by how long it takes a reaction to give a certain amount of product.

However, you know that rate depends on the concentration of reactants. Reactants are used up during a reaction and their concentrations decrease. Therefore, the reaction slows down. That is why the graphs are curves.

You can calculate the rate at any point during the reaction. This is called an **instantaneous reaction rate**. It is given by the gradient of the graph at that point.

For example, to find the rate at 0.25 min in Figure 3:

1. Draw a tangent at this point, and construct a triangle as shown.

2. Find the scale lengths on each axis (0.5 min and 7.4 cm³).

3. Rate = amount of product ÷ time = $\frac{7.4}{0.5}$ = 14.8 cm³ of H_2/min

The gradient of the graph (the rate of reaction) decreases as the reaction proceeds. Finally it becomes horizontal when the reaction is complete and the rate is zero.

Reactions with gases

Increasing the pressure affects the reaction rate only when the reactants include a gas. That is because pressure squeezes gas molecules closer together: there are more in the same volume. Doubling the pressure doubles the concentration of gas molecules: the reaction rate doubles.

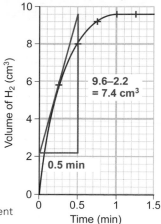

FIGURE 3: Using a tangent to calculate rate.

9.6–2.2 = 7.4 cm³

0.5 min

QUESTIONS

4 Explain the difference between average and instantaneous reaction rates.

5 How does the instantaneous rate vary over the course of a reaction? Explain why.

6 The magnesium–acid reaction produces a gas. Why is the reaction rate not affected by pressure?

Two solutions

The 'Think ink' trick: A magician's fingers click and a colourless mixture suddenly turns blue. How does it happen? At the start he mixes two solutions. They react but do not turn blue until the reaction is complete. By altering the concentrations, the magician can adjust the time delay, so knows when the blue will appear.

Doubling the concentration of one solution will usually double the reaction rate. What happens if both concentrations are doubled? Since the chance of collisions has doubled twice, the reaction rate increases four-fold.

FIGURE 4: Doubling one and both concentrations.

concentration x2

concentration x2 twice

QUESTIONS

7 In the 'Think ink' trick, the 'ink' appears after 20 seconds. How could the magician make it appear after (a) 40 s (b) 5 s?

Particle size

You will find out:
> how surface area affects reaction rate
> surface area also applies to liquids
> some everyday applications of altering the surface area

Danger... bread ingredients!

You probably do not think of bread ingredients as being dangerous, but in April 2010 three people suffered serious burns in a flour mill explosion near St Louis, USA. Similar explosions killed five people in Rawalpindi, Pakistan, in 2003, and ten in Wichita, USA, in 1998. Tiny particles of flour dust are more explosive than coal dust, which occasionally causes explosions in mines. Fine particles react much faster than large particles because they have a greater surface area.

FIGURE 1: Don't try this at home. Flour dust burns almost instantly.

Did you know?

Liquids do not burn – their vapours do. Putting a lighted match into paraffin at room temperature will not ignite it. Petrol will ignite because its lower boiling point produces plenty of vapour above the liquid.

Investigating particle size

Reactions are caused by reactants colliding. Therefore, solids only react at their surface. Breaking a solid into smaller pieces exposes more surface area. That provides more reactant particles to be available for collisions: the reaction rate increases.

Small pieces react faster than a large lump. Powders react faster still.

Marble chips and acid

Marble chips react with hydrochloric acid, giving off carbon dioxide.

$$CaCO_3(s) + 2HCl(aq) \rightarrow CaCl_2(aq) + CO_2(g) + H_2O(\ell)$$

As the gas escapes, the reaction mixture gets lighter. The loss in mass equals the mass of carbon dioxide given off.

Figure 3 shows results for different sized pieces. Notice that the powder reacted fastest (had the highest reaction rate). The small chips were the second fastest. The single large chip was the slowest.

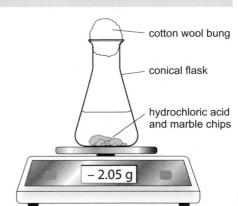

FIGURE 2: Monitoring the reaction of marble chips with hydrochloric acid. Why is the balance showing a negative reading?

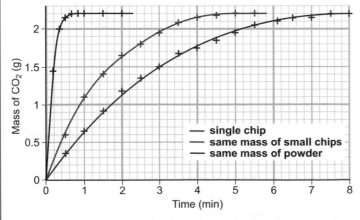

FIGURE 3: Rates of reaction for the same mass of marble chips and powder with hydrochloric acid. Why is the red line much steeper than the other two?

QUESTIONS

Use Figure 3 to answer the following.

1 Why does the reaction mixture lose mass?

2 Which size of marble pieces produced most gas in the first half minute?

3 The same mass of marble was used in each experiment. Explain why the small chips reacted faster than the single chip.

surface area reaction rate

Size and surface area

Cutting a 2 cm cube into eight 1 cm cubes doubles the surface area from 24 to 48 cm². What happens if a 1 cm (= 10 mm) cube is cut into a thousand 1 mm cubes?

> 1 cm cube:

each face = 10 mm × 10 mm = 100 mm²

total surface area = 6 × 100 mm² = 600 mm²

> 1 mm cube surface area:

each face = 1 mm × 1 mm = 1 mm²

total surface area = 6 × 1 mm² = 6 mm²

> A 1 cm cube contains 1000 cubes of 1 mm side length.

Total surface area of 1000 cubes = 6000 mm²

Dividing into a thousand pieces gives ten times the surface area, making the reaction rate ten times higher. Grinding into a fine powder increases the rate hundreds or thousands of times.

Some everyday examples

> Log fires are difficult to light; they burn slowly. Sticks ignite more easily and burn faster, because they have more surface area in contact with the air.

> Sparklers contain metal filings, which burn rapidly. The metal stick does not burn.

> Chewing breaks food into smaller pieces. It is digested more quickly.

> Some power stations use powdered coal, which burns faster than lumps. This releases more energy per minute.

> Acid rain corrodes limestone more quickly than it corrodes marble, although both are calcium carbonate. Marble's surface is smooth. Limestone's is rough, exposing more surface area to the rain.

> The catalyst in catalytic converters has a honeycomb structure. The large area of catalyst allows exhaust gases to react very quickly.

FIGURE 4: How is reaction rate being increased here?

QUESTIONS

4 A 5 mm cube is cut into 1 mm cubes. By how much will the reaction rate increase?

5 Explain why crisps (thinly sliced potatoes) cook in hot fat quicker than chips.

6 Explain why increasing surface area increases reaction rates.

Liquids and gases

Liquids react with gases much faster if sprayed into tiny droplets first. As with powders, this increases the total surface area.

> Oil-fired boilers and furnaces spray in the fuel oil, so that it burns rapidly.

> Aerosol air fresheners quickly react with smelly gases in the air. A bowl of the same liquid would be much less effective.

Even the smallest speck of powder or aerosol droplet contains millions of atoms or molecules. Gases, however, consist of separate molecules – the ultimate small particles.

A mixture of individual gas molecules gives the maximum chance of collisions. Provided collisions are energetic enough to cause the reaction, gases react instantly. Flammable gas mixtures can explode.

QUESTIONS

7 Explain why flammable liquids burn only when vaporised.

acid rain limestone GCSE

Clever catalysis

You will find out:

> what catalysts do and why they are useful

> some examples of catalysts

> how catalysts increase reaction rates

Paved with platinum

City streets are paved with platinum. The platinum was worth about £36 per gram in spring 2011. It comes from catalytic converters in modern car exhausts. 'Cats' contain platinum, rhodium and palladium. They reduce air pollution by catalysing reactions that remove harmful exhaust gases. The metals are slowly lost from the converters and end up on our streets.

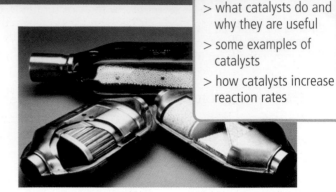

FIGURE 1: Catalytic converters contain precious metals in various forms.

Investigating catalysts

Catalysts are chemicals that speed up reactions, but are not used up in the process. They can be used over and over again. That makes catalysts very useful and important.

Catalysts make reactions happen at lower temperatures than normal. In industrial processes, that saves fuel. Sometimes, catalysts make reactions happen that would not normally occur.

A catalyst experiment

Hydrogen peroxide, (H_2O_2) slowly decomposes into water and oxygen.

hydrogen peroxide → water + oxygen

Add a catalyst such as manganese(IV) oxide and the hydrogen peroxide will decompose very quickly. The catalyst does not dissolve: it can be filtered off, dried and weighed. Its mass is the same as it was at the start, proving that it has not been used up.

Comparing catalysts

Several other transition metal compounds catalyse this reaction. Others do not, although they do catalyse different reactions. Catalysts are 'fussy' – different catalysts work with different reactions. Where several catalyse the same reaction, some work better than others.

FIGURE 2: Reaction rates for decomposing hydrogen peroxide. Which catalyst is most effective?

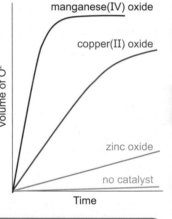

QUESTIONS

1 Give two essential features of a catalyst.

2 Suggest how catalysts can save industry money.

3 Name a catalyst that will decompose hydrogen peroxide.

Using catalysts

How catalytic converters work

Car exhaust fumes contain poisonous carbon monoxide and oxides of nitrogen. Platinum, rhodium and palladium, in the catalytic converter, catalyse reactions between these gases. This converts the gases into less harmful ones. For example:

carbon monoxide + nitrogen monoxide → carbon dioxide + nitrogen

$$2CO(g) + 2NO(g) \rightarrow 2CO_2(g) + N_2(g)$$

Although these metals are very expensive, they continue working for many years.

Catalysts in industry

Many catalysts are transition metals or their compounds. Catalysts can be expensive, but they are economic because they are not consumed. A little catalyst makes a lot of product. Catalysts save time and money by making reactions take place faster at lower temperatures.

TABLE 1: Examples of industrial catalysts

Catalyst	Process	Product	Reaction catalysed
iron	Haber process	ammonia	$N_2(g) + 3H_2(g) \rightleftharpoons 2NH_3(g)$
iron oxide Fe_2O_3	oxidising methane	hydrogen (for ammonia)	$CH_4(g) + H_2O(g) \rightleftharpoons CO(g) + 3H_2(g)$
platinum	oxidising ammonia	nitric acid	$4NH_3(g) + 5O_2(g) \rightleftharpoons 4NO(g) + 6H_2O(g)$
nickel	hydrogenating vegetable oils	spreads (butter substitutes)	adding H_2 across C=C in unsaturated fats
vanadium(V) oxide V_2O_5	Contact process	sulfuric acid	$2SO_2(g) + O_2(g) \rightleftharpoons 2SO_3(g)$
alumina Al_2O_3 and silica SiO_2	cracking petroleum	petrol	breaking down large hydrocarbon molecules into smaller ones

Did you know?

Your body relies on biochemical catalysts called enzymes. Enzymes increase reaction rates, enabling reactions to take place quickly at body temperature.

QUESTIONS

4 Explain how catalytic converters reduce air pollution.

5 Why is it economic to use precious metals, such as platinum, as catalysts?

How catalysts work

A catalyst provides a way for a reaction to occur with lower activation energy. Since less energy is needed, a higher proportion of collisions cause reaction and the reaction goes faster.

The reactants do not react directly. One combines with the catalyst. This combination reacts with the second reactant, forming the normal products and regenerating the catalyst. The catalyst then repeats the process (Figure 3).

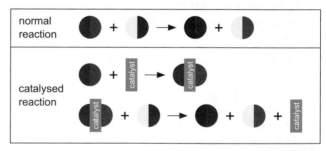

Poisoning catalysts

If a catalyst bonds to a chemical too strongly, it stops working – the catalyst is poisoned. Cars with catalytic converters must use unleaded petrol, because lead poisons the catalyst by coating the active metals.

Biochemical catalysts

Reactions in various body cells, particularly in the liver, produce hydrogen peroxide. As hydrogen peroxide is toxic, it must be destroyed before it damages other cells. It is decomposed by a biochemical catalyst, or enzyme, called catalase (see Figure 4).

Hundreds of different enzymes catalyse other reactions in the body. Each enzyme catalyses one specific reaction.

Enzymes are used, for example, in washing powders and in cheese-making, brewing and baking.

FIGURE 3: Each circle represents a reactant. The catalyst reacts and changes, but is regenerated again.

FIGURE 4: Liver dropped into hydrogen peroxide. What causes this rapid reaction?

QUESTIONS

6 Explain why certain chemicals may poison catalysts.

catalyst poison definition ... enzymes catalysts GCSE

Controlling reactions

You will find out:

> why reaction rates are important in industrial reactions

> how the reaction rates are controlled

> that reaction conditions are often a compromise

Feeding the world

In the last fifteen years, the world's population has increased by about 20%. It is expected to reach seven billion during 2011. About one billion people will be under-nourished. The number would be much higher if it were not for fertilisers made using nitrogen from the air. Nitrogen is very unreactive, so chemical engineers had to greatly increase the reaction rate.

Did you know?

In some processes, reactants are kept in the reaction vessel for minutes or hours. In others, they flow through, reacting in a fraction of a second. It depends on the reaction rate.

FIGURE 1: Spreading fertiliser over a field of rice.

Making ammonia

Ammonia is the starting point for most fertilisers. The **Haber process** combines nitrogen (from the air) with hydrogen (from the reaction of water with natural gas).

nitrogen + hydrogen $\rightleftharpoons$ ammonia

$$N_2(g) + 3H_2(g) \rightleftharpoons 2NH_3(g)$$

The problem is making unreactive nitrogen react fast enough. It took Fritz Haber from 1904 to 1908 to find this answer:

> increase temperature to 450 °C, to get enough successful collisions

> increase pressure to 200 times normal atmospheric pressure, to increase the concentration of the gases

> use a good catalyst, to produce a large increase in reaction rate

> make the catalyst porous, to increase the contact area with the gases.

Worldwide, about 150 million tonnes of ammonia are manufactured every year. About 80% of this is used to make fertilisers.

Remember

Temperature, pressure, surface area of reactants and the catalyst are chosen to give reaction rates that are fast, but safe.

FIGURE 2: Why is this ammonium nitrate fertiliser called 'Extra N'?

QUESTIONS

1 Where does the nitrogen in ammonia come from?

2 Give the reason why (a) high temperature and (b) pressure are needed to make ammonia.

3 Describe how the Haber process ensures good contact between the gases.

Economics and safety

Balancing rate, cost and yield

The ammonia reaction is reversible. The higher the temperature, the more ammonia decomposes again. A temperature of 450 °C is a compromise between a high reaction rate and an economic yield.

High pressure produces more ammonia, but costs more. It requires thick reactor vessels, expensive pumps and energy to power them. The compromise is a pressure of 200 atmospheres.

Haber found the best catalyst was osmium metal, but iron works quite well and is much cheaper – another compromise.

The contact process

Sulfuric acid, H_2SO_4, is needed for detergents, paints, fertilisers, de-rusting steel and many other uses.

> Molten sulfur is sprayed into a furnace. Spraying increases its surface area, so the droplets burn instantly.

> The sulfur dioxide produced is oxidised to sulfur trioxide, using vanadium(V) oxide as catalyst.

$$2SO_2(g) + O_2(g) \rightleftharpoons 2SO_3(g)$$

The reaction is reversible, and 425 °C is again a compromise between increasing the rate and decomposing the product.

Using pure oxygen and higher pressure would increase concentrations, and therefore the rate. However, the process works well using air and normal pressure, so the higher cost is not justified.

Keeping reactions under control

Many reactions are **exothermic** – they give out heat. This raises the temperature, increasing the reaction rate. The reaction then gives out heat more quickly, so the temperature and rate rise faster and faster. The reaction would run out of control.

To avoid disaster, the reaction rate must be kept constant by cooling – for instance, by pumping cold water through pipes in or around the reactor.

FIGURE 3: Sometimes it is worth using an expensive catalyst. To make nitric acid, ammonia and air are passed through several layers of platinum gauze (each worth thousands of pounds).

QUESTIONS

4 Give one advantage and one disadvantage of operating a process at high pressure.

5 Suggest why it is worth using platinum to make nitric acid, but not ammonia.

6 Why are some reaction rates controlled by cooling, not heating?

Catalysts in nature

Living organisms rely on natural catalysts.

Reacting carbon dioxide with water, in a test tube, gives only carbonic acid, but plants convert them into glucose through photosynthesis, using chlorophyll as a catalyst.

Similarly, glucose and oxygen do not react at room temperature, and heating only causes glucose to decompose and char. Plants and animals respire, converting glucose back to carbon dioxide and water, using various enzymes.

$$\begin{array}{c} \text{photosynthesis} \rightarrow \\ 6CO_2(g) + 6H_2O(\ell) \rightleftharpoons C_6H_{12}O_6(aq) + 6O_2(g) \\ \leftarrow \text{respiration} \end{array}$$

QUESTIONS

7 Each enzyme catalyses only one specific reaction. Find out why.

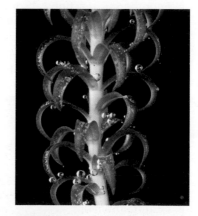

FIGURE 4: Pondweed photosynthesising. What are the bubbles?

The ins and outs of energy

You will find out:
> exothermic reactions release energy
> endothermic reactions take in energy
> how these energy transfers affect reaction rates

Fireworks

What's inside fireworks? Lots of energy waiting to escape. Light a rocket, and the energy lifts it skywards, where it produces a colourful, noisy explosion. Fireworks contain gunpowder and other chemicals that react very quickly when ignited. The reactions do not form just new products, they also release energy.

FIGURE 1: A spectacular release of energy.

Transferring energy

All chemical reactions transfer energy. Energy released in a chemical reaction may be transferred, for example, by

> heating the surroundings, such as combustion (burning)

> electrical current, such as chemical reactions in batteries

> electromagnetic waves such as light, as in glow sticks

> sound, as in explosives.

Energy may be transferred to chemicals, for example by:

> heating them, as in thermal decomposition reactions (such as making lime or cracking hydrocarbons)

> electromagnetic waves such as light, as in photosynthesis

> electric current, as in electrolysis.

Heating up and cooling down

Exothermic reactions transfer energy to the surroundings, heating them.

Mixing acids and alkalis produces no visible change, but the temperature of the solution rises. The chemicals react, transferring energy to the water and heating it. Neutralisation reactions are exothermic.

Endothermic reactions absorb energy transferred to them from the surroundings. This lowers the temperature of the surroundings.

Dissolving ammonium nitrate in water lowers the temperature of the water. The reaction needs energy. The energy is taken from the water, lowering the temperature of the solution. This is an endothermic reaction.

Remember

Think of *exo-* as *exit* = going out, and *endo-* as *into* = coming in.

QUESTIONS

1 Give one example of (a) an exothermic reaction (b) an endothermic reaction.

2 When two solutions mix they get hotter. Have the chemicals taken in heat energy or given it out?

Energy stores

All substances store energy, but different substances store different amounts.

> If reaction products store less energy than the reactants did, the reaction is exothermic. The extra is transferred to the surroundings, heating them up.

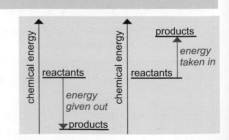

FIGURE 2: Energy transfer during reactions.

exothermic endothermic reactions

> If the products store more energy, they must have absorbed some. The extra energy is transferred from the surroundings in endothermic reactions. The surroundings cool down.

'Surroundings' means whatever is around the reacting chemicals. For reactions in solution, it means the water in which the chemicals are dissolved.

Some applications

> The *Thermit* reaction is spectacularly exothermic. Aluminium powder reduces iron oxide, forming molten iron at 3500 °C. It is used to weld railway lines.

> Self-heating packs provide hot food or drinks anywhere. Water reacts exothermically with calcium oxide in the base of the pack.

> Disposable hand-warmers for outdoor activities contain powdered iron. Unwrapping allows the iron to react with oxygen from the air, gently releasing heat.

> Sports injuries packs give relief by cooling. Breaking a seal mixes chemicals that undergo an endothermic reaction.

FIGURE 3: How was this white-hot iron produced?

FIGURE 4: A self-heating can of coffee. How does it work?

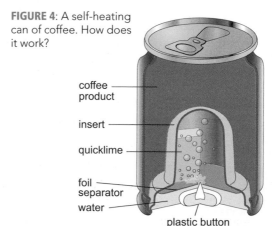

- coffee product
- insert
- quicklime
- foil separator
- water
- plastic button

QUESTIONS

3 Explain how chemicals, reacting at room temperature, can give out heat.

4 Write the word equation for the *Thermit* reaction.

5 Suggest how the iron in hand-warmers is wrapped, so that it reacts when needed, but not before.

Energy and reaction rates

Energy changes affect temperature, which affects reaction rates.

Endothermic reactions need an energy supply. Without continuous heating, the reactants cool, the rate slows down and the reaction stops. In contrast, once started (by providing activation energy) exothermic reactions keep going. If the heat released cannot escape, the temperature rises, increasing the reaction rate. Cooling is needed to prevent a dangerous 'runaway' reaction.

Reversible reactions

If a reversible reaction is endothermic in the forward direction, the reverse reaction is exothermic. Here are two examples.

> Heating blue copper sulfate crystals drives off water, leaving white anhydrous powder. Adding water restores the blue colour, but also gives off heat.

$$\text{endothermic} \rightarrow$$
$$\text{hydrated copper sulfate (blue)} \rightleftharpoons \text{anhydrous copper sulfate (white)} + \text{water}$$
$$\leftarrow \text{exothermic}$$

> Making ammonia is exothermic, but decomposing it is endothermic.

$$\text{exothermic} \rightarrow$$
$$\text{nitrogen} + \text{hydrogen} \rightleftharpoons \text{ammonia}$$
$$\leftarrow \text{endothermic}$$

QUESTIONS

6 Explain why horticulturalists use electric lighting in greenhouses, to extend the hours of daylight.

Acid–base chemistry

The sting cure myth

Folklore says: *put bicarbonate of soda on a bee sting, vinegar on a wasp sting.* This supposedly neutralises the sting, because bee stings are acidic, wasp stings are alkaline. These treatments will not do any harm, but not much good either. Neutralisation means exactly balancing the amounts of acid and alkali to give a neutral solution – not replacing one with the other.

> ### Did you know?
> Universal indicator is a clever mixture of several different indicators. Each changes colour at a different pH value, producing the rainbow colour sequence.

You will find out:
> which ions make a solution acidic or alkaline
> the difference between an alkali and a base
> what happens during neutralisation

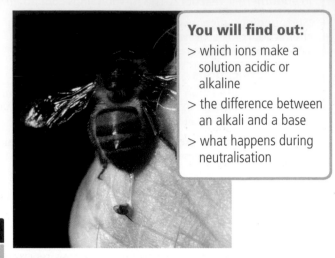

FIGURE 1: When a bee stings, it injects an acid. The stinger and part of its body are torn out, and the bee will die.

Acids and alkalis

Acids dissolve in water to form **acidic** solutions containing **hydrogen ions**, $H^+(aq)$. For example,

$$HCl(g) + water \rightarrow H^+(aq) + Cl^-(aq)$$

Alkalis dissolve in water to form **alkaline** solutions containing **hydroxide ions**, $OH^-(aq)$. For example,

$$NaOH(s) + water \rightarrow Na^+(aq) + OH^-(aq)$$

Acid	Formula	Alkali	Formula
sulfuric acid	H_2SO_4	sodium hydroxide	NaOH
hydrochloric acid	HCl	potassium hydroxide	KOH
nitric acid	HNO_3	calcium hydroxide	$Ca(OH)_2$

| 1 | 2 | 3 | 4 | 5 | 6 | 7 | 8 | 9 | 10 | 11 | 12 | 13 | 14 |

strongly acidic — weakly acidic — neutral — weakly alkaline — strongly alkaline

← increasingly acidic   increasingly alkaline →

FIGURE 2: Universal indicator colours show the acidity or alkalinity of a solution measured on the pH scale.

The pH scale

Acidity or alkalinity depends on the concentration of $H^+(aq)$ or $OH^-(aq)$ ions in the solution. Acidity and alkalinity are measured on the **pH scale**, often by testing with universal indicator.

> Strongly alkaline solutions have high $OH^-(aq)$ concentration, and high pH values.

> Strongly acidic solutions have high $H^+(aq)$ concentration, and low pH values.

> pH 7 means the $H^+(aq)$ and $OH^-(aq)$ concentrations are equal. The solution is neither acidic nor alkaline – it is **neutral**.

Neutralisation

During **neutralisation** reactions between acids and alkalis, the $H^+(aq)$ and $OH^-(aq)$ ions combine to form H_2O, water:

$$H^+(aq) + OH^-(aq) \rightarrow H_2O(\ell)$$

QUESTIONS

1 Ethanol, C_2H_5OH, is neither an acid nor an alkali. Suggest why not.

2 What does the pH scale measure?

3 Universal indicator turns orange in vinegar. What does this show?

Q ph scale GCSE ... acid base indicators

Acid–base reactions

Alkalis and bases

Most metal oxides, hydroxides and carbonates are bases.

A **base** reacts with an acid to form a **salt** and water. Metal carbonates also produce carbon dioxide. For example:

> copper oxide + sulfuric acid → copper sulfate + water

$CuO(s) + H_2SO_4(aq) \rightarrow CuSO_4(aq) + H_2O(\ell)$

> calcium carbonate + hydrochloric acid
>   → calcium chloride + water + carbon dioxide

$CaCO_3(s) + 2HCl(aq) \rightarrow CaCl_2(aq) + H_2O(\ell) + CO_2(g)$

Most bases are insoluble in water. Bases that do dissolve in water are alkalis because their solutions contain hydroxide ions. Alkalis are usually hydroxides of metals in Group 1 and Group 2 in the periodic table.

Alkalis react with acids to give a salt and water only.

Group 1 carbonates and hydrogencarbonates are bases. They dissolve in water, but they are not alkalis because they are not hydroxides – and because they react with acids to give carbon dioxide, as well as a salt and water.

Neutralisation reactions

Acids or alkalis, dissolved in water, split into ions. For example:

> hydrochloric acid contains $H^+(aq)$ and $Cl^-(aq)$ ions

> sodium hydroxide solution contains $Na^+(aq)$ and $OH^-(aq)$ ions.

When they neutralise each other, the sodium chloride produced is also present as ions. The reaction is:

$H^+(aq) + Cl^-(aq) + Na^+(aq) + OH^-(aq) \rightarrow Na^+(aq) + Cl^-(aq) + H_2O(\ell)$

The sodium and chloride ions have not changed. They do not take part in the reaction and are often called 'spectator ions'. The reaction is simply:

$H^+(aq) + OH^-(aq) \rightarrow H_2O(\ell)$

This is an **ionic equation**. It omits anything that was there before reaction and is still there after reaction. It focuses attention on what actually reacts. The ionic equation above is the same for any acid–alkali neutralisation.

Remember

All alkalis are bases, but not all bases are alkalis.

FIGURE 3: This rust treatment "removes rust chemically". What type of reaction occurs?

 QUESTIONS

4 What is the difference between a base and an alkali?

5 Write the word equation for nitric acid reacting with lead oxide.

6 Write a full symbol equation, and an ionic equation, for neutralising nitric acid with potassium hydroxide solution.

Ammonia

Ammonia gas, NH_3, is an unusual base. It reacts with acids to form ammonium salts, but not water. For example:

$2NH_3(g) + H_2SO_4(aq) \rightarrow (NH_4)_2SO_4(aq)$

Ammonia solution is alkaline, because it contains $NH_4^+(aq)$ and $OH^-(aq)$ ions. Ammonia molecules dissolve in water:

$NH_3(g) + aq \rightarrow NH_3(aq)$

Then dissolved ammonia molecules react with water molecules:

$NH_3(aq) + H_2O(\ell) \rightleftharpoons NH_4^+(aq) + OH^-(aq)$

A solution of ammonia consists mainly of dissolved ammonia molecules with some ammonium ions and hydroxide ions. It neutralises acids in the usual way. As hydroxide ions are used up, more ammonia molecules ionise to replace them.

 QUESTIONS

7 Write a balanced symbol equation for the reaction of ammonia solution with sulfuric acid.

Making soluble salts

You will find out:
> three methods for making soluble salts
> these methods involve neutralisation
> how to tell when the reaction is complete

Soluble plant food

Plants make glucose using carbon dioxide from the air, but they must absorb all other nutrients through their roots. These must therefore be soluble compounds. Millions of tonnes of soluble salts are manufactured every year for use as fertilisers. Most are made by reacting acids with ammonia.

FIGURE 1: Slow-release fertilisers gradually dissolve, feeding a plant over several months.

Reactions that form soluble salts

Solutions of soluble salts can be made from acids in three ways.

1. Acids and metals

Any metal higher than hydrogen in the **reactivity series** can react with an acid.

acid + metal → salt + hydrogen

However, very reactive metals, such as sodium, react too violently. They would be dangerous to use.

Metals below hydrogen in the series, such as copper, will not react with most acids.

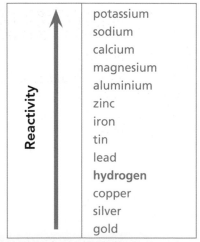

Reactivity
potassium
sodium
calcium
magnesium
aluminium
zinc
iron
tin
lead
hydrogen
copper
silver
gold

TABLE 1: The reactivity series.

2. Acids and insoluble bases

Most bases are insoluble in water, but dissolve in warm acid. They react, neutralising the acid.

acid + base → salt + water

This method is particularly useful for making salts of unreactive metals, which will not react directly.

3. Acids and alkalis

Solutions of alkalis or other soluble bases, such as ammonia, also neutralise acids. This method is limited to making salts of metals in Group 1 or Group 2, and ammonium salts.

Naming salts

The salt formed depends on the metal and acid involved. Here are some examples.

> Magnesium and sulfuric acid give magnesium sulfate.

> Zinc oxide and hydrochloric acid give zinc chloride.

> Potassium hydroxide and nitric acid give potassium nitrate.

Acid		Salt
sulfuric	H_2SO_4	sulfate
hydrochloric	HCl	chloride
nitric	HNO_3	nitrate
phosphoric	H_3PO_4	phosphate

QUESTIONS

1 Give three ways of making salts.

2 Which method is suitable for making copper salts?

3 Suggest two compounds that would react to form sodium phosphate.

Q production salts GCSE ... naming salts

Practical methods

Using a metal

1. Add a little metal (small pieces or powder) to dilute acid. Warm if necessary, to speed up the reaction. When the hydrogen bubbling slows down, add more metal.

2. Continue adding metal until no more will react and dissolve. This ensures that all the acid has been neutralised.

3. Filter off the unreacted metal from the salt solution.

Using an insoluble base

1. Stir small portions of base into warm acid until no more dissolves. Once the reaction is complete the insoluble base remains unreacted.

2. Filter off the unreacted base.

Using an alkali

Because the alkali is in solution, there is no unreacted solid to show when to stop adding. Performing a **titration** determines the volumes of acid and alkali needed. An indicator changes colour when the reaction is complete.

1. Use a **pipette** to measure 25.0 cm³ of alkali into a conical flask. Add phenolphthalein indicator, which turns pink/purple in the alkaline solution, but colourless when neutral.

2. Add acid from a **burette** until the indicator starts to fade. Continue adding slowly, mixing well, until one drop turns the indicator colourless. The solution is now neutral, but contaminated with indicator.

3. Reading the burette scale before and after gives the volume of acid added. In a clean flask, add exactly this volume of acid to another 25.0 cm³ of alkali to make the pure salt solution.

Obtaining the solid salt

Boil the salt solution in an evaporating basin to evaporate most of the water. Stop heating when solid forms around the edge.

On cooling, the salt will **crystallise**.

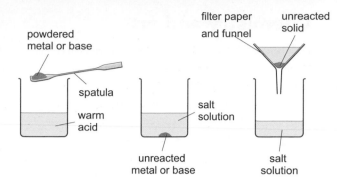

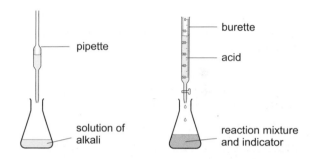

FIGURE 2: Making salts. Which two factors will increase the rate of reaction?

FIGURE 3: It needs two burette readings to measure the volume of acid added. Explain why.

QUESTIONS

4 Explain why the acid is warmed.

5 Suggest two methods, with suitable reactants, for preparing aluminium chloride.

6 In the alkali method, why is it necessary to determine the volume of acid to the nearest drop?

Neutralisation may not work

Not all salts are soluble – the methods described above sometimes fail.

> Lead and sulfuric acid react, forming a layer of insoluble lead sulfate on the lead's surface. This prevents any more acid reaching the lead, so the reaction stops.

> For the same reason, silver oxide will not dissolve in hydrochloric acid. Silver chloride is insoluble.

Insoluble salts must be made by a different method.

QUESTIONS

7 Look up information to find out (a) for which metals all salts are soluble (b) for which acid all metal salts are soluble.

Insoluble salts

Chemistry and art

Art is colourful, thanks to chemistry. Early paints were made from natural minerals, limiting the range of colours. Later, painters began using synthetic pigments. The wide variety of new, bright colours available inspired 19th century artists such as van Gogh. Many of these pigments were insoluble metal salts, such as lead chromate ('chrome yellow').

Did you know?

Older yellow lines along roadsides contain lead chromate – the same pigment as van Gogh used. Alternative pigments are now used because lead chromate is toxic.

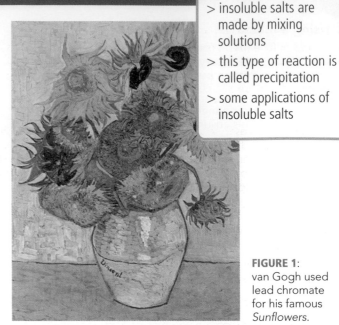

FIGURE 1: van Gogh used lead chromate for his famous *Sunflowers*.

Making insoluble salts

Precipitation

An insoluble salt is one that will not dissolve in water. Like soluble salts, they are made up of metal and non-metal parts. Insoluble salts are made by mixing together two solutions, each containing one part. The metals swap partners. For example:

lead nitrate solution + potassium *chromate* solution
→ solid *lead chromate* + potassium nitrate solution

$$Pb(NO_3)_2(aq) + K_2CrO_4(aq) \rightarrow PbCrO_4(s) + 2KNO_3(aq)$$

A solid product formed by mixing two solutions is called a **precipitate**. The process is **precipitation**, because the solid falls to the bottom of the mixture. Falling rain or snow is also called 'precipitation'.

To obtain the pure insoluble salt:

> filter off the precipitate

> rinse it with distilled water to wash away the other (soluble) product

> dry it in a warm oven.

FIGURE 2: Which insoluble salt is being made here?

Choosing suitable solutions

To make a named insoluble salt, mix solutions of two soluble salts – one of the named metal, the other of the named non-metal (Table 1). Solution B is usually a sodium, potassium or ammonium salt, because these are all soluble.

TABLE 1

Insoluble salt needed	Mix solution A...	...with solution B
barium sulfate	barium chloride	sodium sulfate
copper carbonate	copper sulfate	potassium carbonate
silver bromide	silver nitrate	ammonium bromide

QUESTIONS

1 What does 'insoluble' mean?

2 Name the process used to obtain insoluble salts.

3 Suggest how to make a precipitate of silver chloride.

Q precipitation reaction GCSE

Useful precipitations

Why precipitation happens

Salts are ionic. A mixture of two salts in solution contains positive ions of two metals and negative ions of two non-metal groups. If any pair combines to form an insoluble salt, they produce a solid precipitate (as shown in Figure 3).

Water treatment

> Water supplies in some areas contain dissolved calcium ions. They make the water 'hard' – that is, difficult to form a lather with soap. **Hard water** can be 'softened' by adding a solution, such as sodium carbonate (sometimes known as washing soda) that combines with the calcium ions to form a precipitate.

calcium ions + carbonate ions → calcium carbonate precipitate

$Ca^{2+}(aq) + CO_3^{2-}(aq) \rightarrow CaCO_3(s)$

> Domestic waste water contains phosphate ions from washing powders. They cause pollution, so treatment works remove them before returning the water to a river. Common treatments use calcium, iron or aluminium ions.

phosphate ions + aluminium ions → aluminium phosphate precipitate

$PO_4^{3-}(aq) + Al^{3+}(aq) \rightarrow AlPO_4(s)$

Precipitated chalk

Toothpaste and some medicines contain calcium carbonate. Limestone and chalk are too impure, so the pure product is made by precipitation.

calcium chloride + sodium carbonate → calcium carbonate + sodium chloride

$CaCl_2(aq) + Na_2CO_3(aq) \rightarrow CaCO_3(s) + 2NaCl(aq)$

Photography

Though less used in the digital age, photographic film is still important, for X-ray images in hospitals for instance. Films are coated with light-sensitive silver bromide, precipitated from silver nitrate solution.

FIGURE 3: (a) Mixed ions. (b) Opposite ions join together. (c) Insoluble salt sinks.

QUESTIONS

4 What is left in solution when soda softens hard water?

5 Suggest why ionic equations are used to describe water treatments.

6 Write the balanced ionic equation for removing phosphate ions from waste water using calcium ions.

Chemical analysis

Certain salts are known to be insoluble so, precipitation can be used to identify ions in unknown samples.

TABLE 2

Test solution added	Precipitate formed		Indicates presence of	
	colour	formula	ions	
silver nitrate, $AgNO_3$	white	AgCl	chloride	Cl^-
	cream	AgBr	bromide	Br^-
	yellow	AgI	iodide	I^-
barium chloride, $BaCl_2$	white	$BaSO_4$	sulfate	SO_4^{2-}
sodium hydroxide, NaOH	blue	$Cu(OH)_2$	copper(II)	Cu^{2+}
	green	$Fe(OH)_2$	iron(II)	Fe^{2+}
	brown	$Fe(OH)_3$	iron(III)	Fe^{3+}

FIGURE 4: Results of three silver nitrate tests. What does each result show?

QUESTIONS

7 Suggest how to test for silver ions in a solution.

Ionic liquids

Glass can conduct

Glass is an excellent electrical insulator – it is used to hang high voltage cables from pylons. Yet, if glass is heated until almost molten, it begins to conduct. Glass is an ionic substance, but can conduct electricity only when the ions are free to move, not while solid.

Did you know?

If a cell or battery is cut open, it appears to contain dry powder. In fact, it is damp. If it were dry it could not produce electricity.

FIGURE 1: Glass insulators between overhead power cables and a pylon.

Conduction needs movement

Ionic solids

Ions are atoms, or groups of atoms, that have lost or gained electrons.

> Metal atoms lose electrons, forming positive ions such as Na^+, Cu^{2+} and Fe^{3+}.

> Non-metals gain electrons, forming negative ions such as Cl^-, O^{2-}, NO_3^- and SO_4^{2-}.

Because opposite charges attract, each ion is surrounded by oppositely charged ions. This produces a regular ionic lattice (see pages 114–115). Strong ionic bonding holds the ions in position. Since the ions cannot move, they cannot carry charge through the lattice. Ionic solids cannot conduct electricity.

Aqueous solutions

When a soluble ionic compound dissolves in water, the lattice breaks up. The ions separate and move around at random.

Because the ions are now free to move, they can carry their electrical charges through the solution. In other words, the solution can conduct electricity.

Molten ionic compounds

When an ionic compound melts, the lattice breaks down and becomes liquid. As in a solution, the ions move around, so the molten compound can conduct.

 metal ion ⬤ non-metal ion

FIGURE 2: An ionic lattice. Explain why the ions cannot move around.

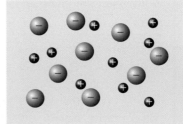

FIGURE 3: Ions in solution. How would the diagram differ if the solution were more concentrated?

QUESTIONS

1 Why do salt (sodium chloride) crystals not conduct electricity?

2 Give two ways to make salt conduct.

3 Does a solution of sugar, $C_{12}H_{22}O_{11}$, conduct electricity? Explain your answer.

Conduction in liquids

Does water conduct electricity?

Water consists of H_2O molecules, not ions. Therefore, pure water cannot conduct. However, safety advice is never to touch electrical equipment with wet hands, and regulations prohibit mains sockets in bathrooms. This is to avoid electric shock, because tap water is not pure water. It contains various dissolved ionic compounds, so tap water does conduct.

How liquids conduct

When a liquid conducts, the electric current must enter and leave the liquid through solid conductors, called **electrodes**. They may be metal or graphite (carbon), the only solid non-metal that conducts. One electrode is connected to the positive of the battery or electricity supply, the other to the negative.

The ions in a solution or molten compound move around randomly. However, when the electricity is switched on, ions in the liquid are attracted towards the oppositely charged electrode.

> Positive metal ions, **cations**, gradually move towards the negative electrode, or **cathode**.

> Negative non-metal ions, **anions**, move towards the positive electrode, or **anode**.

As the ions move, they carry electric charge through the liquid in both directions, from one electrode to the other.

At the electrodes

> At the cathode, cations pick up electrons from the outside circuit. The metal ions become atoms, coating the cathode with metal.

> At the anode, anions release the same number of electrons into the circuit, replacing those removed at the cathode. The ions become atoms, which form non-metal molecules as a gas or in solution.

Chemical changes occur at both electrodes. When an ionic compound conducts electricity, the current decomposes the compound into its separate elements. This process is called electrolysis.

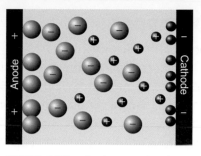

FIGURE 4: Why do the particles that are in contact with the electrodes have no charge?

Remember

Passing an electric current through an ionic liquid decomposes the compound into its elements.

QUESTIONS

4 Suggest some ions present in tap water that make it conduct.

5 What is the difference between an anode and a cathode?

6 Why do metal ions move towards the cathode? Describe what happens when they arrive there.

Direct and alternating currents

To cause electrolysis, **direct current** (d.c.) must be used, so that one electrode is positive, and the other negative. Direct current flows in one direction. It can come from a battery or from a d.c. supply, such as a lab pack.

Mains electricity is an **alternating current** (a.c.), which would make the electrodes alternate rapidly between positive and negative. Ionic liquids do conduct an alternating current, which is why mains electricity and water are a dangerous combination. However, an alternating current does *not* cause electrolysis.

QUESTIONS

7 Explain why a.c. does not cause electrolysis.

 anode vs cathode ... cation anion

Electrolysis

4N copper

About 18 million tonnes of copper are produced each year worldwide. About three-quarters of this is used for wiring in buildings, electric motors, transformers, electromagnets and other devices. To conduct well, the copper must be at least 99.99% pure, known as 4N (4 nines). This can only be achieved by electrolysis.

FIGURE 1: Replacing copper electrodes in an electrolysis tank.

Decomposing ionic compounds

Ionic compounds decompose when they conduct electricity.

Metal cations and non-metal anions are attracted towards electrodes, which are in opposite directions. This causes the cations and anions to separate. Electrolysis takes place. This process is used to extract reactive elements from their compounds.

Sodium and chlorine

Electrolysis decomposes molten sodium chloride:

$2NaCl(\ell) \rightarrow 2Na(\ell) + Cl_2(g)$

> Sodium ions, Na^+, move to the negative cathode. Each gains an electron to become a sodium atom.

> Chloride ions, Cl^-, move to the anode. They each give up an electron to become atoms, which then form molecules of chlorine gas.

Electrolysis of brine

Electrolysis of sodium chloride solution (brine) gives different products from molten sodium chloride.

$2NaCl(aq) + 2H_2O(\ell) \rightarrow 2NaOH(aq) + Cl_2(g) + H_2(g)$

Chlorine is produced, but not sodium. Instead, hydrogen is produced at the negative electrode. This is because hydrogen is lower than sodium in the reactivity series.

In a mixture or solution of ions like this, the reactivity series decides the products. The lowest element in the reactivity series is the one produced. Hydrogen is lower than sodium, so hydrogen gas is made instead of sodium metal. The sodium remains in solution.

Electrolysis of brine is an important industrial process. Its products, sodium hydroxide and chlorine, are starting materials in many chemical processes. Sodium hydroxide for making soap, and chlorine for the production of bleach and plastics are examples.

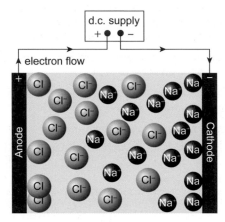

FIGURE 2: Electrolysing molten sodium chloride.

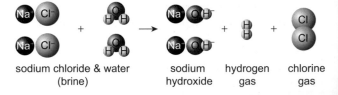

sodium chloride & water (brine)   sodium hydroxide   hydrogen gas   chlorine gas

FIGURE 3: Electrolysing brine. At which electrode is hydrogen formed?

QUESTIONS

1 Explain why electricity decomposes liquid ionic compounds.

2 Suggest how magnesium metal can be manufactured.

3 Describe what happens to chloride ions during electrolysis.

Q chloride ions electrolysis GCSE

More applications of electrolysis

Aluminium production

Aluminium is extracted by electrolysing molten aluminium oxide, Al_2O_3. The aluminium oxide is mixed with cryolite to lower the melting point from 2100 °C to about 900 °C so that less energy is needed to melt it.

Molten aluminium forms at the cathode. Oxygen is given off at the carbon anode and reacts with it, forming carbon dioxide. Therefore the anode must be replaced regularly.

Electroplating

Many metal objects are not what they seem. Thin layers of shiny, expensive metals like chromium or silver are electroplated onto cheaper materials.

Electroplating uses electrolysis. This is the process in Figure 4.

1. The object is attached to the negative terminal. It is now the cathode.

2. Silver ions from the **electrolyte** move towards to the cathode. Each silver ion, $Ag+$, gains an electron and becomes a silver atom. The surface is coated with a layer of silver.

3. Silver atoms in the anode transfer electrons into the circuit. They become silver ions, replacing those used up at the cathode.

Purifying copper

The impure copper is used as anodes in an electrolyte of copper sulfate solution. 99.99% pure copper builds up on the cathodes, while the anodes dissolve away. The impurities form a sludge.

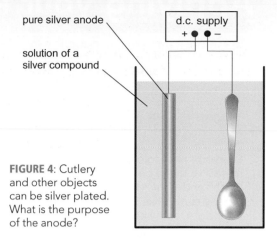

FIGURE 4: Cutlery and other objects can be silver plated. What is the purpose of the anode?

◉ QUESTIONS

4 In electroplating, the anode reaction differs from its reaction during the extraction of a metal or chlorine. Describe the difference.

5 Most 1p and 2p 'copper' coins are steel, coated with copper. Briefly describe how this might be done.

Half equations (Higher tier)

When molten sodium chloride is electrolysed, the reactions that happen at the electrodes are:

> at the cathode (negative electrode) $Na^+ + e^- \rightarrow Na$ (reduction)

> at the anode (positive electrode) $2Cl^- \rightarrow Cl_2 + 2e^-$ (oxidation)

Note: e^- represents an electron.

These are called **half equations**. In one half equation, electrons are lost. This is **oxidation**. In the other half equation, electrons are gained. This is **reduction**.

The number of electrons lost and gained must be the same. This is why the overall equation is:

$2Na^+ + 2e^- + 2Cl^- \rightarrow 2Na + 2e^- + Cl_2$

or $2NaCl \rightarrow 2Na + Cl_2$

Similarly, when molten aluminium oxide is electrolysed, the reactions that happen at the electrodes are:

> at the cathode $Al^{3+} + 3e^- \rightarrow Al$ (reduction)

> at the anode $2O^{2-} \rightarrow O_2 + 2e^-$ (oxidation)

Since electrons lost and gained must be the same, the overall equation is:

$4Al^{3+} + 12e^- + 6O^- \rightarrow 4Al + 12e^- + 3O_2$

or $2Al_2O_3 \rightarrow 4Al + 3O_2$

Remember

OIL RIG:
Oxidation Is Loss,
Reduction Is Gain
(of electrons).

◉ QUESTIONS

6 Write the two half equations for

(a) purifying copper

(b) the electrolysis of molten lead bromide, $PbBr_2$.

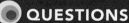

Q electroplating animation ... half equations GCSE

Preparing for assessment: Applying your knowledge

To achieve a good grade in science, you not only have to know and understand scientific ideas, but you need to be able to apply them to other situations and investigations. These tasks will support you in developing these skills.

✸ Make mine shiny!

Everyone likes to see bright and shiny objects. At first people used precious metals to make beautiful objects, as the metals could be polished and made bright and shiny. Unfortunately these metals, such as gold and silver, are usually expensive, so people invented ways of coating cheaper metals with the bright, shiny and expensive ones.

One method was known as Sheffield Plate. It was made by melting a thin layer of silver metal onto a cheaper metal such as copper, then making the object from the sheet of combined metal. Unfortunately, lots of polishing eventually caused the copper to show.

Today, electroplating is used to make objects shiny. Chromium plating, for example, can make items rustproof and good to look at.

To make chromium plated motorcycle handlebars (and other parts of the bike), the handlebars are first made from mild steel (a type of iron).

> The mild steel handlebars are then connected to the negative side of an electric circuit becoming the negative electrode, and placed into a hot solution of chromium sulfate.

> An electric current is passed into the solution. The electric current provides a flow of electrons that cause a thin layer of chromium to cover the handlebars. It makes the handlebars shiny.

> The positive electrode of the circuit is often made from chromium. This dissolves into the solution, maintaining the concentration of chromium ions in the solution.

The electroplating process removes the Cr^{3+} ions from the solution by giving each ion three electrons. The ion becomes a chromium atom and sticks to the negative electrode – the handlebars.

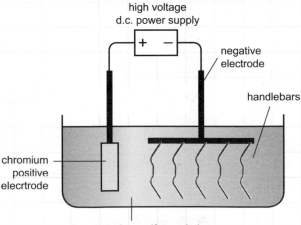

high voltage
d.c. power supply

negative electrode

handlebars

chromium positive elecrtrode

hot chromium sulfate solution

✸ Task 1

(a) Explain why it is necessary for chromium sulfate to be a solution.

(b) Use diagrams to suggest the structure of solid chromium sulfate and also how the ions are in the solution. You could represent sulfate ions using a circle with 'sul' inside and chromium ions as 'Cr^{3+}'.

✸ Task 2

At the negative electrode, a process called reduction is happening. Explain what is meant by reduction in terms of electron transfer.

✸ Task 3

(a) Draw diagrams to show what happens to the chromium ions at the negative electrode.

(b) Use a half equation to represent the reaction.

✸ Task 4

Explain why the positive electrode of the circuit is made from chromium metal. Draw a diagram to show what happens at the positive electrode.

✸ Task 5

When solutions of sodium chloride are electrolysed, the negative electrode does not get covered in sodium metal; instead, hydrogen gas is produced.

(a) Explain why the more reactive sodium cannot be electroplated in this way.

(b) Write an equation to show what happens to the sodium ions at the negative electrode.

✸ Maximise your grade

Answer includes showing that you...	
	know that chromium sulfate needs to be a solution to conduct electricity.
	can draw diagrams to show solid crystal lattice and free moving ions.
	can draw a diagram to show each Cr^{3+} ion gains electrons at the negative electrode.
	can explain reduction in terms of electron transfer.
	understand that a chromium positive electrode is needed to replace Cr^{3+} ions removed from the solution.
	can explain how the reactivity of sodium affects the electrolysis of sodium chloride solutions.
	can write a balanced symbol equation for the reaction of sodium with water.
	can write half equations for the reactions at both electrodes.

Checklist C2.4–2.7

To achieve your forecast grade in the exam you will need to revise

Use this checklist to see what you can do now. Refer back to the relevant topics in this book if you are not sure. Look across the three columns to see how you can progress. **Bold** text means Higher tier only.

Remember that you will need to be able to use these ideas in various ways, such as:

> interpreting pictures, diagrams and graphs

> applying ideas to new situations

> explaining ethical implications

> suggesting some benefits and risks to society

> drawing conclusions from evidence you are given.

Look at pages 278–299 for more information about exams and how you will be assessed.

To aim for a grade E	To aim for a grade C	To aim for a grade A
Know that the rate of a chemical reaction is how quickly a reactant is used up or how quickly a product is formed.	Describe how the rate of a chemical reaction can be found by measuring the amount of a reactant used or the amount of product formed over time.	Use the equations rate of reaction = $\dfrac{\text{amount of reactant used}}{\text{time}}$ rate of reaction = $\dfrac{\text{amount of product formed}}{\text{time}}$
Recall that reactions occur only when particles collide with each other and with sufficient energy.	Explain that particles must have a minimum amount of energy to react and this is called the activation energy.	
Recall that the rate of reaction increases by increasing the temperature, the pressure of reacting gases or the concentration of solutions.	Explain that increasing the temperature increases the speed of particles and hence their energy and the frequency of collisions. Explain that the closer particles are together, through pressure or concentration, the more frequently they collide.	Explain that more frequent collisions and more energetic collisions of particles increases the rate of reaction.
Recall that increasing the surface area of solid reactants makes more of it available for reaction.	Explain that only particles on the surface of a solid can collide with the other reacting substance.	Explain that increasing the surface area increases the rate of reaction.
Recall that catalysts change the rate of chemical reactions but are not used up during the reaction.	Describe why catalysts are important in industrial processes, for example, to reduce costs.	Describe specific ways that costs can be reduced by using catalysts.
Know that, when chemical reactions occur, energy is transferred to or from the surroundings.	Explain that, in exothermic reactions, energy transfers to the surroundings and, in endothermic reactions, energy transfers from the surroundings.	Understand that, if a reversible reaction is exothermic in one direction, it is endothermic in the opposite direction.

To aim for a grade E   To aim for a grade C   To aim for a grade A

To aim for a grade E	To aim for a grade C	To aim for a grade A
Recognise the state symbols in equations: (s), (ℓ), (g) and (aq).	Understand that soluble salts can be made from acids by reacting them with metals, insoluble bases or alkalis.	Write chemical equations for making soluble salts from named reactants.
Recall that salt solutions can be crystallised to produce solid salts.	Understand that insoluble salts can be made by mixing solutions of ions that will form a precipitate.	Write chemical equations for making insoluble salts from named reactants.
Recall that metal oxides and hydroxides are bases.\nRecall that soluble hydroxides are called alkalis.	Understand that the particular salt produced in any reaction between an acid and a base or alkali depends on the reactants used.	Name both reactants needed to make a particular salt, and the method to use.
Know that ammonia dissolves in water to produce an alkaline solution.	Know that ammonia is used to produce ammonium salts. Ammonium salts are important as fertilisers.	
Recall that the pH scale is a measure of the acidity or alkalinity of a solution.\nRecall that the pH scale ranges from 0 to 14, and that pH 7 is a neutral solution.	Know that hydrogen ions, $H^+(aq)$, make solutions acidic and hydroxide ions, $OH^-(aq)$, make solutions alkaline.\nUnderstand that, in neutralisation reactions, hydrogen ions react with hydroxide ions to produce water.	Represent the reaction by the equation:\n$$H^+(aq) + OH^-(aq) \rightarrow H_2O(\ell)$$
When an ionic substance is melted or dissolved in water, the ions become free to move about within the liquid.\nKnow that electrolysis is used to electroplate objects.	Describe electrolysis as passing an electric current through an ionic liquid, to break it down into elements.\nDescribe the movement of ions to electrodes.	Use the term electrolyte.\n**Know how to complete and balance half equations at each electrode during electrolysis.**
Recall that aluminium is manufactured by the electrolysis of a molten mixture of aluminium oxide and cryolite.	Know that cryolite reduces the temperature at which aluminium oxide becomes molten.\nKnow that aluminium forms at the negative electrode and oxygen at the positive electrode.	Know that, in the electrolysis of aluminium, the positive electrode is made of carbon, which reacts with the oxygen to produce carbon dioxide.
Recall that the electrolysis of sodium chloride solution produces hydrogen and chlorine.	Know that sodium hydroxide solution is also produced.\nUnderstand that hydrogen, chlorine and sodium hydroxide are important industrial chemicals.	Know that sodium hydroxide is used in making soap and that chlorine is used in making bleach and plastics.

*In the examination, data and the periodic table will be given on a separate sheet.
You will be expected to select appropriate data from the sheet.*

1. A student investigated the rate of reaction between zinc metal and hydrochloric acid.

The student used an excess of zinc metal.

The reaction can be represented by this equation:

$Zn(s) + 2HCl(aq) \rightarrow ZnCl_2(aq) + H_2(g)$

The student measured the volume of gas given off for six minutes.

The results are shown on this graph. Use the graph to answer the questions.

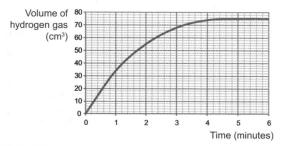

AO2 **(a)** Describe as fully as you can how the rate of the reaction changes over time. [2]

AO2 **(b)** Use the graph to find out how much gas is produced after 2.5 minutes. [1]

AO3 **(c)** Explain why the volume of gas does not change after 4 minutes [2]

AO1 **(d)** The student repeated the investigation using hydrochloric acid at a higher temperature. The reaction was much faster. Explain why. [2]

2. Heating blue copper sulfate crystals turns them white. The reaction is reversible, so adding water to the white crystals makes them become blue again. When the crystals become blue again they get hot. Here is a word equation for the reaction.

blue copper sulfate $\rightleftharpoons$ white copper sulfate

AO3 **(a)** Explain why the reaction is described as reversible. Use information in the question. [2]

AO3 **(b)** The reaction of white copper sulfate crystals to blue crystals is exothermic. Explain how you can tell that this is an exothermic reaction. [2]

AO3 **(c)** The reaction of blue copper sulfate crystals to white crystals is endothermic. Suggest how much energy is needed to make the crystals turn white. [1]

3. Nickel can be electroplated onto objects so that they can conduct electricity. This diagram shows a method for doing this.

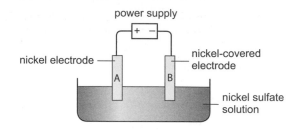

AO2 **(a)** The nickel is plated onto the electrode marked A. Using information on the diagram, explain this process. [2]

AO2 **(b)** The electrode B is made from nickel. This slowly dissolves into the solution to maintain the concentration of nickel sulfate. Explain why it is necessary to maintain the concentration of the solution. [2]

AO3 **(c)** After 5 minutes, 25g of nickel has been deposited on electrode A. If the current in the circuit is doubled, what will happen to the mass of electrode A after a further 5 minutes? Explain your answer. [2]

4. Lead iodide is a salt. It can be made by mixing a solution of lead nitrate with a solution of sodium iodide. A reaction happens and a precipitate of lead iodide forms.

AO2 **(a)** Write a word equation for the reaction showing both products. [2]

AO3 **(b)** Lead iodide is insoluble. Describe how you would obtain a pure, dry sample of lead iodide after mixing the two solutions. [3]

AO2 **(c)** Using the charges on the ions, sodium and iodide, write the correct formula for sodium iodide. [1]

AO2 **(d)** Calculate the formula mass of lead nitrate $Pb(NO_3)_2$ using the mass numbers in the table. [2]

Element	N	O	Pb
Mass number	14	16	207

AO3 **(e)** Suggest how precipitation reactions might remove unwanted ions from drinking water. [2]

AO1 recall the science   AO2 apply your knowledge   AO3 evaluate and analyse the evidence

✱ WORKED EXAMPLE – Foundation tier

In this question you will be assessed on using good English, organising information clearly and using specialist terms where appropriate.

Copper sulfate crystals can be made from copper carbonate and dilute sulfuric acid.

Describe a method to make copper sulfate crystals from copper carbonate and dilute sulfuric acid.

For the method you should include the names of the pieces of apparatus used and a risk assessment.

[6]

I will put some sulfuric acid in a beaker and then add some copper carbonate. I will keep adding it until no more reacts. I will then filter it, and put it into an evaporating basin before heating it up over a Bunsen burner. Allow it to evaporate to about one-third the volume, and cool it.

You should have crystals of copper sulfate in the bottom.

Hazards; none really, only the chance of getting burned so you need to be really careful.

How to raise your grade!
Take note of these comments – they will help you to raise your grade.

The candidate has named some of the apparatus, but not all of it. How big a beaker should he use? Add as much detail as possible. There is no mention of the equipment used for filtration.

Make a list at the top, using a fresh line for everything you remember at first. You then have space to add other items alongside, as you think of them when writing the method.

The method is good, but needs more detail. For example, how much acid should be used; when would you know that the reaction is complete?

The risk assessment is very weak. It makes no mention that hot acid is dangerous, nor that copper compounds are toxic, or at least should be checked.

If you do not know about a chemical, say that you need to check what hazard(s) it may present.

This answer is not well organised.

The candidate would receive only two marks.

In the examination, data and the periodic table will be given on a separate sheet.
You will be expected to select appropriate data from the sheet.

1. A student investigated the rate of reaction between zinc metal and hydrochloric acid using several different concentrations of the acid.

The student used an excess of hydrochloric acid.

The reaction can be represented by this equation:

$Zn(s) + 2HCl(aq) \rightarrow ZnCl_2(aq) + H_2(g)$

The student measured the volume of gas given off for nine minutes.

The results are shown on this graph. Use the graph to answer the questions.

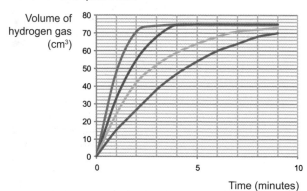

AO2 **(a)** Describe as fully as you can how the rate of the reaction changes over time, for 2 mol/dm³ [2]

AO2 **(b)** Use the graph to find out the maximum volume of gas that can be made from the reaction. [1]

AO3 **(c)** What conclusion can you make from the graphs about the concentration of the acid and the rate of reaction? [2]

AO1 **(d)** Explain why changing the concentration of the hydrochloric acid changes the rate of reaction. [2]

2. Lead iodide is an insoluble salt that is coloured bright yellow. It is made like this:

• Compound A, which is soluble and contains potassium ions, is dissolved in water.

• The solution is added to a solution of compound B which contains nitrate ions.

AO2 **(a)** Use information from the table to suggest a suitable compound for:

(i) compound A **(ii)** compound B. [2]

The table gives information about the solubility of some compounds.

Soluble	Insoluble
all nitrates	
all Group 1 salts	silver , lead chlorides, bromides and iodides
sodium, potassium and ammonium carbonates.	most other carbonates
most sulfates	barium sulfate and lead sulfate

AO2 **(b)** Use the data sheet to write the formula of lead iodide. [1]

In part (c) of this question you will be assessed on using good English, organising information clearly and using specialist terms where appropriate.

AO3 **(c)** Describe how you would obtain a pure and dry sample of lead iodide after the reaction. [3]

3. Soldiers in war zones use self heating packs of food.

Rations to go!

CHICKEN CURRY
88-09-914

squeeze here

calcium oxide mass = 75g

The pack contains three compartments: one contains the food to be heated, the second water, and the third contains a chemical such as calcium oxide. Squeezing the corner of the pack mixes the water and calcium oxide together. The reaction is exothermic.

AO1 **(a)** Explain how the reaction provides energy to heat the food. [2]

AO3 **(b)** Some students wanted to know how effective the pack was. They squeezed the pack and measured the temperature change with a thermometer. It was 15 °C. One of the students tasted the curry and said that it was not hot enough. He suggested a 30 °C temperature rise was needed, and that the pack should have 150 g of calcium oxide.

Suggest reasons why simply doubling the mass of calcium oxide may not double the temperature rise. In your answer you should consider other variables that might have an effect. [3]

AO1 recall the science   AO2 apply your knowledge   AO3 evaluate and analyse the evidence

✴ WORKED EXAMPLE – Higher tier

Some students carried out an investigation into the products formed when sodium chloride (Na^+Cl^-) solution is electrolysed. They used the apparatus shown below.

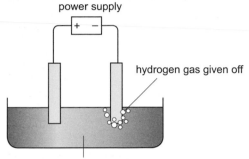

power supply

hydrogen gas given off

sodium chloride with universal indicator

Two gases, hydrogen and chlorine, which is a bleach, were produced, together with an alkali, sodium hydroxide.

(a) What is the purpose of the universal indicator in the sodium chloride solution? [1]

to identify if the solution is acid or alkali

(b) After five minutes, the students noticed that the universal indicator was blue around the negative electrode.

Explain what was happening. [2]

an alkali was being produced called sodium hydroxide

(c) They also noticed that the positive electrode smelt, and the solution was turning clear around the electrode.

Explain what was happening. [2]

chlorine gas is bleaching the universal indicator.

(d) The chloride ions reacted like this:

$2Cl^-(aq) \rightarrow Cl_2(g) + e^-$

Write a half equation for the reaction of the sodium ions. [1]

$Na^+(aq) + e^- \rightarrow Na(s)$

(e) Explain why hydrogen gas and not sodium metal was produced at the negative electrode. [3]

Sodium is a very reactive metal and will react with the water of the sodium chloride solution to produce sodium hydroxide and hydrogen gas instead of sodium metal.

How to raise your grade!

Take note of these comments – they will help you to raise your grade.

This is correct, but the indicator will monitor the pH, not just whether the solution is acid or alkali. The candidate will not receive the mark.

Two marks would be given for interpreting the observation and drawing a conclusion based on the ions present in the solution.

The candidate receives one mark only. This is a weak answer as it fails to mention that the chlorine gas is being produced at the positive electrode. The candidate recognises that the chlorine is bleaching the indicator, but this is in the question. Repeating the question does not gain any marks.

One mark is given for writing the correct half equation.

This is a good answer that covers the three marks well.

The candidate could also have made reference to sodium being more reactive than hydrogen, which is the reason why hydrogen gas is produced.

Physics P2.1–2.3

What you should know

Forces

Forces are measured in newtons using a newtonmeter.

The resultant force is calculated by combining the forces acting on the same object.

Balanced forces do not change motion but unbalanced forces make objects change speed or direction.

 Draw a force diagram showing the forces acting on a boat floating in water.

Speed

Speed = $\dfrac{\text{distance}}{\text{time}}$

Distance–time graphs show the changes in distance of an object, from its starting point, over time.

 A person travels 100 metres in 40 seconds. What is their speed?

Energy

Energy transfers include kinetic energy, sound, light and electricity.

Energy stores include fuels, lifted objects and deformed elastic objects.

The total energy before and after any energy transfer is the same.

 Describe the energy transfers when an electric light is switched on.

Electricity

Current is measured using an ammeter in series in a circuit.

Potential difference is measured using a voltmeter in parallel in a circuit.

When you draw a circuit diagram, symbols represent the components in the circuit.

Resistance limits current flow in a circuit.

A charge flow model explains how current flows in a circuit.

 Draw a series circuit that includes a bulb and two cells.

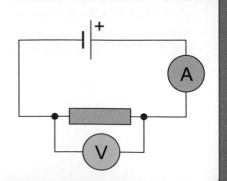

You will find out

Forces and their effects

> Several forces acting on an item can be combined into one resultant force.

> When forces on an item are unbalanced, the item changes speed or direction.

> Forces can change the shape of objects, and this stores energy in elastic objects.

> Distance–time graphs and velocity–time graphs show an object's motion over a period of time.

> The stopping distance and the terminal velocity of a vehicle depend on different factors.

The kinetic energy of objects speeding up or slowing down

> Energy is transferred when a force does work on an object.

> Power measures the rate at which energy is transferred.

> The kinetic energy and gravitational potential energy of an object can be calculated.

> Moving objects have momentum, and this can be used to analyse collisions and explosions.

Currents in electrical circuits

> Electric charge builds up in electrical insulators, but moves through circuits.

> Circuit diagrams show how components are connected together.

> Since different components in a circuit behave in various ways, they have different uses.

> The current, potential difference and resistance in series and parallel circuits can be calculated.

> The current in a circuit depends on several factors.

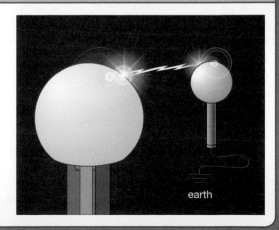

earth

See how it moves

You will find out:

> how to represent high speed and low speed movement on a distance–time graph

> why the gradient (slope) of a distance–time graph represents speed

> how to construct distance–time graphs

Who will win?

A bicycle can travel faster than a runner and a car can go faster still. Yet, in a race between a car, a bicycle and a runner, the runner might win. Why? Distance–time graphs can help – they show that the runner reaches top speed sooner than the car or the bicycle, so the runner will win very short races.

Did you know?

The current land-speed record for a wheeled vehicle is over 1200 km/h. That is travelling one kilometre in less than three seconds.

FIGURE 1: If the bus and the bike are going to the same place, who will get there first?

Distance–time graphs

You can use **distance–time** graphs to 'picture' journeys. A 'journey story' tells how something moves. For example:

"I started by travelling very slowly because I was walking to the bus stop. Then I stopped and waited for the bus. The bus travelled quickly, but it slowed down when it got near to school. Then it stopped and I got off."

Figure 2 shows the distance–time graph for this same journey.

A: The person is walking. The very shallow gradient means the **speed** is very slow. The person does not move very far each second, so the distance does not increase much.

B and E: The person has stopped. The horizontal line shows that the time is changing but the distance is not.

C: The bus is travelling quickly. A steep gradient means it travels a long way in each unit of time.

D: The bus is travelling slowly. The gradient is shallower than when the bus was travelling fast but is steeper than for the person walking.

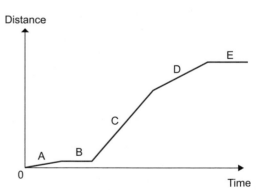

FIGURE 2: A journey to school. How would this distance–time graph change if the bus went more slowly all the way to school?

Drawing distance–time graphs

Graphs must always be labelled with the quantity and unit for each axis. For a distance–time graph:

Axis	Quantity	Unit
y	distance	metre (m) – sometimes kilometre (km) or mile is used
x	time	second (s) – sometimes minute or hour is used

> If distance in measured in metres and time in seconds, the unit for speed in metres per second (m/s).

> If distance in measured in kilometres and time in hours, the unit for speed in kilometres per hour, km/h (often written as kph).

Q distance time speed

QUESTIONS

1 Describe the journey in each of graphs P, Q and R.

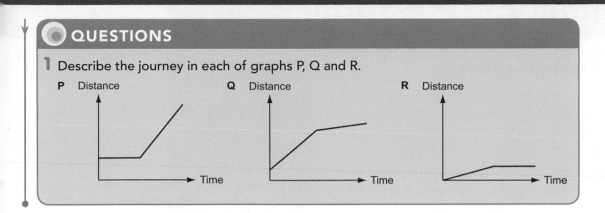

Speed and average speed (Higher tier)

If a distance–time graph is a straight line, its gradient tells you the speed that an object is moving:

$$speed = \frac{distance\ travelled}{time\ taken}$$

The graph in Figure 3 consists of two straight lines A and B.

A: The car travels 80 km in 1 hour. Its speed is 80 km/h.

B: The distance travelled by the car = (160 − 80) = 80 km. It takes (4 − 1) = 3 hours. Its speed is 80 ÷ 3 = 26.7 km/h.

A + B: Altogether the car travelled 160 km in 4 hours, so its **average speed** = 160 ÷ 4 = 40 km/h.

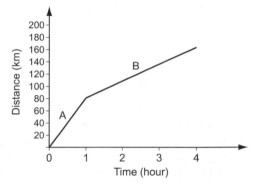

FIGURE 3: The car's average speed was 40 km/h. If the speed limit is 50 km/h, why did the driver get a speeding ticket?

QUESTIONS

2 Draw a distance–time graph for a vehicle which changes speed, but which has an average speed of 20 m/s.

Speed from the graph (Higher tier)

The gradient of a distance–time graph at any point gives the speed of the object at that time. In real life, speeds rarely change suddenly, so many real life distance-time graphs have curved, not straight lines.

The speed of the object at any point can be found by:

> drawing the tangent to the curve at that point

> then calculating the gradient of the tangent.

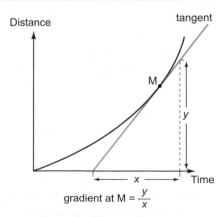

gradient at M = $\frac{y}{x}$

QUESTIONS

3 Draw a distance–time graph for a car that speeds up gradually to a maximum speed, then brakes hard in an emergency stop.

FIGURE 4: A 'real life' distance–time graph. Suggest why it is curved.

Speed is not everything

You will find out:
> the difference between speed and velocity
> how to find the acceleration of an object
> how to interpret velocity-time graphs

Boldly going where?

Scientists use space shuttles to take astronauts into space and to launch probes to explore the planets in our solar system. The shuttle moves fast to escape Earth's gravity, but high speed alone is not enough. What else do scientists have to think about, if a spacecraft is to reach Venus, for example?

Did you know?

The acceleration that you experience on a big roller-coaster may be five times the maximum acceleration of a sports car.

FIGURE 1: What else matters for this space shuttle, besides speed?

Velocity and speed

When a footballer kicks the ball, the speed alone will not tell you whether or not he will score a goal. You need to know the ball's **velocity**.

Velocity tells you about an object's speed *and* direction.

The symbol for velocity may be *v* or *u*.

Velocity can be positive (+) or negative (-). It depends on the direction of travel. In Figure 2, all the velocities are measured upwards, away from the ground.

The positive velocity of the rocket shows that it is moving upwards, away from the ground, at 12 000 m/s.

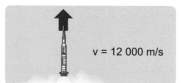

$v = 12\,000$ m/s

The parachute has a velocity of –2 m/s. It is negative because it is travelling downwards.

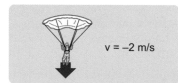

$v = -2$ m/s

The aeroplane has a very high speed, but its velocity upwards is 0 m/s. It is not moving towards the ground or away from the ground.

$v = 0$ m/s

FIGURE 2: Velocity shows speed and direction. Here, upwards is positive.

Velocity–time graphs

Velocity–time graphs show how velocity changes with time. Figure 3 shows the velocity–time graph for a cyclist. The velocity is measured along the road in the direction the cyclist is travelling.

A: The cyclist is speeding up steadily from stationary.

B: The cyclist is travelling at a high constant velocity.

C: The cyclist is slowing down.

D: The cyclist is travelling at a low constant velocity.

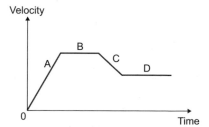

FIGURE 3: A cyclist's velocity–time graph.

QUESTIONS

1 What two things does the velocity of an object tell you?

2 Two runners both have the same velocity. What direction are they each moving in?

3 A ball is dropped. It hits the ground when its velocity is 5 m/s. It bounces back up with a speed of 3 m/s. What is its velocity?

Q velocity time graphs GCSE

Velocity and acceleration

Acceleration is the change in an object's velocity. Like velocity, acceleration has a direction. You can calculate acceleration using the equation:

$$a = \frac{v - u}{t}$$

where
a is acceleration in metres per second squared (m/s²)
v is final velocity in metres per second (m/s)
u is initial velocity in metres per second (m/s)
t is time taken in seconds (s)

Suppose an athlete is running at 4 m/s. She speeds up to pass another runner. It takes her 10 seconds to pass the other runner, at which point she is running at 6 m/s. What was her acceleration?

$v = 6$ m/s $u = 4$ m/s $t = 10$ s

Therefore, her acceleration was $\frac{(6 - 4)}{10} = 0.2$ m/s²

She slows down again to 4 m/s in another 20 seconds.

$v = 4$ m/s $u = 6$ m/s $t = 20$ s

Therefore, her acceleration was $\frac{(4 - 6)}{20} = -0.1$ m/s²

It is negative (−) because she is slowing down. The acceleration is in the opposite direction – she is **decelerating**.

> If an object is getting faster, it is accelerating.

> When it is slowing down, it is decelerating.

The direction of acceleration

In section C of Figure 3, the acceleration is negative, because the final velocity (v) for that section is smaller than the initial velocity (u).

In Figure 4:

> The ball is moving upwards. Its velocity is positive.

> The ball is slowing down. Its acceleration is acting downwards, towards the ground.

> The ball's acceleration is negative because it is slowing down *and* in the opposite direction to its velocity.

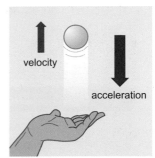

FIGURE 4: This ball has been thrown upwards. What will happen to it next?

 QUESTIONS

4 Which way does the acceleration act when a car is braking?

5 Use acceleration to explain why an arrow, fired horizontally, falls downwards, as well as moving forwards.

Using velocity–time graphs (Higher tier)

Acceleration

The gradient of a velocity–time graph shows how velocity changes with time. Hence, the gradient of a velocity–time graph is the acceleration of the object.

$$\text{gradient} = \frac{\text{change in } y}{\text{change in } x}$$

$$= \frac{\text{final velocity } (v) - \text{initial velocity } (u)}{\text{time } (t)}$$

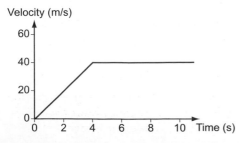

FIGURE 5: The motion of an object falling under gravity.

Distance

The distance travelled by an object can be calculated from:

distance travelled = velocity × time

This is the area underneath the velocity–time graph. Add together the areas of the triangles and rectangles under the graph.

 QUESTIONS

These questions refer to the object in Figure 5.

6 Describe the motion of the object.

7 Calculate (a) the acceleration while the object is accelerating at a constant rate (b) the average acceleration (c) the distance fallen during the 10 seconds.

Forcing it

Forces everywhere

In most situations, objects have lots of forces acting on them, all having different effects. Some of the forces act together, making the effect larger; some forces act against each other, making the effect smaller. Fortunately there is a way to find out the overall effect of lots of forces acting together – it is the resultant force.

You will find out:

> how the resultant force affects the way an object moves

> how to find the size of the resultant force

> how to resolve forces

Did you know?

An empty roof rack on a car increases the car's fuel consumption by about 10%.

FIGURE 1: What forces are acting here?

Adding up or cancelling out

Many forces on an object will affect the way it moves. To find out how, imagine that all the forces are replaced by one single force – the **resultant force**.

The resultant force on an object is the single force that would make the object move in exactly the same way as all the original forces acting together.

This resultant force would cause the object to speed up, slow down or change direction in exactly the same way as all the original forces put together.

> Forces that act in the same direction as each other add up.

In Figure 2, the total force backwards on the car is:

air resistance + friction = 400 N + 400 N = 800N

> Forces that act in opposite directions cancel out. Subtract them.

The total forwards force on the car in Figure 2
= forward force from engine – total backward force

= 1000 N – 800 N = 200 N

There is a resultant force of 200 N acting forwards on the car.

Remember

A single force on an object makes the object move in the same direction as the force. A resultant force gives an object a resultant direction.

○ QUESTIONS

1 Describe two forces that might act on the person abseiling in Figure 1.

2 Describe what a resultant force is.

3 How do you find the resultant force when two forces act in the same direction?

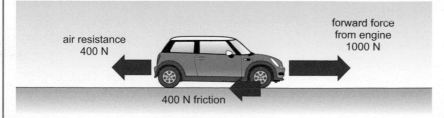

air resistance 400 N

forward force from engine 1000 N

400 N friction

FIGURE 2: The three main forces acting on a car.

Q forces GCSE ... friction activity

How things move

If the resultant force on an object is zero:

> stationary objects stay stationary

> moving objects stay moving at the same speed and in the same direction.

If the resultant force on an object is not zero:

> stationary objects start to move and accelerate (speed up) in the direction of the resultant force

> moving objects start to accelerate (speed up) in the direction of the resultant force.

FIGURE 3: Forces on a skier.

For an example, look at Figure 3.

The air resistance and friction both act in the same direction, up the slope. The total backward force on the skier equals the air resistance plus the friction, added together.

Gravity tends to 'pull' the skier down the slope. One component (a part) of his weight acts down the slope. The steeper the slope, the greater this forward force will be.

> If the forward force is greater than the total backward force, the skier will accelerate down the slope.

> If the total backward force is greater than the forward force, the skier will slow down, and eventually stop.

> If the forward force and the total backward force are the same size, the skier will continue to slide down the slope at a constant speed.

QUESTIONS

4 What happens to an object when the resultant force on it is not zero?

5 In Figure 3, how do the forces affect the way the skier moves?

6 What might change the size of the total backward (resistive) force on a skier?

Using resultant forces

A roof rack increases a car's fuel consumption. At any particular speed, it uses more fuel than without the roof rack.

> The roof rack increases the car's air resistance *at any particular speed*, so increases the backwards force *at that speed*.

> The forwards force *at that speed* must increase as well, to keep the car moving at a constant speed. (Otherwise the resultant backwards force would make the car slow down.)

> To provide a larger forwards force, the engine must work harder. It therefore uses more fuel.

QUESTIONS

7 Use energy transfers to suggest why an engine uses more fuel when it increases the forward force.

Q resultant force … energy speed forces AND motion

Force and acceleration

You will find out:
> a resultant force on an object causes an acceleration
> the relationship between acceleration, force and mass

High performance cars

People looking for high performance cars usually want to know how quickly the car can accelerate from 0 to 100 km/h. High performance sports cars may do so in four seconds or less, small family cars may take more than ten seconds.

FIGURE 1: How do designers make high performance cars that accelerate quickly?

How quickly

Unbalanced forces

When the forces on an object are not balanced, there is a resultant force. A resultant force changes the way an object moves.

When the car in Figure 2 travels at a constant speed, along a straight road, the resultant force is zero. The car is not speeding up, slowing down or changing direction. The driving force from the engine is the same size as the resistive force (the sum of air resistance and the friction between tyres and road).

Pushing on the accelerator increases the driving force from the engine. The resultant force is forwards and the car speeds up. The acceleration is in the direction of the resultant force.

Braking adds to the friction force between the brake and the wheels. This increases the resistive force. The resultant force is backwards and the car slows down. The acceleration is in the direction of the resultant force (the car decelerates).

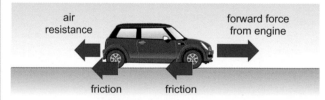

FIGURE 2: The forces acting on a moving car.

How big is the acceleration?

> The greater the acceleration, the more quickly something speeds up.

> The greater the deceleration, the more quickly something slows down.

In Figure 3, the car will be moved more quickly if more people help push it. They will provide a larger force. A larger force increases the acceleration.

You can move a broken down bicycle more quickly than a broken down car. The acceleration is greater if the **mass** being moved is smaller.

FIGURE 3: What affects how quickly the car can be pushed?

QUESTIONS

1 How can you tell if the resultant force on a car is zero?

2 Use forces to describe why the car speeds up when the driver presses the accelerator.

3 How can you tell the direction in which an object will accelerate?

science friction air resistance

Investigating acceleration

Acceleration increases when force increases and decreases when mass increases, but what is the precise relationship?

The apparatus in Figure 4 can be used to make a velocity–time graph. The gradient of this graph gives the acceleration.

Using different forces, and plotting a graph of acceleration against force used, shows that acceleration is directly proportional to force.

Changing the mass of the trolley, while keeping the force constant, shows that acceleration is inversely proportional to mass.

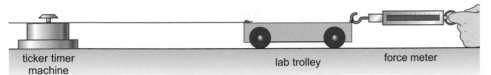

ticker timer machine   lab trolley   force meter

FIGURE 4: Apparatus to investigate how acceleration changes.

Force, mass and acceleration

The unit of force is the **newton** (N). One newton is required to give a mass of one kilogram an acceleration of one m/s^2.

The equation for the relationship between force, mass and acceleration is:

$F = m \times a$

where
F is the resultant force in newtons (N)
m is mass in kilograms (kg)
a is acceleration in metres per second squared (m/s^2)

> If the mass stays the same, the greater the force, the greater the acceleration.

Acceleration is proportional to force: $a \propto F$

> If the force stays the same, the greater the mass, the lower the acceleration.

Therefore acceleration is inversely proportional to mass: $a \propto \frac{1}{m}$

QUESTIONS

4 Describe an example of your own to show that acceleration decreases when mass increases.

5 What graph would you plot to show $a \propto \frac{1}{m}$? What would you expect the graph to look like?

Mass in space

Imagine that scientists want to find out about a rock on another planet, using a remote-controlled robot. To calculate the rock's density, they need to find its mass. Usually mass is found by measuring the pull of gravity on an object (its weight), but that does not work in space: weight changes when gravity changes. Instead scientists use a force of a known size to push the rock and then measure the rock's acceleration.

QUESTIONS

6 Explain how scientists calculate the mass of an object in space using the equation $F = m \times a$

FIGURE 5: This Mars explorer is analysing rocks on Mars.

Balanced forces

You will find out:

> about equal and opposite forces
> how Newton's Third Law applies to stationary and moving objects

Where are the forces?

Good ice skaters have no problems skating where they want to. Learner ice skaters often find it very hard to move how and where they want.

FIGURE 1: What forces are acting on this skater?

 Balanced forces

In a tug-of-war teams, if neither team moves, they must both be pulling as hard as each other. You can show this using two **force meters** joined together so that they pull on each other. Hold one force meter still and pull on the other one. The readings on both force meters are always the same, no matter how hard or how gently you pull.

Finding pairs of forces

Newton said that, for objects that are joined together, or in contact with each other, both objects exert exactly the same size force on each other. This is his **Third Law**.

For every force there is an equal and opposite force.

Equal and opposite forces are two forces that are the same size as each other, but they act in opposite directions. There are balanced forces in lots of situations.

When you sit on a chair:

> Your **weight** pushes down on the chair seat. The chair seat pushes up on you just as hard. That is why you do not sink through the chair.

> The chair legs push down on the floor, but the floor pushes up on the chair legs as well.

When you push down on a floating tray:

> The **upthrust** of the water pushes back.

> The harder you push, the harder the water pushes back.

FIGURE 2: Which team is pulling harder?

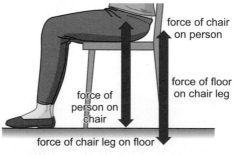

force of chair on person

force of floor on chair leg

force of person on chair

force of chair leg on floor

FIGURE 3: Equal and opposite forces.

QUESTIONS

 1 If two tug-of-war teams are not moving, what does that tell you about the forces?

2 Explain what 'equal and opposite forces' means.

3 The force of a person down on the ground is 500 N. (i) What other force acts? (ii) What size is it?

Forces on moving objects

Newton's Third Law works for moving objects as well as for stationary objects.

A parachutist falling downwards has two main pairs of equal and opposite forces acting.

> The weight of the parachutist pulling downwards on the parachute is the same size as the force from the parachute pulling upwards on the parachutist. This is due to the tension in the parachute strings.

> The force of Earth pulling downwards on the parachutist is the same size as the force of the parachutist pulling upwards on Earth. This is because gravity is a two-way force that acts to pull masses together.

When a car is pulling a caravan, the forward force of the car on the caravan is the same size as the backward force of the caravan on the car.

> As a car moves forwards, the backward force of the road on the wheels is the same size as the forward forces of the wheels on the road.

> The 'work done to overcome friction' is the work done to provide this force of the wheels pushing against the road.

FIGURE 5: This balloon will fly around, if it is released without tying the neck. Do you know why?

FIGURE 4: If the kayaks were inside the caravan, would it change any of the forces?

QUESTIONS

 4 Draw an arrow diagram to show the forces acting on the car and caravan.

5 Use Newton's Third Law to explain why an untied, blown-up balloon flies around the room when it is let go and the air comes out.

That lifting feeling

 When a lift starts accelerating upwards, the force of the floor pushing upwards on someone inside is greater than their weight. The resultant upward force makes them accelerate upwards too. The larger-than-usual upward force of the floor on them causes a larger-than-usual downward reaction force: they feel heavier than usual.

When the lift slows down, the floor pushes less hard than usual on their feet. The smaller-than-usual force of the floor causes a smaller-than-usual reaction force: the person feels lighter than usual, as the lift slows down.

QUESTIONS

 6 Explain what people inside a lift would feel when a lift (a) starts to accelerate downwards (b) stops.

conservation of momentum gases

Stop!

Keep your distance

In moving traffic, if one car brakes suddenly the car behind may not be able to stop soon enough to avoid an accident. How can drivers tell what a safe distance is, between them and the car ahead? It certainly helps to know how far your car will travel while you are trying to stop.

You will find out:
> about factors affecting the distance that a moving car takes to stop
> the difference between thinking distance, braking distance and stopping distance

Did you know?

Antilock Braking Systems (ABS) adjust a car's braking force, so that the driver can brake as hard as possible without the wheels locking and the car skidding.

FIGURE 1: When is traffic like this dangerous? Why?

Force it to stop

Kinetic energy

When a car travels at a constant speed, the driving force balances the resistive forces. Resistive forces on a moving car are air resistance (the main factor) and friction. The resultant force is zero.

When the driver brakes, friction between the brakes and the wheels increases the resistive force. The resultant force is negative. The car slows down.

The faster a car is moving, the more **kinetic energy** it has. It needs a larger force to make it stop in the same distance. The driver has to brake harder.

Work done by friction between the wheels and brake pads transfers this kinetic energy to the brakes by heating up the pads. When a car brakes, its brakes become warmer.

FIGURE 2: What is the link between this damage and the energy of the car?

Stopping distance

The **stopping distance** is how far a vehicle travels before it stops.

Stopping distance has two parts:

> **Thinking distance** – the distance a vehicle travels while a message transfers from the driver's eyes to the brain and from the brain to the foot on the brake pedal.

> **Braking distance** – the distance a vehicle travels while it is slowing down, after the brake pedal has been pressed.

stopping distance = thinking distance + braking distance

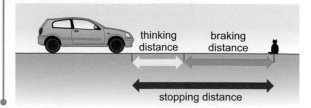

FIGURE 3: Stopping distance has two parts.

QUESTIONS

1 Why does braking harder make a car slow down more quickly?

2 Discuss why accidents at high speeds do much more damage to a car and its passengers than accidents at slower speeds.

Q friction car brakes ... car stopping distance

Thinking and braking

Thinking distance and reaction time

Thinking distance depends on the car's speed and on its driver's **reaction time**. You can calculate thinking distance using the equation:

thinking distance = speed × reaction time

A driver's reaction time will be slower than usual if they are tired or have taken alcohol or drugs (including many prescription drugs). Accidents are more likely to happen.

Braking distance

The braking distance of a car depends on its speed – doubling the speed increases the braking distance by four times.

Braking distance also depends on:

> the mass of the car. Heavier cars travel further before stopping because they have more kinetic energy to transfer.

> the condition of the tyres and brakes. Badly worn tyres or brakes give a smaller friction force and, therefore, a longer braking distance. That is why brakes are tested in an MOT test and why tyres are illegal if they have too little tread.

> the weather and road conditions. Wet or icy road surfaces decrease the friction force. On a wet road the braking distance can be doubled; on an icy or snowy road it can be up to 10 times as far as on a dry road.

Remember

Distractions may affect a driver's ability to react quickly.

 QUESTIONS

3 Look at the data in Table 1. Is the thinking distance directly proportional to speed?

4 Draw a graph of stopping distance against speed.

5 At low speed, does thinking distance or braking distance have most effect on the total stopping distance?

6 At high speed, does thinking distance or braking distance have most effect on the total stopping distance?

Speed (mph)	Speed (km/h)	Thinking distance (m)	Braking distance (m)	Stopping distance (m)
20	32	6	6	12
30	48	9	14	23
40	64	12	24	36
50	80	15	38	53
60	96	18	55	73
70	112	21	75	96

TABLE 1: Stopping distance.

Aquaplaning

On a wet road, tyre treads channel water out from between the tyre and the road. Bald tyres allow a thin layer of water to stay between the tyre and the road making the vehicle aquaplane. When a vehicle aquaplanes, it is sliding across the surface of the water, rather than gripping the road.

The danger of aquaplaning is one of the reasons why car tyres must have a tread depth of at least 3 mm.

 QUESTIONS

7 Formula 1 cars have 'slick' tyres; they are smooth. Explain why the driver has the tyres changed, if the race track is wet.

FIGURE 4: Both of these types of tyre have their advantages. What are they?

Q aquaplaning GCSE

Terminal velocity

Free fall

Skydivers enjoy the feeling of free fall, when they are falling downwards without a parachute, but they all open their parachutes before they land. Without a parachute they would hit the ground at about 200 km/h, which would be fatal.

FIGURE 1: Forces explain how a parachute helps provide a safe landing.

You will find out:

> why accelerating objects reach a terminal velocity

> the connection between mass and weight

> how to draw and interpret velocity–time graphs involving terminal velocity

Faster and faster

However hard you pedal, a bike always reach a top speed. You cannot go faster. Your bike has three main forces acting on it: the driving force of you pedalling, the resistive force of friction and the resistive force of air resistance.

> To begin with, air resistance is very small. There is a resultant force forwards and the bike speeds up.

> As you get faster, air resistance increases because you have to push more air particles out of the way each second.

> Eventually, air resistance and friction added together equal the forward force of you pedalling. There is no resultant force on the bike. It does not speed up any more.

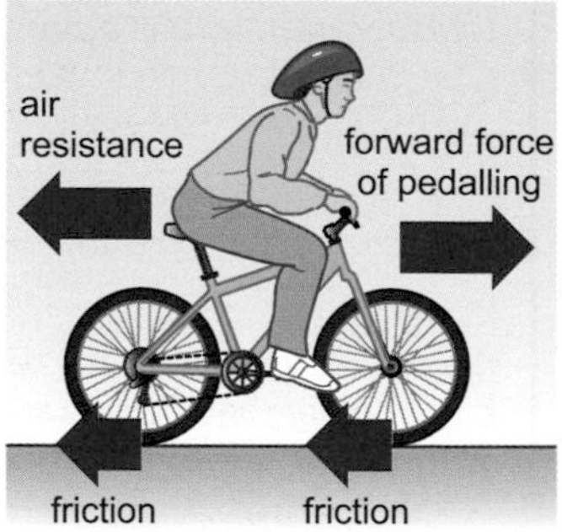

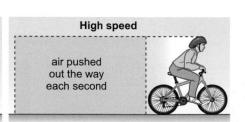

FIGURE 2: Forces acting on a cyclist.

FIGURE 3: Air resistance and cycling speed.

QUESTIONS

1 How could you change the size of any of the forces shown on Figure 2?

2 Racing cyclists wear streamlined clothing to reduce their air resistance. How does this affect their top speed?

Terminal velocity

Any object moving through a liquid or a gas experiences a force that opposes the motion. This force increases as the object gets faster. There is a maximum possible top speed – its **terminal velocity**.

For an object falling through air, its terminal velocity depends on the size of the upward force (the air resistance) and the size of the downward force (its weight).

$W = m \times g$

where
W is weight in newtons (N)
m is mass in kilograms (kg)
g is gravitational field strength in newtons per kilogram (N/kg)
On Earth, the value of g, is 9.8 N/kg (often rounded to 10 N/kg).

As an object falls faster, air resistance increases. The object stops accelerating when the air resistance equals the weight. This is when the object is falling at its terminal velocity.

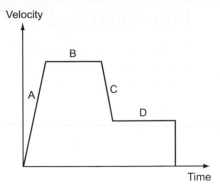

FIGURE 4: A velocity–time graph for a skydiver.

Graphs to show terminal velocity

Figure 4 shows how the velocity towards the ground changes for a skydiver before and after their parachute opens.

A: Weight is greater than air resistance. The resultant force towards the ground makes the skydiver accelerate downwards. As he accelerates, air resistance increases.

B: Weight and air resistance are equal. He has reached terminal velocity.

C: The parachute is open and air resistance increases. Air resistance is much greater than weight. The skydiver slows down. His air resistance decreases.

D: Air resistance and weight are again equal. The skydiver is again travelling at terminal velocity, but a slower terminal velocity than at B.

Weight also affects terminal velocity. There is a point at which air resistance and weight are balanced for a skydiver. At this point, another, heavier skydiver with the same air resistance will still have a resultant force downwards. The heavier skydiver accelerates for longer and reaches a higher terminal velocity.

QUESTIONS

3 Discuss which changes on the Moon, mass or weight. Do you know why?

4 Copy the velocity–time graph for the skydiver. Add an extra line to show what you would see if he had a larger parachute.

5 A spider does not fall like you would, but floats gently downwards. Use forces and terminal velocity to explain why.

Birdmen

Between 1930 and 1961, 75 'birdmen' tried to build wing suits to enable them to fly like birds – 72 of them died. Now, this dream is almost possible. Modern wing suits are strengthened with struts, to create a mini parachute. The terminal velocity can be as low as 40 km/h. A skilled birdman jumping from a plane at 4000 metres can glide for over eight kilometres.

QUESTIONS

6 Use your understanding of forces to suggest how birdmen might change direction by adjusting their wing suits.

FIGURE 5: This is a modern wing suit.

Forces and elasticity

You will find out:
> how forces may change the shape of an object
> how energy can be stored in a stretched elastic object
> about the factors affecting the extension of an elastic object

Pulling strings

Toy bows and arrows are fun toys and mostly harmless. Yet in Olympic archery, competitors fire at a target 70 metres away and the arrows travel at speeds of over 240 km/h. The difference is due to the materials that are used to make bows and arrows at competition level, and the forces involved.

Did you know?

Normally, dry human hair can be stretched by 20% before breaking. Wet hair can stretch 40–50% before it breaks.

FIGURE 1: This bow and arrow is not very dangerous. Can you use forces to explain why?

 ## Changing shape

Forces do not always make objects move. Sometimes a resultant force makes an object change shape, or break. An object that returns to its original shape when the force stops acting is an **elastic** object.

When a force stretches an elastic object, such as an elastic band or a spring, the force does work. It changes the shape of the object. The work done by the force transfers energy to the elastic object. This energy is stored by the object's new shape. It is called **elastic potential energy**.

If a force stops stretching an elastic object, the object will return to its original shape. The energy stored as elastic potential energy will be transferred away. It may do work to make something else move, such as a catapult pellet.

FIGURE 2: What would happen if the girl let go of the chewing gum?

QUESTIONS

1 Name three effects that a force can have when it acts on an object.

2 Give an example of an elastic object.

3 Give two examples of situations where elastic potential energy is stored.

4 Give one example where elastic potential energy is transferred as work to make an object move.

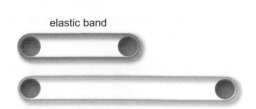
elastic band

FIGURE 3: The more work done, the greater the change in shape and the more elastic potential energy stored.

Q elastic potential energy GCSE

How far does it stretch?

An elastic object, such as a spring or elastic band does not just keep stretching further as you apply more and more force. Eventually it breaks.

Figure 4 shows an experiment to investigate how the extension of an elastic object is related to the force used to stretch it. Work is done to stretch the spring by the weight of the mass acting downwards.

The graph in Figure 5 shows:

> a linear section. The extension is directly proportional to the force applied.

> a non-linear section. Here, the **limit of proportionality** has been exceeded. The spring or other elastic object stretches non-linearly, then breaks.

Hooke's Law describes how force and extension are related for the linear part of the graph, where the limit of proportionality is not exceeded. Hooke's Law says that:

$F = k \times e$ where
F is force in newtons (N)
k is the spring constant in newtons per metre (N/m)
e is extension in metres (m)

> **Remember**
>
> Weight is a force and is measured in newtons.

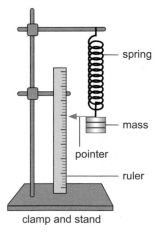

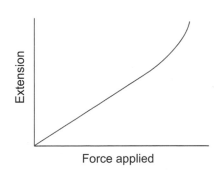

FIGURE 4: An experiment to test Hooke's law.

FIGURE 5: The straight line part of this graph shows Hooke's Law.

QUESTIONS

5 A strong spring does not stretch as easily as a weaker spring. Does a strong spring have a high or a low value of k?

6 How could you use a graph for Hooke's Law to find out the size of an unknown mass?

Using elastic potential energy

Whenever an elastic object is stretched or compressed, work is done to change its shape. The energy transferred by that work is stored as elastic potential energy. The work is done to pull the particles of the material further apart – or to push them closer together – than normal.

When the force stops acting, the particles return to their original positions and the extra energy they stored is transferred. This energy can be usefully transferred.

> In a catapult, energy is transferred to do work, making a projectile move.

> In a tightly wound clockwork spring, the stored energy is released slowly as work done, making a wind-up toy move.

> In devices such as clockwork radios, the energy is transferred, turning a generator to generate electricity.

QUESTIONS

7 In a set of scales, springs are compressed to find an unknown mass. Explain what happens to the force and energy.

8 How could you compare the amount of elastic energy stored by two different catapults? Explain what you would do to ensure valid and reliable results.

Q Hooke's law spring constant

Preparing for assessment: Planning an investigation

To achieve a good grade in science, you not only have to know and understand scientific ideas, but you need to be able to apply them to other situations and investigations. These tasks will support you in developing these skills.

✱ Investigating terminal velocity

Confectionery is often made by pouring sugary liquids into moulds where they cool and set hard.

Food manufacturers can compare how easily fluids flow by measuring the terminal velocity (steady speed) of a marble falling through the fluid. Fluids such as syrup, honey and chocolate flow more easily when they are warm.

A group of students visited a factory that uses syrup to make sweets.

After the visit, the students decided to investigate how the steady speed (terminal velocity) of a marble falling through syrup and the temperature of the syrup were linked.

✱ Planning

1. Devise a hypothesis linking the terminal speed of a marble falling in syrup, and the temperature of the syrup.

A hypothesis should state how you think an independent variable that you can control is linked with a dependent variable.

2. Outline a method for the investigation.

An outline method does not need to give details of how to carry out measurements. When any hypothesis is tested, the experiment must be designed as a fair test or the data will not be valid.

3. List the equipment needed to carry out the investigation.

When choosing equipment, think about which variables will be measured or changed, and how to calculate the top speed (terminal velocity).

4. Write instructions for the investigation to measure the time taken for a marble to fall 10 cm through syrup at 10, 20, 30, 40 and 50 °C.

A good method explains how to use equipment to take readings and how many times a measurement should be repeated. A preliminary experiment to check that the marble has reached terminal velocity should be included.

✳ Processing data

5. Prepare a table for the results. ●

6. Describe how to calculate the mean (average) time for tests carried out at a given temperature. ●

7. Describe how to calculate the mean speed for tests carried out at a given temperature. ●

Your table needs headings for all measurements, including their units. It needs space for the readings that you plan to take.

Check for anomalous data and ignore these when calculating the mean.

For each temperature, use the mean time for the marble to fall.

✳ Reviewing the method

8. List two hazards in your method. Describe changes that you could make that would reduce the risks from these hazards. ●

9. Two groups of students carried out initial experiments. Their results did not always agree.

Suggest one way of improving their investigation, to obtain reliable results. ●

Hazards are things that could harm a person or the surroundings. Include the things that you think are most likely to go wrong.

There are several things that the students can do. For example, they could make sure that all details of the experiments are carried out in the same way, by both groups.

✳ Connections

How Science Works

- Plan practical ways to develop and test scientific ideas
- Assess and manage risks when carrying out practical work
- Collect primary and secondary data
- Select and process primary and secondary data
- Analyse and interpret primary and secondary data

Science ideas

P2.1.2 Forces and motion

P2.1.4 Forces and terminal velocity

Energy to move

Exciting energy

Most exciting sports transfer lots of energy. Imagine yourself on a quad bike. Think about all the ways you would notice energy being transferred. There are useful and wasted energy transfers. Could you draw an energy diagram? Where has the quad bike's energy come from, originally?

FIGURE 1: In what ways is this quad bike transferring energy?

What is kinetic energy?

The energy stored in a moving object is called kinetic energy. Energy cannot be created or destroyed – this is the **Law of Conservation of Energy**.

Energy stored in the movement of an object has been transferred there from somewhere else. For example:

> The body's kinetic energy has been transferred from energy in food.

> The kinetic energy of a moving car has been transferred from energy in fuel.

> The kinetic energy of water in a solar fountain has been transferred from the Sun, by electromagnetic radiation.

The kinetic energy stored in the movement of an object can also be transferred away from the object. For example:

> When a car crashes, its kinetic energy is transferred as work done to damage the car.

> The kinetic energy of a moving loudspeaker is transferred as sound waves.

 QUESTIONS

1 Describe how energy is transferred when you kick a ball.

2 Describe ways in which energy is transferred by a quad bike.

3 Describe another example of energy transfer involving kinetic energy.

FIGURE 2: How does this fountain transfer energy?

Kinetic energy transfers

Friction and heating

You need fuel to keep a car moving. The energy used in its movement (kinetic energy) is being continually transferred mechanically, by moving parts of the car.

However, wherever moving parts are in contact with something else, there is friction. Energy transferred from the fuel is used to do useful work (to make the car move) but much is **'wasted'** heating up the car and the air around it as frictional forces are overcome.

Did you know?

Your hands become warm if your rub them hard together because work done against friction is transferred by heating.

Q kinetic energy GCSE

Examples of the heating effect of friction:

> Car tyres become warm. The correct tyre pressure for a car should be checked when the tyres are warm.

> Aircraft heat up and expand as they do work against air resistance. A passenger jet may be up to 20 cm longer in the air due to heating.

> Very fast moving objects, such as spacecraft or meteorites heat up as they do work to move through Earth's atmosphere. Shooting stars are meteors that are glowing white hot because of friction.

Kinetic energy and cars

If a car crashes, its kinetic energy is transferred as work done to damage the car. **Crumple zones** are safety features designed to reduce injuries to passengers. They are in places such as the front or rear of the car, so that work is done to crumple these parts of the car, leaving less kinetic energy to be transferred to the passengers.

When a car slows down, it transfers kinetic energy. **Regenerative braking** uses some of this energy – as work done to put the brakes on. The car can brake harder and the brakes heat up less.

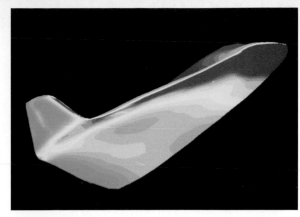

FIGURE 3: This computer-generated image shows how the space shuttle heats up as it re-enters Earth's atmosphere.

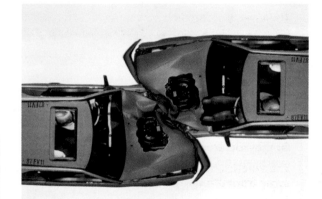

FIGURE 4: Crumple zones absorb energy in a crash. This test shows how a car's occupants are protected.

QUESTIONS

4 High speed aircraft tend to fly at high altitude, where the air is thinner. Suggest some reasons why.

5 Describe one example of how kinetic energy is transferred away from a stopping car.

Storing kinetic energy

Kinetic energy is difficult to store. Moving objects stop eventually unless energy is transferred to them.

A flywheel is a heavy, fast-spinning wheel used to store kinetic energy for short periods. Heavy flywheels store more energy than light ones. However, heavy wheels have the same air resistance as lighter wheels of the same diameter. So, heavy flywheels transfer (and therefore waste) a smaller proportion of their energy as work done against air resistance.

QUESTIONS

6 Discuss situations where it might be useful to store kinetic energy.

FIGURE 5: What design features of this flywheel make it good at storing kinetic energy?

Working hard

Floating around

Canals have locks to lift the boats up or down hills. One person can easily pull a canal boat into a lock using ropes, but it would certainly be impossible for one person to lift the canal boat without using the lock. Where is the energy transferred from, to lift the boat in the lock?

You will find out:

> how work, energy transferred and forces are related

> how to calculate the work done and the energy transferred by a force

FIGURE 1: Do you know how canal locks work?

Working

Work and energy

If you do lots of **work** lifting or moving things, you 'use up' lots of energy and feel tired and hungry. Of course, you do not really 'use up' energy. You transfer energy from you to the object that you move or lift. You do work to overcome gravity or frictional forces. The more work you do, the more energy you transfer.

work done = energy transferred

Pushing, pulling, lifting

Work is done and energy transferred every time a force makes something move.

Look at the tasks in Figure 2.

> The person with the umbrella is doing work against air resistance, transferring energy from the air particles to make them move slower.

> The gardener is doing work against frictional forces. Energy is transferred by heating and, probably, making sound.

> The librarian is doing work against gravity. Energy is transferred to the books where it is stored as gravitational potential energy. **Gravitational potential energy** is a way of storing energy in an object's position.

FIGURE 2: Which person is doing the most work?

QUESTIONS

1 Give an example of (a) something doing a lot of work (b) something doing a little work.

2 Describe the connection between using a force and transferring energy.

3 Describe how you are doing work when you climb a ladder.

Q work force energy

How much and how far

Calculating how much work is done

The amount of work done, and energy transferred, depends on:

> the force used. A large force does a lot of work and transfers a lot of energy.

> the distance the force moves an object. More work is done when an object is moved further.

The work done, and energy transferred, can be calculated from:

$W = F \times d$

where
W is work done in joules (J)
F is force applied in newtons (N)
d is the distance moved in the direction of the force in metres (m)

FIGURE 3: Would this person do more work carrying the a rucksack upstairs or up a mountain?

Direction matters

The distance an object moves must always be measured in the direction of the force.

Gravity acts vertically. Two people lifting equal masses through the same vertical distance:

> do the same amount of work against gravity

> transfer the same amount of energy, stored as extra potential energy in the mass they lift.

Using a ramp to lift a mass actually does more work and transfers more energy in total – work also has to be done against friction as the mass slides. This extra work is transferred by heating. However, using a slope feels easier, even though it does more work.

FIGURE 4: Would this cyclist do more work riding, carrying or pushing his bike up the hill?

QUESTIONS

4 Calculate the work done when a force of 40 N moves a block 2 m.

5 Estimate the energy transferred when you climb one flight of stairs. (Your weight is probably in the range 500 N to 750 N.)

6 Discuss why astronauts are able to jump higher on the Moon than on Earth.

The clock has stopped

Battery-powered wall clocks almost always stop with the minute or second hand at the 'quarter to' position, when the battery goes flat. In this position:

> the hand moves the greatest vertical distance against gravity for each 'tick'

> the battery has to transfer most energy per 'tick'.

FIGURE 5: Why did this clock stop at this time?

QUESTIONS

7 Use your knowledge of forces and energy transferred to explain why steep hills are harder to climb than shallower hills.

calculate work by force

Energy in quantity

You will find out:
> how to calculate the gravitational potential energy stored in an object's position
> how to calculate the kinetic energy stored in an object's movement

Potentially spectacular

When water tumbles over the Niagara Falls, it drops 52 metres. In every kilogram of that water, 500 joules of gravitational potential energy transfer into kinetic energy as it falls. In fact, three million kilograms of water tumble over every second, and all that energy makes a spectacular splash.

Did you know?

A typical rifle bullet has about the same amount of kinetic energy as you have when you jog slowly.

FIGURE 1: Some of the energy in this water is diverted to generate hydroelectricity.

Gravitational potential energy and kinetic energy

Gravitational potential energy

Gravitational potential energy is energy stored in an object because of its position.

To 'picture' this energy, think about how much work you would do to move an object from one place to another.

To lift an object higher, you must do more work against gravity and transfer more energy. The gravitational potential energy increases for objects that are further from the ground.

Kinetic energy

Kinetic energy is a way of storing energy in an object's movement.

To 'picture' how much energy different moving objects store, think about how much work the object would do, if it transferred its kinetic energy elsewhere. For example, when you throw a stone into a pond, it will make a bigger splash if:

> its mass is larger

> it is moving faster.

Kinetic energy increases when the mass of an object increases and its speed increases.

QUESTIONS

1 Write down the two factors that affect how much kinetic energy an object has.

2 Give one example of an object with (a) high kinetic energy (b) low kinetic energy.

FIGURE 2: How much kinetic energy will this shot transfer? How could you tell?

Calculating gravitational potential energy

You can calculate the gravitational potential energy in an object by finding the work done to move it to its position.

The gravitational potential energy gained when an object is lifted through a height, h, is the work done against force due to gravity to move it that distance.

The force due to gravity on an object is its weight, which is $m \times g$.

gravitational potential energy gained = weight × height lifted

$E_p = m \times g \times h$

where E_p is the change in gravitational potential energy in joules (J)
m is mass in kilograms (kg)
g is gravitational field strength in newtons per kilogram (N/kg)
h is the change in height in metres (m)

Example

How much does the gravitational potential energy stored by a box increase, if the box is lifted from the floor onto a shelf, 75 cm from the floor? The mass of the box is 4 kg.

$E_p = m \times g \times h$

$E_p = 4 \times 10 \times 0.75 = 30$ J

Note:

(i) The value of g is actually 9.8 N/kg, but it is often rounded to 10 N/kg for calculations.

(ii) The value of h used is 0.75, not 75, because h must be in metres, not centimetres.

The work done to lift the box is equal to the gravitational potential energy transferred to the box.

Therefore, work done = 30 J.

QUESTIONS

3 Calculate the change in gravitational potential energy for an object of mass 200 g, lifted though a height of 6 m.

4 An object of mass 1 kg rolls down a slope, moving through a vertical distance of 5 m.

(i) Describe and calculate the change in gravitational potential energy for the object.

(ii) What other energy transfers will there be?

Investigating kinetic energy

The energy stored by the pendulum, in Figure 3, changes from energy stored in its position (at A and C), to energy stored in its movement (at B), and back again. Ignoring work done against air resistance and friction, the gravitational potential energy 'lost' moving from A to B is the same as the kinetic energy gained at B.

Accurate measurements, using light gates and 'friction-free' pendulums shows that the kinetic energy can be found from:

$E_K = \frac{1}{2} \times m \times v^2$

where E_K is kinetic energy in joules (J)
m is mass in kilograms (kg)
v is speed in metres per second (m/s)

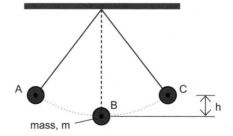

FIGURE 3: Energy changes can be used to investigate the kinetic energy of this pendulum.

QUESTIONS

5 Calculate the kinetic energy of an object of mass 100 g, travelling at a speed of 20 m/s.

6 Discuss how you could find the elastic potential energy in a stretched elastic band using (a) the height reached by a pellet fired vertically (b) the speed of a pellet fired horizontally.

Energy, work and power

You will find out:
> the scientific meaning of power
> how to calculate the power of a device

Mighty muscles

Words such as 'strong' or 'powerful' might describe a weightlifter. Female champions will lift a mass of more than 150 kilograms. For a successful lift, the athlete must hold the bar steady above their head, with arms and legs straight and motionless. 'Being powerful' is related to the work that she does and the energy she transfers.

FIGURE 1: Is there any scientific way to find out just how powerful this athlete is?

Work and power

Whenever something does work, it is transferring energy. However, **power** is not the same as work.

The fork lift truck Figure 2 is much more powerful than you. It can lift and move the loads of bricks in minutes.
Something is described as powerful if:

> it can do a lot of work in a short time

> it can transfer energy at a high rate.

Power is calculated by finding out how much energy something transfers each second. Power is calculated using the equation:

$$P = \frac{E}{t}$$

where
P is power in watts (W)
E is energy transferred in joules (J)
t is time taken in seconds (s)

One **watt** is the same as one joule per second (1 W = 1 J/s)

FIGURE 2: You could do the work needed to lift all these bricks onto the back of a lorry, but it might take you all day.

Did you know?

The average power of humans is about the same as a laptop and a printer together, about 100–120 W.

QUESTIONS

1 When a machine is called 'powerful', what does it mean?

2 What is the unit of power?

3 A crane moves the same number of bricks as you can, but in less time. Does it do more work, less work or the same work as you?

Q power GCSE

Finding the power

Method one

You can calculate the power of a machine, or of a person, by measuring out how much work it can do in a given time using the equation:

$W = F \times d$

where
W = work done (energy transferred) in joules (J)
F = force applied in newtons (N)
d = distance moved in the direction of the force in metres (m)

For example, a force of 15 N moves an object 3 m, in the direction of the force, in 5 s.

$W = 15 \times 3 = 45$ J

Work done (W) is the amount of energy transferred (E)

so you can use $P = \dfrac{E}{t}$ to calculate power.

$P = 45 \div 5 = 9$ W

Method two

Suppose the person in Figure 3, mass 50 kg, takes 10 s to climb the 2 m up the flight of stairs. The energy they transfer is the work they do against the force of gravity.

Energy transferred is the gravitational potential energy gained.

$E_p = m \times g \times h$

Taking g as 10 N/kg

$E_p = 50 \times 10 \times 2 = 1000$ J

Then using $P = \dfrac{E}{t}$

$P = \dfrac{1000}{10} = 100$ W

FIGURE 3: How could you find out your maximum power?

 QUESTIONS

4 (i) A crane lifts a mass of 60 kg to a height of 5 m. How much work does it do?

(ii) The crane takes 1 minute to do this. Calculate the power of the crane.

5 A crane has a power of 100 kW. How far can it lift a 10 t mass in 10 s?

The same work, but easier

Dragging blocks up slopes feels easier than lifting them the same height, even though friction means you do more work in total. Gravity acts straight downwards, but it can be imagined as two forces instead: one pulling down the slope and one pulling at right angles to the slope.

The 'component of gravity' acting down a slope is smaller than the real force due to gravity. Even when the frictional force is added, the force down the slope – that the people in Figure 4 are pulling against – is still smaller than the force of gravity acting straight downwards.

The shallower the slope, the smaller the force you have to pull with, and the easier it feels. You have to pull for longer though.

FIGURE 4: What forces are acting here?

 QUESTIONS

6 A pyramid is 40 m high and a 1000 kg block needs to be taken to the top. If the block can be dragged up a slope 500 m long, how many men, each pulling with a force of 80 N, would it need? Ignore frictional forces.

Momentum

You will find out:
> what momentum is and how to calculate it
> how to investigate the momentum of colliding or exploding objects

The hose is alive

Have you ever dropped a garden hosepipe and seen it flip round soaking everyone nearby? It is more likely to happen if the tap is on full. A fire hose, with very high-pressure water, sometimes needs two firemen to hold the hose steady and point it in the right direction. It is all to do with momentum.

FIGURE 1: Can you explain why fire hoses sometimes need two people to hold them?

What is momentum?

Momentum affects many everyday things. Momentum describes how hard it would be to stop a moving object. Objects with a greater momentum are harder to stop. The momentum of an object is found from the equation:

$p = m \times v$

where
p is momentum in kilogram metres per second (kg m/s)
m is mass in kilograms (kg)
v is velocity in metres per second (m/s)

Momentum increases when mass increases or velocity increases. Heavier, faster objects are harder to stop.

The momentum is always in the same direction as the velocity.

Momentum is not the same as kinetic energy.

> Energy is a **scalar quantity** – it only has size not direction; you cannot have negative energy.

> Momentum is a **vector quantity** – it has size and direction.

FIGURE 2: Momentum has size and direction.

QUESTIONS

1 Give an example of an object with (a) high momentum (b) low momentum.

2 Describe the difference between momentum and kinetic energy.

Collisions and explosions

Investigating momentum

There are two different types of collision:

> **elastic collisions** – two objects collide and bounce apart again

> **inelastic collisions** – two objects collide, stick together and remain together after the collision.

Newton showed that, whenever two objects collide, or an object explodes, the total momentum does not change, provided that it is a closed system, with no other forces acting on the objects.

This is the **Law of Conservation of Momentum**:

The total momentum before an event is equal to the total momentum after the event.

Did you know?

Safety features, such as sports mats and crash helmets, are designed so that momentum changes more slowly, helping to limit damage.

The Law of Conservation of Momentum can be used to find the speeds of objects after a collision.

Look at the inelastic collision in Figure 3. If both trolleys have the same mass, 1 kg, and the first trolley had a velocity of 5 m/s before the collision, what is the velocity of the two trolleys together after collision?

momentum before collision = momentum after collision

$(1 \times 5) + (1 \times 0) = 2 \times v$

$5 + 0 = 2 \times v$

$v = 5 \div 2 = 2.5$ m/s

in the same direction as the first trolley was moving originally.

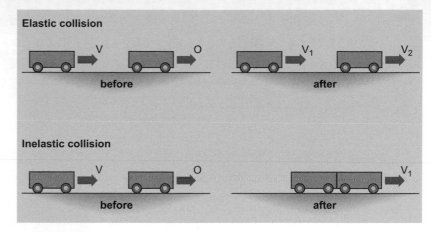

FIGURE 3: Elastic and inelastic collisions.

Explosion in a gun

When a gun is fired, an explosion inside the gun propels the bullet in one direction and the gun in the opposite direction. Before the collision, nothing was moving, so the total momentum was zero. The momentum after the gun fires is also zero. The momentum of the gun in one direction is the same size as the momentum of the bullet in the opposite direction. Because the bullet has a much smaller mass than the gun, it moves off with a much greater velocity.

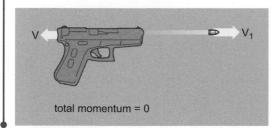

total momentum = 0

FIGURE 4: Momentum is zero before and after firing a gun.

QUESTIONS

3 Describe what is meant by 'inelastic collision'.

4 Two identical trolleys, travelling in opposite directions at equal speeds, collide with each other then bounce apart. Discuss the total momentum and how the momentum of each trolley changes.

5 What are the effects of making a gun lighter?

Jet engines

Jet engines work in a similar way to the balloon in Figure 5. Hot gas rushing out the back of the jet engine at very high speed has a high momentum. This is equal to the size of the momentum of the aeroplane in the opposite direction.

Because the mass of the aeroplane is much greater than the mass of the exhaust gases, the aeroplane moves more slowly than the exhaust gases do.

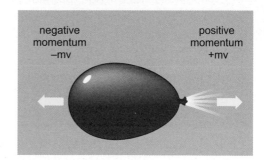

FIGURE 5: Air rushing out of the balloon makes it move.

QUESTIONS

6 (i) What would happen to the total momentum if jet engine exhaust gases went faster?

(ii) What would happen to the velocity of the aeroplane?

Preparing for assessment: Applying your knowledge

To achieve a good grade in science, you not only have to know and understand scientific ideas, but you need to be able to apply them to other situations and investigations. These tasks will support you in developing these skills.

✸ At the circus

Circuses often include a trapeze act. A trapeze is a swing hanging high above the ground. Once the trapeze is swinging, the performer carries out tricks by turning, jumping, and somersaulting. They also jump through the air to other trapezes or change places with other performers while the trapeze is still swinging.

An exciting act uses a trapeze and a swinging platform, many metres apart and many metres above the ground. Trapeze artists jump further than seems possible, from the platform to the trapeze, catching it using their hands or working with partners to perform even more daring tricks.

Success depends on exact timing. Each performer must control their momentum and speed totally, to avoid disaster. When the act starts with two performers on the platform, the trapeze artist can jump even faster and further.

Maxine works as a trapeze artist with partners Molly who shares the platform with her, and Toni who catches her after she jumps.

> At the start of the act, Molly and Maxine step onto the platform, and start it swinging faster and higher.

> At the same time, Toni starts the trapeze swinging, carefully timed to match the moving platform.

> When she is ready and the platform is moving fast towards Toni, Maxine leans forward, bracing herself.

> When it reaches the right position, Maxine jumps, pushing hard against the edge of the platform.

By timing this exactly, Maxine can push Molly and the platform away, giving Maxine enough momentum to fly metres through the air and still reach Toni in time to finish the act safely and with style.

The whole act depends on conservation of momentum. The combined momentum of Maxine and Molly stays the same during the jump. Because Maxine gives Molly extra momentum, Maxine (who is travelling in the opposite direction by now) gains momentum herself, which means faster speeds.

✵ Task 1

(a) A performer stands on the platform. How can you tell if the forces are balanced?

(b) The performer starts the trapeze swinging. How can you tell that the forces are not balanced?

✵ Task 2

The performer jumps from the platform and catches a trapeze several metres away.

(a) What forces act on the performer as they travel through the air?

(b) When they catch the trapeze and slow down, explain the direction in which the forces from the trapeze act.

✵ Task 3

(a) Two performers and the platform have a mass of 140 kg, and move at 2 m/s. What is their total momentum?

(b) One performer, mass 60 kg, jumps off, moving in the same direction at a new speed of 6 m/s. What is her momentum now?

(c) Use your answers to parts (a) and (b) to calculate the new momentum and speed of the platform and the other performer.

✵ Task 4

(a) Using a second person on the platform helps the first performer to jump forward faster. Explain why.

(b) Momentum is only conserved if no external forces are acting. How can you tell that external forces are acting, when the trapeze artists are swinging backwards and forwards?

✵ Maximise your grade

	Answer includes showing that you...
	know that moving objects have momentum.
	know that unbalanced forces make objects change speed or direction.
E	know that balanced forces do not change motion.
	can calculate momentum for moving objects.
	know what is meant by conservation of momentum.
C	can calculate the resultant of forces acting on an object.
	can predict changes in the speed when moving objects separate.
A	can describe the effect of a resultant force on motion.
	know that conservation of momentum only applies if no external forces act.

Static electricity

You will find out:
> how and why some insulating materials gain a static electric charge
> how different electric charges affect each other

Lightning

A flash of lightning is a spark of electricity moving between Earth and the clouds. Benjamin Franklin worked out that lightning was a brief flow of electric current. He flew a kite early in a storm (many early scientists were killed doing this) and watched the sparks from a metal key threaded on the kite string.

Did you know?

The energy transferred by a lightning strike can make the sap in a tree boil. This splits open the tree.

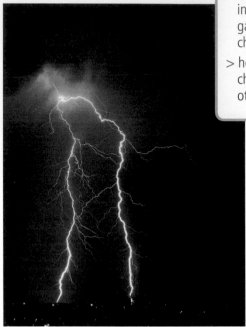

FIGURE 1: What should you do if you are caught in a thunderstorm?

Static electricity

You may see **static electricity** in several places. Rubbing a balloon on a jumper charges it with static electricity, which makes it stick to walls or ceiling. Rubbing a plastic ruler with a duster charges it with static electricity, which:

> makes a trickle of water change direction

> makes tiny pieces of paper or polystyrene beads jump about.

All materials are made from atoms, and atoms contain **electrons** with a negative electric charge. The electrons are much smaller than **protons** and **neutrons**. When some insulating materials are rubbed against each other, negatively charged electrons move from one material to the other.

> The material that gains electrons ends up with a negative electric charge.

> The material that loses electrons ends up with an equal positive electric charge.

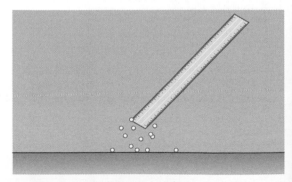

FIGURE 2: The ruler has been charged with static electricity. The trickle of water deflects away from the ruler.

QUESTIONS

1 What charge does an electron have?

2 Describe how a plastic ruler gains a static electric charge when you rub it.

3 What happens, if the water, in Figure 2, touches the ruler?

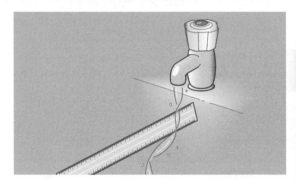

FIGURE 3: The ruler has been charged with static electricity. The polystyrene beads jump about. Why?

Investigating charges

Some types of atoms lose electrons more easily than others. Some materials gain a static electric charge when they are rubbed and others do not. Whether a material becomes positively or negatively charged (loses or gains electrons) depends on what type of atoms the material is made from.

Figure 4 shows how to investigate the charge on different materials. Place a rubbed, electrically charged, cellulose acetate strip in a paper cradle. In turn, hold near it – but not touching – another rubbed cellulose acetate strip and a rubbed poly(ethene) strip.

> Objects that carry the same type of charge repel.

> Objects that carry different types of charge attract.

1. The two cellulose acetate strips move apart. They must carry the same type of electric charge, because they are the same type of material.

2. The cellulose acetate strip and the poly(ethene) strip move together. Poly(ethene) and cellulose acetate must have different electric charges, because they behave differently from the two cellulose acetate strips.

By connecting the charged strips to very sensitive ammeters and looking at which way the needle deflects, scientists have found that:

> cellulose acetate carries a positive electric charge

> poly(ethene) carries a negative electric charge.

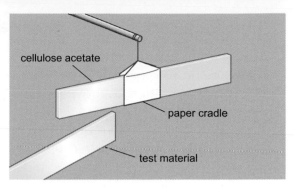

FIGURE 4: Investigating static electric charge.

QUESTIONS

4 Name the two types of charged particles in atoms.

5 What would happen if you held two charged poly(ethene) strips close together?

6 Two charged strips repel. Suggest how they might behave if you rubbed each one for longer.

Identifying unknown charges

The gold leaf electroscope, shown in Figure 5, has a metal stem and a strip of gold leaf.

A: Wiping a negatively charged strip across the cap makes it negatively charged too.

B: The stem and the gold leaf move apart.

C: If another negatively charged strip is held near, the electrons will be repelled further down the electroscope and the gold leaf will rise more.

D: If a positively charged strip is held near, the gold leaf will rise less.

QUESTIONS

7 Suggest what would happen if the top of the gold leaf electroscope was touched by
(a) the negatively charged strip
(b) the positively charged strip.

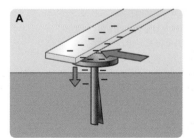

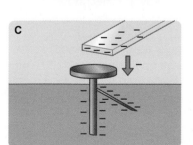
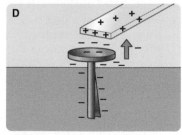

FIGURE 5: A charged strip of poly(ethene) is wiped across the cap of a gold leaf electroscope.

Moving charges

You will find out:
> electric current is a flow of electric charge
> how electric charges move in a range of materials and situations

The Van der Graaff generator

The Van der Graaff generator demonstrates some of the ways in which electric charges can move around. It can give off sparks several centimetres long, make a person's hair stand on end, and many other things. Yet, it is not really as dangerous as it looks.

Did you know?

Aircraft can be struck by lightning without harming the passengers because the electric charge flows round the outside of the aircraft.

FIGURE 1: Is this as hair-raising as it looks?

 ## Electric charge and movement

Forces and movement

Two electrically charged objects exert a force on each other.

This force will make the electrons carrying the charge move closer together or further apart, if they are able to.

Metals

You cannot put a static electricity charge on a metal strip. Metal is an electrical conductor, therefore electricity flows through it easily. As fast as rubbing transfers electrons on to, or off, the metal, they flow away to earth through the metal and your hand. The flow of electrons (electric charge) is an **electric current**.

Van der Graaff generator

The **Van der Graaff generator** is a way of collecting a lot of electric charge in one place.

A motor turns a belt over a roller. As the belt and roller rub together, the belt gains a static electric charge. The belt carries the charge up to the metal dome. The charge cannot flow away from the dome because there are no electrical conductors connected to the dome.

Sparks jump from a Van der Graaff generator to an **earthed** electrical conductor held close to the dome. There is so much charge on the dome that it 'jumps' across to a conductor and flows away. It is not dangerous, because the spark only transfers a tiny electric current and a tiny amount of energy.

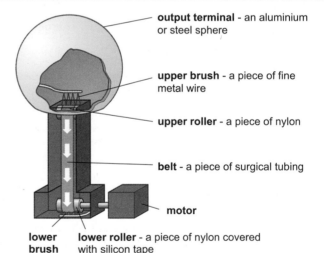

output terminal - an aluminium or steel sphere

upper brush - a piece of fine metal wire

upper roller - a piece of nylon

belt - a piece of surgical tubing

motor

lower brush **lower roller** - a piece of nylon covered with silicon tape

FIGURE 2: The roller of the Van der Graaff generator carries electric charge up to the metal dome.

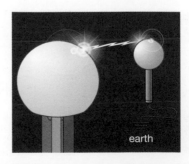

earth

FIGURE 3: What is happening when you see this spark?

Q Van der Graaff generator experiments

If you stand on an insulating mat and touch the dome of the Van der Graaff generator, some of the electric charge from the dome spreads out over you. All your hairs get the same type of electric charge, so they all repel each other, and your hair stands on end.

QUESTIONS

1 Two objects carry the same type of charge. How do they affect each other?

2 Describe what an electric current is.

3 Describe the spark from a Van der Graaff generator.

Electrostatic induction

When an electrically charged balloon is held near the ceiling, the negative charges on the balloon repel the negative charges on the ceiling, leaving the surface of the ceiling with a positive electric charge. The negatively charged balloon and the positively charged ceiling surface attract each other and the balloon sticks. This is **electrostatic induction**.

The balloon is more likely to stick on dry days than on damp days. Water is a good electrical conductor; on damp days, the static electric charge flows away through the moisture in the air, as a tiny electric current. The balloon does not stay charged for long enough to stick.

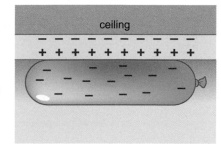

FIGURE 4: The negative charges on the balloon repel the negative charges on the ceiling.

QUESTIONS

4 Why might rubbing a balloon for longer make it more likely to stick to a surface?

5 Suggest why you cannot stick a balloon with a static electric charge to a metal surface, such as a fridge.

6 Discuss how electrostatic induction causes soap bubbles to be attracted to the dome of a Van der Graaff generator.

Lightning conductors

Static electric charges concentrate at points. The sharper the point, the more charge it will have.

A lightning conductor uses a sharp point to attract lightning. The static electric charge on the thunderclouds attracts a high charge up to the point of the lightning conductor. This makes a spark of electricity (the lightning) more likely to jump to this point. From there, the current can be conducted safely to earth.

The noise of the thunder and the crackle of smaller sparks is the sound of the air heating up and expanding very rapidly.

FIGURE 5: The static electricity charge on a thundercloud attracts a high charge up to the point of the lightning conductor.

QUESTIONS

7 Suggest suitable materials for a lightning conductor. Give your reasons.

Q electrostatic induction animation

Circuit diagrams

You will find out:
> the standard symbols used to draw circuit diagrams
> about series and parallel circuits

Know what it does

Engineers use circuit diagrams to build appliances such as washing machines. Without the circuit diagram it would be very hard to know what components were where, and what they did. Complex devices such as aircraft controls may have several pages of circuit diagrams.

Did you know?

Old textbooks have different symbols from those that you use. Some symbols were changed when an international standard was introduced.

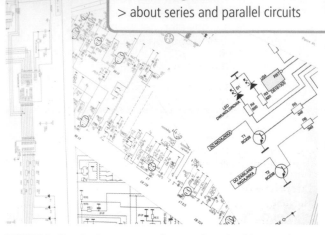

FIGURE 1: Circuit diagrams can be simple or highly complex.

Symbols

You need to know all of the following symbols, and how you might use the different components.

 Open switch

 Closed switch – used to make and break electrical circuits

 Cell – a single unit producing electricity

 Battery – two or more cells connected together

 **Resistor** – a component that slows down the flow of current in a circuit

 Variable resistor – a component in which the resistance can be varied

 Lamp – a device that can give out light

 **Fuse** – a component that protects equipment from electrical surges

 Voltmeter – an instrument that measures the potential difference across a component

 **Ammeter** – an instrument that measures the current flowing in a circuit

 Diode – a component that makes sure that current flows only in one direction

 Thermistor – a component whose resistance varies with temperature

 LDR (light dependent resistor) – a component whose resistance is affected by the amount of light shining on it

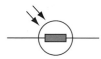 **LED** (light emitting diode) – a component through which current can only flow in one direction and which lights up when the current flows

QUESTIONS

1 What is the difference between a cell and a battery?

2 Which instruments measure (a) current (b) potential difference?

3 Make a list of (a) components that you know how to use (b) components that you do not know how to use. Check at the end of this unit that you know how to use all the components in your list b.

Q battery GCSE ... measure current potential difference

Series and parallel

Figure 2 shows the two types of circuit that you need to know about.

Connections

> **Series circuits** are circuits where all the components are connected in just one 'loop'.

The voltmeter is not in the 'loop'. A voltmeter is a measuring device that connects 'across' a component – ignore it when deciding if a circuit is a series circuit.

> **Parallel circuits** are circuits where the components are connected in more than one 'loop'. There is more than one possible path for the current to take.

Current

> In a series circuit, the current stays the same all the way round the circuit.

> In a parallel circuit, the current divides at a junction. The total current flowing into any junction is always equal to the total current flowing out of the junction.

Series circuit

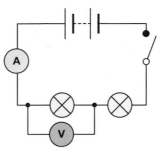

Parallel circuit

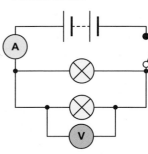

QUESTIONS

4 (i) In a series circuit, how is the brightness of a lamp related to the current flowing through it?

(ii) What effect does changing the number of lamps or cells have?

5 In a parallel circuit, discuss the factors that would affect the brightness of one or more lamps.

FIGURE 2: Connections in a series circuit and a parallel circuit.

Supplying or using energy?

> Components in a circuit that 'supply' energy, transfer energy from elsewhere to energy carried by the electric current in the circuit. For example, cells transfer energy stored in the chemicals in the battery to energy carried by the electric current.

> Components in a circuit that 'use' energy, transfer energy carried by the electric current to energy elsewhere. For example, a lamp transfers energy carried by the electric current to energy transferred away by light radiation.

Potential difference (p.d.) is related to the energy being transferred by each unit of electrical charge.

> In series circuits, the total potential difference across components that 'supply' energy equals the total potential difference across components that 'use' energy.

> In parallel circuits, the total potential difference across components that 'supply' energy equals the total potential difference across components that 'use' energy *in each 'loop' of the parallel circuit*.

QUESTIONS

6 The potential difference across the battery in Figure 2 is 12 V. If all the lamps are identical, what is the potential difference across each lamp in (a) the series circuit (b) the parallel circuit?

Ohm's Law

Torches going flat

Perhaps you would expect a torch to be either 'on' or 'off'. Yet it does not matter whether a torch is wind-up or battery-powered, they never suddenly stop working. They just fade out – the light becomes dimmer and dimmer until there is nothing to see. Ohm's Law helps to explain why this happens.

FIGURE 1: Why does this torch 'fade-out' as its battery goes flat?

 Resistance

Electric current flows more easily through some components and some materials than through others. That is why:

> the connecting wires for circuits are made from metal, most often copper

> some lamps are brighter than others, when connected to the same battery or power supply.

The **resistance** of a component indicates how easy it is for electric current to flow through that component. Figure 2 shows a circuit to find the resistance of a component.

If the resistance is:

> high it is hard for current to flow. The current will be small for any given potential difference

> low it is easy for current to flow. The current will be large for any given potential difference

Ohm's Law

The current flowing through a component, the potential difference across it and the resistance of the component are related by the equation:

$V = I \times R$

where
V is potential difference in volts (V)
I is current in amperes, or amps (A)
R is resistance in ohms (Ω)

This is **Ohm's Law**, named after Georg Ohm, a German physicist.

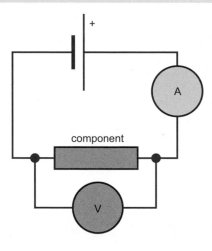

FIGURE 2: A circuit to find resistance.

QUESTIONS

1 Describe what is meant by 'resistance'.

2 Use the Ohm's Law equation to calculate the potential difference across a 5 Ω resistor which has a current of 2 A flowing though it.

3 A lamp has a potential difference. of 3.0 V across it and a current of 0.5 A through it. What is its resistance?

4 Torches become dimmer because the current flowing through the lightbulb decreases. What does this tell you about the potential difference of a battery that is 'going flat'?

Factors affecting resistance

An electric current is a flow of moving electrical charges. In a conducting wire made of metal, these moving charges are **'free' electrons**, held only loosely to the metal atoms. They are able to move through the wire, attracted to the positive potential of the battery (opposite charges attract).

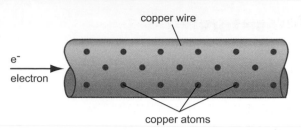

FIGURE 3: The current is a flow of charged electrons.

The resistance is a measure of how easily these electrons can move. Interactions between the moving electrons and the stationary atoms make it harder for the electrons to move.

Resistance of a wire, at constant temperature, is affected by three factors.

1. Type of metal: copper has a low resistivity – electrons can move through it easily.

2. Length: longer wires have a higher resistance – there are more interactions slowing down the electrons.

3. Thickness: thinner wires have a higher resistance – there are fewer 'gaps' between atoms for the electrons to flow through.

FIGURE 4: Some devices have resistances that you can vary. The controls on this guitar are attached to variable resistors. By changing the resistance, you change the sound.

 QUESTIONS

5 For a wire at constant temperature, list the factors that would give it a high resistance.

6 Increasing the temperature of a wire makes the atoms vibrate more. Suggest a reason why this makes the resistance of the wire increase.

Current and potential difference for a resistor

Ohm used a circuit similar that in Figure 2 to investigate how the current through a resistor, at a constant temperature, varied when the potential difference across it was changed. Figure 5 shows the graph he obtained. He investigated negative values of potential difference by connecting the battery or power supply the opposite way round.

The straight line shows that the resistance of a resistor is a constant, at a constant temperature.

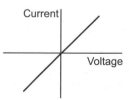

FIGURE 5: Current–potential difference graph for a resistor at constant temperature.

 QUESTIONS

7 The resistance of a wire increases when the temperature increases. How would this cause the shape of the graph to change? Explain why.

Did you know?

Copper is over six times better than tin at conducting electricity. Choosing the right materials for components is very important.

Non-ohmic devices

Transistors

Transistors are made of semiconductors – materials that are neither conductors nor insulators. They were invented in 1947. They work as tiny, very fast switches and as amplifiers, using a small current to turn on a large current elsewhere. Without transistors, all devices using computer chips – from laptops to high-speed Maglev trains – would be impossible.

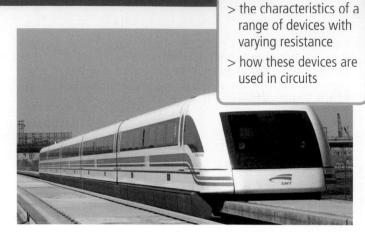

FIGURE 1: Maglev trains are very high speed trains, floating on computer-controlled magnetic fields.

Non-ohmic devices

Non-ohmic devices are devices where the resistance varies, even at constant temperature.

The graph of current against potential difference for a non-ohmic device is not a straight line.

Filament bulb

As the current through a **filament bulb** increases:

> the thin filament wire heats up

> as it becomes hotter, its resistance increases

> it needs a larger and larger potential difference to make the current increase a small amount.

This happens whichever way the current flows – that is, whether the potential difference is positive or negative.

FIGURE 2: Current–potential difference graph for a filament bulb.

QUESTIONS

1 What is the difference between a non-ohmic device and an ohmic device?

2 It does not make any difference which way round a filament bulb is connected. How does the shape of a current–potential difference graph show this?

Direction, light and temperature

Diode

A **diode** is a semiconductor device that only allows current to flow through it in one direction, called the forward direction. The arrow on the circuit symbol shows the allowed direction.

In the forward direction, the current is proportional to the potential difference (the diode obeys Ohm's Law) provided that the potential difference is greater than about 0.6 V. In the reverse direction, the diode has a very high resistance, so no current flows.

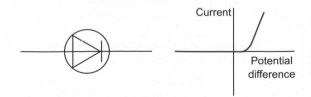

FIGURE 3: Current–potential difference graph for a diode.

Q filament lamp current potential

Light emitting diode

The large arrow in the symbol for a **Light Emitting Diode (LED)** shows the forward direction, the direction in which current will flow. The small arrows show that it gives out light.

LEDs are becoming common in lighting as they give a bright light using a very small current, usually around 25 mA.

FIGURE 4: An LED is a special type of diode that emits light.

Light dependent resistor

The circuit symbol for a **Light Dependent Resistor (LDR)** shows that it is a type of resistor affected by light shining on it. As the light shining on the LDR becomes brighter, the resistance decreases, so the current increases.

LDRs are used in:

> smoke alarms – smoke stops light getting to the LDR and the change in current switches on an alarm

> security lighting – when it gets dark the resistance of the LDR increases, the change in current switches on lights.

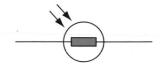

FIGURE 5: This circuit symbol shows that the LDR is a type of resistor.

Thermistor

The circuit symbol for a **thermistor** also shows it is a type of resistor. Heating the thermistor makes it resistance decrease, so the current through the thermistor increases.

Thermistors are used in:

> electronic thermometers – where the temperature can be found because it is a measure of the resistance of the thermistor. These thermometers are very small and do not transfer much energy from whatever they are measuring

> thermostats – where something needs to be switched on or off when temperature changes, such as in fire alarms, heating systems or freezers.

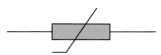

FIGURE 6: This circuit symbol shows that the thermistor is also a type of resistor.

QUESTIONS

3 (i) Which part of the current–potential difference graph for a diode shows the region where the diode acts like an ordinary resistor?
(ii) Is this when the diode is connected in the forward direction or the reverse direction?

4 An LED is a type of diode. What does this tell you about (a) the way it must be connected in a circuit (b) the potential difference across it?

5 Describe how an LDR might be used in a burglar alarm.

6 The current through a thermistor decreases. What does this tell you about the temperature?

How do they work?

Diodes, LEDs, LDRs and thermistors are made from semiconductor materials. Heating a thermistor or shining light on an LDR transfers more energy to the material. More electrons have enough energy to become 'free electrons', able to move through the material as an electric current, so the resistance decreases.

The diode and the LED are made from 'sandwiches' of semiconductor material, where electrons have to cross 'barriers' between one layer and the next. Increasing the potential difference gives these electrons enough energy to cross the barrier. A current can flow once the potential difference is great enough.

QUESTIONS

7 Do you think it would be possible to make a thermistor or an LDR which uses microwave radiation or ultraviolet radiation? Explain your answer.

 LED diode circuit … thermistor LDR

Components in series

Fuses in series

Fuses are safety devices that switch off the current through a device, if the current gets too large. They are always connected in series. This is because it is important that all the current going though the device goes through the fuse as well. Houses have a mains fuse that all the current into the house goes through.

Did you know?
Long ago, Christmas tree lights used to be connected in series. If one bulb broke, all the lights went out.

FIGURE 1: What are the disadvantages of connecting these lights in series?

Series circuits

In a **series circuit** the components are connected in a single 'loop'. There is only one path for the current.

Note that voltmeters are ignored when deciding if a circuit is a series circuit.

Figure 2 shows a series circuit. The 'dotted' ammeters A_2, A_3, A_4 are not extra ammeters; they are alternative positions to put the ammeter A.

By connecting the ammeter in turn at each of the positions, you can show that the current is the same each position.

Adding more identical bulbs to a series circuit makes each bulb dimmer. By measuring the current, you can show that if you increase the number of bulbs connected in series, this decreases the size of the current.

Current flow
Using a centre-zero meter shows:

> The current flows from the positive terminal of the battery, round the circuit to the negative terminal. This is called **conventional current flow**. It is in the opposite direction to the flow of electrons.

Remember
For components in series, the same current flows through each component.

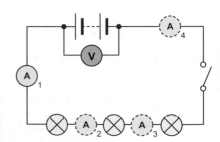

FIGURE 2: Circuit to investigate components in series.

QUESTIONS

1 Describe a series circuit.

2 What can you say about the current flowing in a series circuit?

3 Explain what happens to the brightness when more bulbs are connected in a series circuit.

Q series circuit disadvantages

Potential difference and resistance in series

Potential difference in a series circuit

By connecting the voltmeter in turn across (a) each component in a series circuit, (b) all the components together and (c) the battery, you can find the potential difference across each.

Comparing these values shows:

> The total potential difference across the supply is equal to the sum of the potential difference across each of the components.

For identical bulbs in a series circuit, each bulb has the same the potential difference across it. Adding more identical bulbs to a series circuit decreases the potential difference across each bulb.

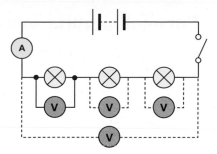

FIGURE 3: Investigating potential difference in a series circuit.

Resistance in a series circuit

Measurements of current and potential difference show how these vary around a series circuit. Putting these values into the equation for Ohm's Law shows how the resistance varies. Ohm's Law says:

$V = I \times R$

Rearranging this gives:

$R = \dfrac{V}{I}$

The current is the same through all the components in a series circuit. Using this value of current, you can calculate the resistance for each bulb on its own and the potential difference across each bulb on its own.

You can calculate the total resistance in the series circuit from the value of the current and the total potential difference across all the bulbs.

Comparing these values shows:

> In a series circuit, the total resistance is the sum of the resistances of each component.

QUESTIONS

4 Three bulbs, with resistances 2 Ω, 3 Ω and 5 Ω, are connected in series. Calculate the total resistance.

5 Two resistors are connected in series. The potential difference of the supply is 12 V and the current around the circuit is 3 A. If the resistance of one resistor is 1Ω, what is the resistance of the other resistor?

Cells in series

The circuit symbol for a battery shows several cells in series. Current flows from the positive terminal of each cell, so the cells in a battery must be connected 'the right way round'.

Connecting a voltmeter across cells connected in series, shows that the total potential difference is the sum of the potential difference of each cell. It can be positive or negative.

The potential differences of two cells connected:

> the same way round as each other, add up

> the opposite way round, cancel out.

QUESTIONS

6 Draw all the possible arrangements of four 1.5 V cells connected in series. What total potential differences are possible?

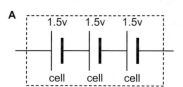

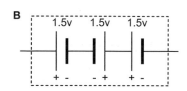

FIGURE 4: In battery A the cells 'push' in the same direction. The total potential difference is 4.5 V. In battery B two cells 'push' one way and one cell 'pushes' the opposite way. The total potential difference is 1.5 V.

Components in parallel

You will find out:
> how the current, potential difference and resistance varies for components connected in parallel

Parallel circuits everywhere

Most electric fan heaters will have several settings: the fan blowing cold air, a warm heater setting and a hot heater setting. This means that the fan and two different heater elements must be arranged in a parallel circuit so that they can each be controlled by different switch positions.

Did you know?

Homes in the UK are wired up with a ring main, which is a parallel circuit, connecting sockets together.

FIGURE 1: Would this fan heater work properly, if the fan and the heater are connected in series?

Parallel circuits

In a parallel circuit there is more than one path for the current to flow along.

In Figure 2, an ammeter can be connected in turn at each of the positions shown. This will show that the current at A_1 and at A_5 is equal to the sum of the current at A_2, A_3 and A_4.

For components in parallel:

> The total current through the whole circuit is the sum of the currents through the separate components.

> At each junction where the current splits, the total current flowing into the junction is the same as the total current flowing out of the junction.

Adding more identical bulbs to a parallel circuit does not change the brightness of each bulb. Measuring the current shows:

> The current through each bulb remains the same and the total current increases.

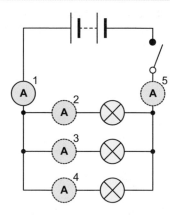

FIGURE 2: A circuit to investigate components in parallel

QUESTIONS

1 Describe a parallel circuit.

2 How does the current vary around a parallel circuit?

3 If you could only see one of the bulbs in a parallel circuit, would you be able to tell how many bulbs there were in the whole circuit? Explain your answer.

Q parallel circuit GCSE ... parallel circuit current split

Potential difference and resistance in parallel

Potential difference in a parallel circuit

In a parallel circuit, connecting the voltmeter in turn (a) across each of the components and then (b) across the battery, shows:

> The total potential difference across the supply is the same as the potential difference across each of the components of a parallel circuit.

Adding more identical bulbs to a parallel circuit does not change the potential difference across each bulb or across the supply.

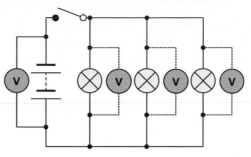

FIGURE 3: Investigating potential difference in a parallel circuit.

Resistance in parallel circuits

To see what happens to the total resistance in parallel circuits, use Ohm's Law again.

If there is one bulb connected to a battery, its resistance can be calculated using:

$$R = \frac{V}{I}$$

When more bulbs are added, the potential difference does not change, but the total current increases.

The equation shows that, if the total current increases, the total resistance must decrease.

> The total resistance of identical bulbs in parallel is less than the resistance of each bulb alone.

 QUESTIONS

4 Two bulbs are connected in parallel. The total current is 6 A. The current through one bulb is 2 A.

(i) What is the current though the other bulb?

(ii) If the total potential difference across the supply is 12 V, what is the potential difference across each bulb?

(iii) Calculate the resistance of the bulb with a current of 2 A flowing through it.

(iv) If the bulbs were identical, what could you say about the total resistance, compared with the resistance of each bulb alone?

Series and parallel

Circuits can have components in series and in parallel, such as in Figure 4.

> The total current is the sum of the currents going along each possible path around the parallel circuit.

> The potential difference across each 'loop' is the same as the potential difference across the supply.

> The potential difference across the single bulb B1 is the same as the potential difference across the two bulbs B2 and B3 together.

 QUESTIONS

5 If the potential difference across the supply in Figure 4 is 12 V and all the bulbs are identical, what will be the potential difference across B1, across B2 and across B3?

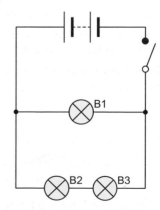

FIGURE 4: Series and parallel in one circuit.

🔍 bulb resistance calculator

Checklist P2.1–2.3

To achieve your forecast grade in the exam you will need to revise

Use this checklist to see what you can do now. Refer back to the relevant topics in this book if you are not sure. Look across the three columns to see how you can progress. **Bold** text means Higher tier only.

Remember that you will need to be able to use these ideas in various ways, such as:

> interpreting pictures, diagrams and graphs

> applying ideas to new situations

> explaining ethical implications

> suggesting some benefits and risks to society

> drawing conclusions from evidence you are given.

Look at pages 278–299 for more information about exams and how you will be assessed.

To aim for a grade E	To aim for a grade C	To aim for a grade A
Recall that unbalanced forces make an object change speed or direction, and that balanced forces do not change its motion.	Calculate the resultant of several forces acting on an object.	Describe the effect of a resultant force on motion.
Describe an object's motion using a distance–time graph.	Use distance–time graphs or velocity–time graphs to describe motion. Carry out calculations involving weight, acceleration or the force causing acceleration.	**Calculate speed from distance–time graphs.** **Calculate acceleration and distance travelled from velocity–time graphs.** Rearrange equations to calculate weight, acceleration or force causing acceleration.
Recall that stopping distance = thinking distance + braking distance	Describe how different factors, including braking force and speed, affect stopping distance. Recall that friction heats up brakes as they slow down a vehicle.	Explain the effects of different factors on stopping distance.
Recall that falling objects reach a top speed (terminal velocity).	Draw velocity–time graphs for objects reaching terminal velocity.	Interpret velocity–time graphs for objects reaching terminal velocity.

To aim for a grade E	To aim for a grade C	To aim for a grade A
Recall that forces can change the shape of objects.	Recall that stretched elastic objects store energy. Know that that a larger force will extend an elastic object more.	Use equations to calculate the energy stored in a stretched elastic object. Interpret force–extension graphs.
Recall that work is done when a force makes an object move.	Use equations to calculate work done and power during energy transfers.	Use equations to calculate work done, power, kinetic energy and gravitational potential energy.
Recall that moving objects have momentum. Describe some of the safety features in cars.	Use equations to calculate momentum for one or for two objects. Explain how some of the safety features in cars work.	Use momentum calculations to predict speed changes or mass of objects involved in collisions or explosions. Evaluate the benefits of safety features in cars.
Recall that rubbing insulators together leaves them electrically charged and thus exerting a force on each other.	Understand that insulators, when rubbed together, become charged by gaining or losing electrons. Recall the effect of forces between charged insulators.	
Recall the names and symbols of electrical components.	Know graphs that relate current–potential difference for different components.	Interpret graphs that relate current–potential difference. **Describe resistance changes in terms of ions and electrons.**
Know how current and potential difference are measured.	Use equations to calculate current, potential difference and resistance.	Use and rearrange equations to calculate current, potential difference and resistance.
Know that some electrical circuit components have practical uses.	Describe the practical applications of some components in electrical circuits.	Explain the practical applications of some components in electrical circuits.
Describe a series circuit.	Recall differences between series and parallel circuits.	Explain differences between current and potential difference in series and parallel circuits.

In the examination, equations will be given on a separate equation sheet.
Write down the equation that you will use. Show clearly how you work out your answer.

1. The diagram shows a skydiver falling to the earth.

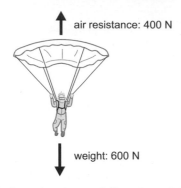

air resistance: 400 N

weight: 600 N

AO2 **(a)** Calculate the size and direction of the resultant force on the skydiver.
Show clearly how you work out your answer. [2]

AO1 **(b)** The skydiver will stop accelerating and reach his terminal velocity after a short time.
What is the resultant force when he reaches his terminal velocity? [1]

AO2 **(c)** Write down the size of the two forces (weight and air resistance) acting on the skydiver when he falls at his terminal velocity. [2]

2. A boy kicks a football lying on the ground. The football has a mass of 0.8 kg. After he kicks it, the ball moves with a velocity of 3 m/s.

AO1 **(a)** What is the momentum of the ball when it is lying on the ground? [1]

AO2 **(b)** What is the momentum of the ball when it starts moving? Show all your working. [3]

3. A student is designing a circuit, to turn on a light when the temperature increases.

AO2 **(a)** Name one component that she could use in the circuit to sense changes in temperature. [1]

AO1 **(b)** Explain why she should choose this component. [1]

AO3 **(c)** Write down one practical use in the home for this circuit. [1]

4. A toy car is powered using a stretched elastic band. The manufacturers tested the force needed to stretch the elastic band and their results are shown in the table below.

Force used (N)	Extension (cm)
0	0
2	3
4	6
6	7
8	12
10	15
12	18

AO2 **(a)** One of these results was written down incorrectly.
Draw a circle around the incorrect result. [1]

AO2 **(b)** Describe the pattern shown by the results. [2]

AO2 **(c)** The car moves further when the elastic band is stretched more, before it is used. Explain why. [3]

5. When a balloon is rubbed on a piece of cloth, the balloon becomes charged.

AO1 **(a)** Which particles have a negative charge? [1]

AO2 **(b)** Explain why rubbing the balloon made it negatively charged. [2]

AO2 **(c)** Describe what happens when the charged balloon is held near another negatively charged balloon. [1]

AO2 **(d)** State one use of static electricity. [2]

AO2 **6.** Two friends are sitting in a boat floating on a river.

(a) Write down one pair of balanced forces acting on the boat. [2]

(b) Another person gets in the boat. The boat floats lower in the water.

Describe how the forces acting on the boat change. [3]

AO1 recall the science   AO2 apply your knowledge   AO3 evaluate and analyse the evidence

✹ WORKED EXAMPLE – Foundation tier

Here is some information from a police report that was prepared after a car accident.

How to raise your grade!
Take note of these comments – they will help you to raise your grade.

• Weather conditions: Dry but cloudy. Temperature: 10 °C

• Road conditions: Wet surface due to earlier rainfall. No puddles. The accident took place on a series of sharp bends.

• Car condition: Brake pads worn down and in a bad condition. Three tyres in bad condition with a worn tread pattern.

• Driver condition: The driver had been using his mobile phone. He was not over the legal limit for alcohol.

(a) Explain what is meant by thinking distance. [2]

Thinking distance is the time taken before the driver starts to brake.

This is worth one mark, because the candidate explained thinking time, not thinking distance.

(b) Use the report to write down three factors that increased the driver's braking distance. [3]

Wet road, worn brake pads, using a mobile phone

Make sure that you only include the factors that affect the braking distance. Using a mobile phone affects thinking distance.

Only two marks would be given for this answer.

(c) The police must decide whether to prosecute the driver. Explain whether you think the driver was responsible for causing the accident. [3]

Yes he was because he had not kept the car in good condition and was using the mobile phone.

This is a good response as it answers the question and gives reasons from the information. If you say that the driver was not responsible, you must be able to give reasons for your answer.

In the examination, equations will be given on a separate equation sheet.
Write down the equation that you will use. Show clearly how you work out your answer.

1. This table gives information of the thinking distance and braking distance of a vehicle.

Speed (m/s)	Thinking distance (m)	Braking distance (m)	Stopping distance (m)
9.0	6	6	12
13.5	9	14	23
18.0	12	24	36
22.5	15	38	53
27.0	18	56	74
31.5	21	75	96

AO2 **(a)** Using the data in the table, describe how thinking distance and speed are related. [2]

AO2 **(b)** A car is travelling along a road when a child steps out into the road.

(i) The driver sees the child when the child is 12 m away. The driver stops just in time.

How fast was the car travelling? [1]

(ii) If the car was travelling at 18 m/s, what speed would it be travelling at, when it reached the child? Explain your answer. [2]

AO3 **(c)** One researcher found that younger drivers react more quickly than older drivers, but tend to drive faster. Use this information to explain whether younger drivers are more likely to stop in a shorter distance than older drivers. [4]

2. Bicycles are often painted using an electrostatic paint spray. A paint gun gives paint droplets an electrostatic charge. The bicycle has the opposite charge to the paint drops.

AO2 **(a)** The paint droplets spread apart as they leave the paint gun. Why does this happen? [2]

AO2 **(b)** Using this method means that the whole bicycle will be covered with a layer of paint. Explain why. [3]

AO3 **(c)** Once the bicycle is painted, excess charge must be removed.

Say whether the bicycle frame should be discharged using a metal wire or a plastic cable. Explain your answer. [2]

3. The graph shows a potential difference–current curve for a filament bulb.

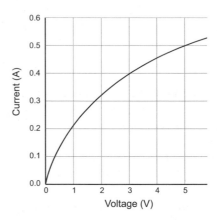

AO2 **(a)** Explain how the current varies with the potential difference. [2]

AO2 **(b)** Calculate the resistance when the potential difference is 3.0 V. [3]

AO2 **(c)** While a filament bulb is switched on, the resistance changes. Explain why. [3]

AO1 recall the science   AO2 apply your knowledge   AO3 evaluate and analyse the evidence

✴ WORKED EXAMPLE – Higher tier

The diagram shows two cars travelling towards each other.

mass: 750 kg
speed: 20 m/s

mass: 800 kg
speed: -25 m/s

(a) Calculate the total momentum of the two cars before the collision. [5]

Momentum = mass × velocity = 15 000 kg m/s for car 1 and 20 000 kg m/s for car 2.
Total momentum = 35 000 kg m/s

The two cars collide and stick together

(b) What is the momentum of the two vehicles after the collision? [1]

The same as before, 35 000 kg m/s

(c) Calculate the speed and direction of the cars after the collision. [3]

35 000 ÷ (750 + 800) = 22.6

In part (d) of this question you will be assessed on using good English, organising information clearly and using specialist terms where appropriate.

(d) Explain whether a crumple zone protects passengers better in a side-on collision or head-on collision. [6]

The crumple zone is at the front of the car. It crumples in a collision and increases the time of collision. This reduces the forces during a head-on collision.

How to raise your grade!
Take note of these comments – they will help you to raise your grade.

Since momentum has direction, the candidate should subtract the momentum of one car from the other. The candidate will lose one mark.

The candidate uses the idea that momentum is conserved correctly, so receives the mark.

The candidate shows correct working so gains two marks, even though the wrong calculation from part (a) was used. The candidate did not show a unit for speed, so loses a mark.

This candidate explains the idea of crumple zones well. However, the candidate did not answer the question. There is no description or comparison of the protection in a head-on collision with the protection in a side-on collision.

The candidate would receive only one mark.

Physics P2.4–2.6

What you should know

Energy transfer and electricity

Different forms of stored energy are used to generate electricity.

Electrical circuits are used to transfer energy.

Measurements are needed to compare energy transfers.

 Write down four forms of stored energy.

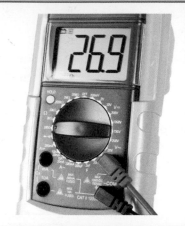

Atoms and electromagnetic radiation

Atoms consist of protons, neutrons and electrons.

Atoms are characterised by their atomic number and mass number.

Gamma rays are part of the electromagnetic spectrum.

 Describe the nature and properties of gamma rays.

Changing Earth

In our solar system, the moon orbits Earth and Earth orbits the Sun.

The Sun is a star.

Planets and their moons have different compositions from one another.

Stars, planets and moons are different in size from one another.

Space science and astronomy are used to find out about the solar system and galaxies.

 Draw a diagram showing the arrangement of objects the solar system.

You will find out

Using mains electricity safely and the power of electrical appliances

> Electric plugs and cables have different parts that are designed to carry out different jobs.

> There are differences between alternating current and direct current electricity supplies.

> Earthing, fuses and circuit breakers are designed to keep people and equipment safe from electrocution and overheating.

> The rate and amount of energy transferred by electrical appliances can be calculated.

What happens when radioactive substances decay, and the uses and dangers of their emissions

> Background radiation surrounds us and has many different causes.

> Alpha particle scattering experiments helped scientists to develop the model of the atom: a central nucleus orbited by electrons.

> There are three main types of nuclear radiation: alpha, beta and gamma. They have different properties and uses, and are affected differently by electric and magnetic fields.

> Radioactive materials become less radioactive over time.

Nuclear fission and nuclear fusion

> Nuclear fission is used in nuclear power stations to generate electricity.

> Nuclear fusion takes place in stars, forming new elements and releasing energy.

> All stars move through different stages in their life cycle.

Household electricity

You will find out:
> about the UK mains electricity supply
> the difference between a.c. and d.c. supply

Different countries, different plugs

Mains electricity is not the same across the world. UK mains plugs have three pins, but plugs for many countries only have two. The size of the mains voltage varies too, so even if you could use the plugs from home, the appliance might not work. Electric shavers and laptop computers are usually manufactured so that they work on different values of mains electricity.

Did you know?

Power stations coordinate the electricity they send out, so that they do not work against each other. Otherwise it would be like connecting batteries the opposite way round.

FIGURE 1: If you want to recharge your phone abroad, you may need a mains travel adaptor.

Direct current and alternating current

Cells and batteries produce electricity as **direct current (d.c.)**. The current always flows in the same direction – from the positive terminal of the battery to the negative terminal. The potential difference does not change (unless the battery is going flat).

Domestic electricity is generated as **alternating current (a.c.)**. The direction of the current flow is continually swapping round – that is why it is called 'alternating'. The potential difference changes from positive to negative and back again in a continual cycle.

Figure 2 shows that, for a.c. supply, the size of the potential difference changes.

The potential difference goes through **cycles** of positive and negative. The number of complete cycles each second is the **frequency**.

> The 'voltage' of the **mains supply** in the UK is 230 V. It really means that the *average* potential difference of the mains supply is 230 V.

> The frequency of the mains supply in the UK is 50 cycles per second – 50 hertz (Hz).

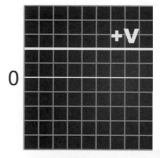
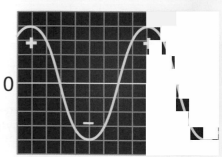

FIGURE 2: Oscilloscope traces show how potential difference stays the same for d.c. and changes for a.c. electricity.

QUESTIONS

1 (i) What do the abbreviations d.c. and a.c. stand for?
(ii) Describe the difference between d.c. and a.c.

2 What is the 'potential difference' of mains electricity in the UK?

3 (i) For an electricity supply, what does frequency mean?
(ii) What is the frequency of electricity in the UK?

Q electricity direct alternating current

Using a cathode ray oscilloscope

The screen display

A cathode ray **oscilloscope** – usually just called an oscilloscope – has a screen that displays how an input potential difference varies with time.

The vertical axis shows the size of the potential difference: the bigger the height of the wave on the screen, the larger the potential difference. Dials on the oscilloscope show how many volts each division on the vertical grid represents.

The horizontal axis shows time: the more cycles of the wave you can see across the screen, the higher the frequency. Dials on the oscilloscope show how many seconds each division on the horizontal grid represents.

Calculating potential difference

The **amplitude** (maximum height) of the wave on the oscilloscope screen shows the peak value of the mains potential difference.

This enables you to calculate potential differences of d.c. supplies and the peak potential differences of a.c. supplies

Note that the peak is not the same as the 'mains voltage' of 230 V, which is the average value of the potential difference.

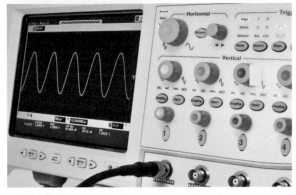

FIGURE 3: Cathode ray oscilloscope.

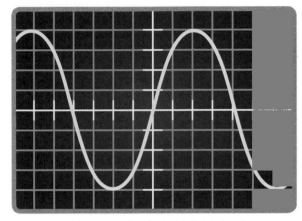

FIGURE 4: Oscilloscope trace showing how potential difference varies for alternating current.

> ## QUESTIONS
>
> **4** Describe what an oscilloscope screen shows.
>
> **5** What can you measure using the vertical grid of the oscilloscope?
>
> **6** If each vertical division in Figure 4 represents 80 V, what is the peak potential difference for the trace?

Calculating period and frequency (Higher tier)

The horizontal wavelength of an oscilloscope trace shows how long it takes for one complete cycle of an alternating current. From this, you can find out how many cycles there would be in one second. This is the frequency of the signal.

In Figure 4, suppose each horizontal division represents 2.5 milliseconds (2.5 ms). Then each complete wave takes 2.5 × 8 = 20 ms. The **period** for the wave is 20 ms.

In 1 second there are 1000 ms, therefore in 1 second there will be 100 ÷ 20 cycles.

The wave has a **frequency** of 50 Hz.

> ## QUESTIONS
>
> **7** What can you measure using the horizontal grid of the oscilloscope?
>
> **8** Describe the oscilloscope trace that you would see for a 12 V d.c. signal.
>
> **9** How would the oscilloscope trace in Figure 4 change, for a wave with a frequency of 100 Hz?

Plugs and cables

You will find out:
> what is inside a three-pin plug
> what is inside electrical cables

Connecting to the mains

Mains leads for electrical appliances must be sold with moulded plugs that you cannot open. A three-pin plug connects to the mains socket. Many years ago, people fitted plugs to appliances themselves – a mistake could result in someone being electrocuted.

Did you know?

Electricians have to know the right type of cable to put in different places in a house. An electric cooker requires very different wiring from an outside light.

FIGURE 1: It is a legal requirement for all new electrical appliances in the UK to have moulded plugs.

Making the connection

Sockets

In your home, the mains electricity sockets have three holes connecting to three different wires.

> The two holes at the bottom connect to the live wire and to the neutral wire. Together these provide the very high potential difference of the mains. It is similar to the two terminals of a battery.

> The third wire hole (at the top) connects to an earth wire, which is a safety feature. The earth pin on a **three-pin plug** is longer than the other two, so that the plug always connects to the earth wire first.

The three-pin plug

A three-pin plug connects the wires from your electrical appliance to the wires in the mains supply.

The wires inside the mains cable from the appliance have colour-coded plastic round them: it is vital that each wire connects to the right place.

> The live wire is **brown**. It connects to the right hand pin – usually also labelled L for Live.

> The neutral wire is blue. It connects to the left hand pin – labelled N for Neutral.

> The **earth wire** is green and yellow. It connects to the top pin – labelled E for Earth.

Parts of the plug not intended to conduct electricity are made from electrical insulators, usually plastic or rubber.

Wires are usually copper because it is a good electrical conductor. The connecting pins are usually brass (a copper alloy) because it conducts electricity well, but is cheaper than copper.

The plug also has a cable grip. This is a strip that is fastened tightly onto the cable where it enters the plug, so that if the cable is pulled, the connections inside the plug will not come undone.

FIGURE 2: Inside a three-pin plug.

QUESTIONS

1 Why is the earth pin on a plug longer than the other two pins?

2 Why is it important that the wires are colour-coded?

3 What do the live and neutral wires connect to? What colour are they?

4 Describe what the cable grip is for.

Q GCSE mains electricity ... wires plug

Cables and fuses

Three core cable

Three core cable has a live, a neutral and an earth wire.

> Each 'wire' may be a bundle of finer wires.

> Each wire is encased in plastic insulation – brown, blue and green/yellow.

> All three wires are covered in an insulating sheath.

If the cable has to carry a large current, the cable should have a larger diameter. For example, the cable for a tumble dryer will be thicker than the cable for a desk lamp.

The earth wire connects to the mains wiring earth. This wire connects to a metal rod in the ground. It is a safety device.

FIGURE 3: What is the earth cable for, in this three core cable?

Two core cable

Two core cable does not have an earth wire. It is used only for appliances that have a plastic case.

If something goes wrong and one of the wires touches the inside of the case, the case itself is an insulator. Current will not flow through a person touching the outside of an insulated case.

Fuses

The **fuse** protects the circuit, if too much current flows through an electrical appliance.

A fuse is always connected in series with the live wire.

FIGURE 4: The fuse in a moulded plug can be changed without opening the plug itself. Why?

Shocking

Electric shock is caused by the current as it passes through the human body.

These are the effects of an alternating current:

> about 1 mA (0.001 A) – an unpleasant tingling.

> above 10 mA – severe muscle pain. The victim cannot let go of the conductor due to muscle spasm.

> between 100 mA and 200 mA – ventricular fibrillation of the heart. It will probably be lethal.

Electric current will be lethal not just because of its size but because of the length of time that it is flowing.

Current is more dangerous when skin is wet. This is because the skin's resistance changes dramatically. The resistance of the human body varies enormously between different people. However, the difference between dry skin (over 15 000 Ω) and wet skin (500 Ω) is even greater.

It is why shaver points in bathrooms are at 115 V instead of 230 V, and why other electrical appliances should not be taken into a bathroom.

FIGURE 5: Why does the plug on this razor have only two pins?

Electrical safety

You will find out:
> how fuses and earth wires keep us safe
> about more modern safety devices

Zero potential

Planet Earth is electrically neutral. This means that it is not electrically charged – its electric potential is zero. Any electrical charge flowing down the earth wire from appliances simply spreads out over the ground. Earth is so big that charge just moves until it is cancelled out by an opposite charge.

Did you know?

Birds can perch safely on very high potential power cables. They are not touching anything else and so no current can flow through them.

FIGURE 1: Our neutral Earth.

Protection

Earthed appliances

All electrical appliances that have a metal case are connected to the mains using three core cable. The earth wire, in the cable, connects the metal casing of the appliance to earth.

If a fault develops in an appliance and the live wire touches its case, current will flow through the earth wire more easily than through a person touching the appliance case. This protects the person from an electric shock.

Remember

Earthing is designed to protect you. Fuses are designed protect electrical appliances.

Fuses

Fuses stop the current flowing when the current becomes dangerously high.

Problems caused by large currents:

> The live wire may touch the casing of an appliance. A large current will flow to earth.

> If larger currents flow than the appliance is designed for, the current can destroy the appliance.

Fuses are thin pieces of wire enclosed in a ceramic case. They are designed to melt (or 'blow') if too large a current flows. This breaks the circuit, stopping the current flowing and protecting the appliance.

It is important to use the right size fuse. If the fuse rating is:

> too low, the normal operating current of the appliance will 'blow' the fuse – the appliance will not work

> too high, the current that can flow may be large enough to damage the appliance.

FIGURE 2: Standard fuses are available in 3 A, 5 A and 13 A ratings. The rating indicates the maximum current that can flow before the fuse 'blows'.

QUESTIONS

1 What types of appliances do not need an earth wire?

2 (i) What is a fuse?

(ii) How does a fuse protect an appliance?

3 Why is it potentially dangerous to use the wrong fuse in an appliance?

Q earth wire GCSE ... wrong fuse danger

Residual current circuit breakers (RCCBs)

Residual current circuit breakers (RCCBs) do the same job as fuses, but they work much faster. They are designed to stop the current flowing in less than 0.05 seconds. RCCBs should always be used with appliances such as lawn mowers, hedge trimmers or power tools.

In an appliance that is working correctly, all the current that flows into the appliance also flows out again. The current in the live wire is exactly the same size as the current in the neutral wire. When a fault causes current to flow to earth, there is a difference in the size of these two currents.

The RCCB works by detecting this difference and turning off the current.

Unlike a fuse, the RCCB does not need replacing after there has been a fault. Once the fault has been repaired. the current can be switched on again by pressing the reset button on the RCCB.

FIGURE 3: Residual Current Circuit Breaker.

QUESTIONS

4 Give two advantages that an RCCB has over a fuse.

5 Suggest why you should use an RCCB with appliances such as lawn mowers.

Choosing the right fuse

Fitting a fuse with the correct current rating is important. To choose the appropriate fuse, you need to know how much current should flow through an appliance when it is working correctly. The thickness of the cable gives the first clue. Large currents need thick cables to carry them safely: so the thicker the cable the larger the current and the larger the value of the fuse.

The exact value of the current can be calculated from the power rating of the appliance. This is always given on a label on the back of the appliance. Current, potential difference and power are related by the equation:

$P = I \times V$

where
P is power in watts (W)
I is current in amperes, amps (A)
V is potential difference in volts (V)

The correct fuse should have a current rating a little higher than the current that the appliance is designed to use.

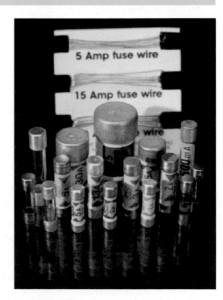

FIGURE 4: How can you tell which fuse to use?

QUESTIONS

6 A lawn mower has a power rating of 1400 W. The potential difference of the mains is 230 V. Rearrange $P = I \times V$ to calculate the approximate current. Explain what size standard fuse you would use.

Q residual current circuit breakers

Current, charge and power

You will find out:
> the relationship between electric charge and electric current
> about the energy transferred by electrical appliances

Danger

Every year, hundreds of people in the UK are injured in domestic fires caused by electrical faults. Wiring that is incorrect or damaged, appliances that are faulty or too many appliances plugged into one socket can all cause electrical fires. Schools and businesses must have all their portable electrical appliances checked for safety by an expert once a year.

FIGURE 1: Do you know how to prevent something like this happening?

Energy transfer

Current and heating

Whenever an electric current flows, energy carried by the current is transferred. When a current flows through a resistor, energy is transferred by heating the resistor.

Wires and electrical components all have resistance. They all act as resistors and all heat up, some much more than others. The heating effect of current is used in hairdryers, electric fires, aquarium heaters and many other places.

Heating and lighting

The popular name for **Compact Fluorescent Lamps (CFLs)** is still 'energy-saving bulbs'. What does this mean?

An old filament lamp becomes very hot when it is switched on; CFLs do not.

> Energy is transferred to an old filament lamp by the current flowing through it. Most of this energy is transferred away from the bulb by heating the surroundings.

> The CFL is much more **efficient**. It transfers most of the energy to give light, not heat. The CFL 'uses' less energy to produce the same amount of light.

FIGURE 2: CFLs come in many shapes and sizes, but they all need less energy than old filament bulbs.

Energy and power

The **power rating** of an appliance is given on the back of the appliance. Power rating is the rate at which an appliance transfers energy. It is measured in joules per second, called watts.

One watt is one joule per second (1 W = 1 J/s).

$$P = \frac{E}{t}$$

where
P is power in watts (W)
E is energy in joules (J)
t is time in seconds (s)

QUESTIONS

1 A filament bulb with a power rating of 60 W and a CFL with a power rating of 14 W both give the same amount of light. How many joules of energy are transferred per second by (a) the filament bulb (b) the CFL?

2 A portable fan transfers 1800 joules of energy in 60 seconds. What is its power rating?

3 A mains-powered hairdryer is rated at 500 W. How much energy will it transfer in 1 minute?

heat transfer and efficiency bulbs

Current, charge and work done

Current is a flow of tiny charged particles, called electrons.

The size of the electric current indicates the rate at which electric charge is flowing. The size of the current is given by:

$$I = \frac{Q}{t}$$

where
I is current in amperes, amps (A)
Q is charge in coulombs (C)
t is time in seconds (s)

The potential difference of a battery or a power supply is a measure of how hard the battery 'pushes' the current around the circuit. Higher voltages 'push' harder and so 'push' more current through the same resistor.

Thus, potential difference is a measure of work done on each unit of charge. It is written as:

$$V = \frac{W}{Q}$$

where
V is potential difference in volts (V)
W is work done in joules (J)
Q is charge in coulombs (C)

FIGURE 3: Recharging a 12 V motorcycle battery.

QUESTIONS

4 How much work is done on a charge of 150 C, when it moves through a potential difference of 4 V?

5 A current of 5 A flows for 30 minutes. How much charge has passed any given point in the circuit?

6 A charge of 50 C moves through a potential difference of 12 V. How much energy is transferred to the charge?

Did you know?

Laptop computers transfer enough energy by heating that, if the cooling fan is dusty or blocked, they can become a fire risk, if left running unattended.

Calculating the energy transferred (Higher tier)

The energy transferred to a unit of charge by a battery or power supply is the same as the work done on the unit of charge.

Replacing W (work done) with E (energy transferred) in $V = \dfrac{W}{Q}$

and then rearranging, gives:

$$E = V \times Q$$

where
E is energy transferred in joules (J)
V is potential difference in volts (V)
Q is charge in coulombs (C)

QUESTIONS

7 The potential difference across a bulb is 5 V and the current through the bulb is 3 A. In one minute, how much energy will be transferred by charge passing through the bulb?

Preparing for assessment: Applying your knowledge

To achieve a good grade in science, you not only have to know and understand scientific ideas, but you need to be able to apply them to other situations and investigations. These tasks will support you in developing these skills.

✳ Using electricity safely abroad

Electrical sockets in other countries are different from electrical sockets in the UK. Tourists use adaptors when they use electrical equipment from the UK in other countries. One side of the adaptor looks like a socket in the UK and equipment plugs straight in. The other side of the adaptor has a different arrangement of pins, so that it can plug straight into the sockets in hotel rooms. Different adaptors are needed in different countries.

Many adaptors work safely with equipment using currents of up to eight amps, but tourists can also buy cheaper adaptors. These are made from low quality plastic, with thin wiring inside. They can only cope with small currents. If the current is too large, the plastic melts and scorches. Before the fuse has a chance to work, an electrical fire can start.

Jodie had a lucky escape on a recent holiday to South America. When she turned on her hairdryer, the adaptor that she was using turned black and started to smell of burning, even though the hairdryer was not damaged.

Hairdryers are very powerful because they contain heating elements. Jodie's hairdryer used a current of 8 A normally. In the UK, she had fitted a 13 A fuse inside the plug. This only melts if the current is more than 13 A. The adaptor Jodie was using was designed only for currents of up to 2 A, and it did not include safety devices to stop the larger currents automatically. Luckily, Jodie realised in time to turn off the hairdryer and switch off the electricity. In some parts of South America, mains voltage (potential difference) is supplied at 120 V. In the UK, mains voltage is supplied at 230 V.

✳ Task 1

(a) Explain why the adaptor turned black and started to smell of burning.

(b) Why would it be safer if the adaptor had contained a 2 A fuse?

✳ Task 2

There were two types of socket in Jodie's hotel bedroom. One type of socket was earthed and the other was not. Equipment from the UK can plug into adaptors which fit either socket.

(a) How does earthing protect the user from electrocution?

(b) If Jodie wanted to use all the sockets in her room, she would have to buy two adaptors to fit both types of socket. She did not want to buy both types of adaptor.

Which type of adaptor would you recommend?

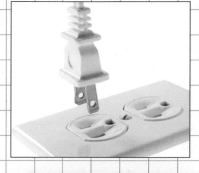

✳ Task 3

(a) Suggest some safety checks that a tourist could make in their hotel room, before using electrical equipment.

(b) A leaflet published by the Electricity Safety Council says that electrical safety standards in the UK are higher than in many other countries.

In the UK, certain types of sockets and light switches are not allowed in bathrooms. This prevents people from touching electrical appliances while in the bath or shower. These regulations do not apply in all countries.

Explain why this increases the risk of electrocution in these countries.

✳ Maximise your grade

Answer includes showing that you...	
	know that a fuse and earthing are safety features.
	recognise dangerous uses of mains electricity.
E	know that current has a heating effect.
	can describe how these safety features work.
	can describe ways to reduce the risks of using mains electricity.
C	can use equations to calculate electrical power.
	can explain how safety features protect the equipment and the user.
A	can evaluate the most suitable fuse to use in a circuit.

✳ Task 4

(a) The power of the hairdryer is 1200 W.

Calculate the current flowing through the hairdryer when it is used in the UK.

(b) Explain why the current in the hairdryer is larger when it is used in South America. Use calculations to illustrate your answer.

Structure of atoms

You will find out:
> about the masses and electric charges in an atom and how they are arranged

Bright ideas

Ernest Rutherford fired some positively charged particles at gold leaf and found that most went straight through and that some bounced off at many different angles. He concluded that most of the atom is empty space with the positive charges gathered together in small, heavy clumps – the model used today.

Did you know?

Physicists have discovered that protons and neutrons are made from even smaller particles – electricity and magnetism are to do with how these particles behave.

FIGURE 1: Ernest Rutherford suggested the basic model of the atom.

Atoms

What is inside an atom?

Atoms are made from smaller particles called **neutrons**, **protons** and **electrons**. Figure 2 shows how these particles are arranged in atoms.

> The **relative mass** of a particle is the mass it has compared with the mass of a neutron.

> Protons and electrons have opposite charges, but the same size. A neutron has no charge.

> Atoms are electrically neutral.

Atoms have the same number of electrons as protons. The effect of the positive and negative charges cancel out, so atoms have no electric charge.

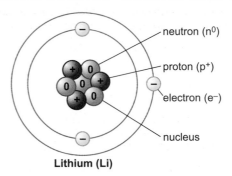

neutron (n^0)
proton (p^+)
electron (e^-)
nucleus

Lithium (Li)

FIGURE 2: An atom of lithium.

Particle	Charge	Relative mass
neutron	0	1
proton	1	1
electron	−1	0

TABLE 1: Charge and mass of sub-atomic particles.

Atoms and elements

> The **atomic number** of an element is the number of protons in the atom. This is the same as the number of electrons.

> The **mass number** of an element is the number of protons and neutrons added together.

QUESTIONS

1 (i) Name the particles in an atom.

(ii) Which particles are electrically charged?

2 What does 'an atom is neutral' mean? What does this tell you about the numbers of protons and electrons?

3 Explain what is meant by atomic number and mass number.

The mass number and the atomic number are written to the top left and bottom left of an element's symbol, for example, lithium is written as $_3^7\text{Li}$

Element	Notation	Atomic number	Number of protons	Number of neutrons	Number of electrons
hydrogen	$_1^1\text{H}$	1	1	0	1
helium	$_2^4\text{He}$	2	2	2	2
lithium	$_3^7\text{Li}$	3	3	4	3
beryllium	$_4^9\text{Be}$	4	4	5	4
boron	$_5^{11}\text{B}$	5	5	6	5
carbon	$_6^{12}\text{C}$	6	6	6	6

TABLE 2: The first six elements in the periodic table.

Electrons, protons and neutrons

Ions

Electrons orbit around the nucleus of an atom, in shells or layers.

When atoms combine, electrons can be transferred to an atom of a different element. The atoms form **ions** which have an electric charge.

> An atom that gains an electron has a negative charge – it becomes a negative ion.

> An atom that loses an electron has a positive charge – it becomes a positive ion.

Isotopes

All atoms of the same element have the same number of protons. For example, all carbon atoms have six protons.

However, the number of neutrons in atoms of an element can vary.

> Atoms of the same element with different numbers of neutrons are called **isotopes**.

Figure 3 shows two isotopes of carbon: a carbon-12 isotope with 6 protons and 6 neutrons, and a carbon-14 isotope with 6 protons and 8 neutrons.

Atoms with more neutrons than protons tend to be unstable. Unstable atoms decay, emitting radiation – they are radioactive.

The greater the difference between protons and neutrons, the more likely an isotope is to be unstable. This is why most radioactive substances are among the heavier elements.

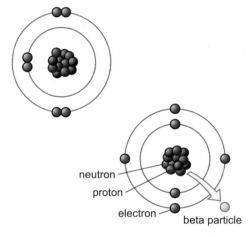

neutron
proton
electron
beta particle

FIGURE 3: Carbon-12 (top) and carbon-14. The carbon-14 isotope is emitting a beta particle.

QUESTIONS

4 How does an atom become an ion?

5 What is the difference between two or more isotopes of an element?

Sub-atomic particles

At one time, atoms were thought to be indivisible. Over time, scientists have developed a more sophisticated model – an atom consists of a nucleus, made up of protons and neutrons, with electrons moving around it. The model is still evolving as scientists probe further.

QUESTIONS

6 Can protons, neutrons and electrons be split into smaller particles? Research on the internet to find out.

 how ions form ... isotope GCSE

Radioactivity

Radiotherapy

Most people have heard of radioactivity. It is often called radiation, but more accurately it is nuclear radiation. Many people are scared by it because they have heard about nuclear bombs or nuclear accidents. Yet, there is lots of good news about radioactivity that does not make the headlines. For example, nuclear radiation is used in the treatment of cancer.

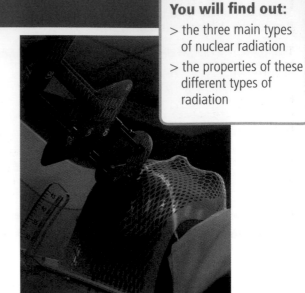

FIGURE 1: In radiotherapy, the radiation is targeted accurately so that it destroys only the cancer cells.

 ## Nuclear radiation

What is nuclear radiation?

Some isotopes of some elements are **radioactive**. They are the isotopes that are unstable and decay naturally, emitting radiation.

There are three types of **nuclear radiation** given out when radioactive substances decay.

> **Alpha** (α) radiation – These are alpha particles made up of two neutrons and two protons (the same as a helium nucleus).

> **Beta** (β) radiation – These are beta particles. They are electrons.

> **Gamma** (γ) radiation – This is high energy electromagnetic radiation emitted from the nucleus. It does not consist of particles.

The nuclei of isotope atoms decay randomly – you cannot tell exactly when each will emit radiation. However, there is an average decay time. Scientists can predict how much radiation will be emitted by a given mass of a particular radioactive substance.

One way to measure radiation is with a Geiger–Muller tube. Radiation entering the tube causes a small electrical pulse, which is counted as a click.

Properties of nuclear radiation

Alpha particles are heavy, positively charged and relatively slow moving (about 5% speed of light). They only travel a few centimetres through air and are stopped very easily, by a sheet of paper. They are strongly ionising – when they collide with atoms, they knock electrons away, making the atom into a positive ion.

Beta particles are small, light and fast moving (about half the speed of light). They will travel a few metres through air but are stopped by 3 mm aluminium. They are weakly ionising – they may cause some ionisation.

Gamma radiation is electromagnetic radiation, so it travels at the speed of light. It is very penetrating – even a thick sheet of lead only reduces its intensity. It is a very poor ioniser.

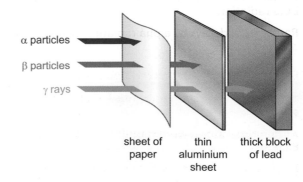

α particles
β particles
γ rays

sheet of paper | thin aluminium sheet | thick block of lead

FIGURE 2: Penetrating power of α, β and γ radiation.

QUESTIONS

1 (i) Name the three types of nuclear radiation.

(ii) Which types are particles?

2 Describe the properties of beta radiation.

3 What do you understand by ionisation?

Q nuclear radiation types ... ionisation

Explaining the properties

The properties of the types of radiation can be explained by their size, mass and charge.

> Alpha particles are large and so have lots of collisions with atoms. They have a large mass, knocking electrons away from atoms that they collide with. All these collisions mean that they rapidly lose their energy and do not travel far.

> Beta particles are small and so do not interact with many atoms. However, they are light and are easily stopped by atoms that they do collide with.

> Gamma rays are short wavelength, high energy electromagnetic radiation.

> Because they do not have an electric charge, they are not attracted or repelled by electrons or ions.

> Because they are electromagnetic radiation, not particles, they do not interact by 'colliding'.

Gamma rays pass easily through materials and are very penetrating.

 QUESTIONS

4 'Radioactive isotopes decay randomly.' What does this mean?

5 What makes alpha particles particularly strong ionisers?

Radiation	alpha (α) particle	beta (β) particle	gamma (γ) ray
Nature	positive helium nucleus $^{4}_{2}$He about 7000 times the mass of an electron	electron – negative charge	electromagnetic waves with very short wavelength and high energy
Charge	positive (+)	negative (−)	neutral
Range in air	a few centimetres	a few metres	very penetrating
Penetration	stopped by skin or a sheet of paper	stopped by 3 mm aluminium	intensity reduced, but not stopped by lead
Ionisation	strong	weak	very weak
Speed	about 5% speed of light	about 50% speed of light	speed of light
Dangerous?	yes	yes	yes

TABLE 1: Properties of the three types of nuclear radiation.

Nuclear equations (Higher tier)

Nuclear equations show how isotopes decay by the emission of alpha and beta radiation.

Radium-224 decays by alpha radiation to become radon-220.

$$^{224}_{88}\text{Ra} \rightarrow {}^{220}_{86}\text{Ra} + {}^{4}_{2}\text{He}$$

Each side of the equation must balance.

> The total mass number (protons + neutrons) on both sides must be the same: 224 = 220 + 4.

> The total atomic number (protons) on both sides must be the same: 88 = 86 + 2.

Bismuth-212 can decay by beta radiation to become polonium-212

$$^{212}_{83}\text{Bi} \rightarrow {}^{212}_{84}\text{Po} + {}^{0}_{-1}\text{e}$$

Again the equation must balance. The polonium-212 is formed by one of the neutrons in the nucleus of the bismuth changing to a proton as it gives out a beta particle (an electron).

 QUESTIONS

6 Radon-220 is radioactive. It decays to polonium-212. Write the nuclear equation for its decay by alpha radiation.

7 Carbon-14 decays by beta radiation. Write the nuclear equation.

Particles in atoms

You will find out:
> how nuclear radiation is affected by electric and magnetic fields

Pioneers

Marie and Pierre Curie were among the early scientists to study radioactive substances. Many of them, including Marie Curie, died of various forms of cancer because they did not realise how harmful nuclear radiation can be. One charity caring for cancer patients – Marie Curie Cancer Care – is named after Marie Curie. Radiation is now used in the treatment of cancer.

Did you know?

There are nine naturally occurring radioactive elements.

FIGURE 1: Marie Curie.

A model of the atom

It took many scientists working together to work out the model for the structure of an atom. Nuclear radiation played an important role in its development.

The plum pudding

In 1897, J J Thompson discovered negatively charged electrons.

In 1904, because he knew that atoms were electrically neutral, he suggested atoms as a positive 'jelly' with electrons dotted through it. This was nicknamed the **'plum pudding' model**.

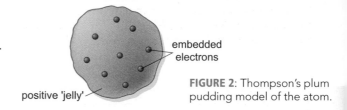

FIGURE 2: Thompson's plum pudding model of the atom.

The scattering experiment

In 1909, Rutherford and two of his students, Marsden and Geiger, carried out the now famous 'Rutherford scattering experiment'.

The scattering experiment fired a narrow beam of alpha particles at a very thin sheet of gold foil in a vacuum. They used a film that would show dots of light where the alpha particles hit it. Using a microscope, they were able to see where the alpha particles went. This is what they discovered:

> Most alpha particles passed straight through the foil.

> A few particles were deflected back, some through quite wide angles.

> A very few bounced straight back towards the source.

These results did not fit with what they expected if the 'plum pudding' model were correct.

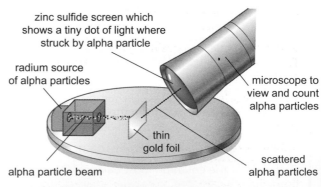

FIGURE 3: Model of scattering apparatus.

QUESTIONS

1 Discuss why Rutherford's results are not what was expected for the 'plum pudding' model.

2 Suggest why this experiment was carried out in a vacuum.

Q Rutherford vs plum pudding

Explaining the scattering experiment results (Higher tier)

Figure 4 shows the alpha particle tracks that Rutherford found.

Rutherford used the experimental evidence to draw the following conclusions.

1. Evidence: Most of the alpha particles passed straight through without being deflected.

Conclusion:

> The atoms in the gold foil contained a lot of empty space.

2. Evidence: Alpha particles are positively charged. When they are deflected back at wide angles they must have been deflected by a force of repulsion.

Conclusion:

> The atom must contain a small positively charged core, the nucleus.

3. Evidence: Alpha particles are quite heavy. Yet some where bounced straight back. They must have collided with something even heavier.

Conclusion:

> The nucleus of the gold atoms must be very small but very dense.

4. Evidence: Negative charges did not seem to be having any effect.

Conclusion:

> Electrons are not part of the inner atom (nucleus) but orbit in the empty space around the nucleus.

Rutherford's model of the atom was accepted by the scientific community in 1911. Later, Niels Bohr used the model to work out how electrons were able to orbit the nucleus without falling into it.

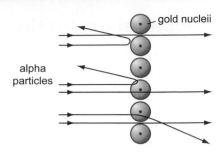

FIGURE 4: The alpha particle tracks that Rutherford had to explain.

QUESTIONS

3 Why do results from the scattering experiment suggest that electrons are not part of the central nucleus?

4 Why do the results also suggest that electrons are very small?

More deflections (Higher tier)

Experiments have shown that nuclear radiation is affected by electric and magnetic fields.

> Alpha particles are positively charged – They are attracted by the negative charge and are deflected towards it.

> Beta particles are negatively charged – They are deflected away from the negative charge.

> Gamma rays are not electrically charged – They are not affected by an electric field.

Both alpha particles and beta particles act like a very tiny electric current. Electric currents and magnetic fields interact, with the charged particles moving if they can. Alpha particles and beta particles are deflected in different directions.

Alpha particles are much heavier than beta particles, so they need a much greater force to make them change direction. Beta particles are always deflected more than alpha particles.

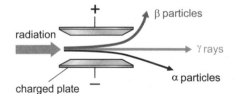

FIGURE 5: The effect of electric and magnetic fields on nuclear radiation.

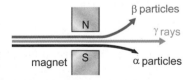

QUESTIONS

5 A laboratory technician has three radioactive sources: cobalt-60 which is a source of gamma rays, strontium-90 which is a source of beta particles and americium-241 which is a source of alpha particles. Suggest how the technician can tell which is which.

Background radiation

Natural protection

Space is a dangerous place to be. There is a lot of radiation out there. Earth's atmosphere protects us from the dangerous nuclear radiation that damages living cells. This is one very good reason to take care of our atmosphere.

Did you know?

Earth's atmosphere extends almost 700 km, but anything over 85 km away is generally considered to be outer space because the outer layer is so thin.

FIGURE 1: Earth's atmosphere is a protection from harmful radiation, including the radiation from the Sun that causes the northern lights.

 ## Radiation everywhere

What is background radiation?

A Geiger counter clicks at a slow rate even when it is a long way from all radioactive sources. The Geiger counter is not faulty: it is recording the **background radiation**.

Background radiation is a low level of radiation that is present everywhere.

> Some of it comes from natural sources – from soil and rocks and from space.

> Some of it is a result of human activity.

Humans have evolved on a planet with background radiation. There is no evidence that natural background radiation is harmful.

Sources of background radiation

The pie chart in Figure 3 shows the proportion of the background radiation coming from different sources. It will not be exactly the same for everyone because:

> the amount of background radiation from different types of rocks varies – it depends on where you live

> to a slight extent, it depends on your lifestyle.

The pie chart does shows that almost 90% of the background radiation is from natural sources.

FIGURE 2: A Geiger counter in use.

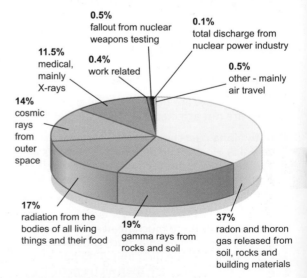

0.5% fallout from nuclear weapons testing
0.1% total discharge from nuclear power industry
11.5% medical, mainly X-rays
0.4% work related
0.5% other - mainly air travel
14% cosmic rays from outer space
17% radiation from the bodies of all living things and their food
19% gamma rays from rocks and soil
37% radon and thoron gas released from soil, rocks and building materials

FIGURE 3: Sources of background radiation.

QUESTIONS

1 What is background radiation?

2 Is background radiation harmful?

Q background radiation GCSE

Natural and lifestyle

Natural sources of radiation

Granite rock emits gamma rays. Consequently, areas with large amounts of granite rock have a higher level of background radiation. Some types of soil and rock naturally give off radon gas.

In 2004, research suggested that high levels of radon gas, seeping into homes, might be linked to an increased risk of lung cancer. Houses in affected areas can have the levels of radon gas measured. If necessary, special vents will be cut into house foundations, to prevent the gas levels building up.

All living things contain some radioactive materials. Bones are slightly radioactive. Food is slightly radioactive, too. How much radioactive material people absorb from food will also depend on where their food grows.

Lifestyle related background radiation

Earth is constantly bombarded by cosmic rays from outer space. Fortunately, the atmosphere absorbs much of this radiation.

However, modern long-haul aircraft fly very high, where the atmosphere is thinner and the protection less. Pilots and aircrew on these aircraft wear radiation badges to measure the radiation that they are exposed to. People who work with medical X-rays or at nuclear power stations also wear these badges. If the badge records that they have received above the 'safe' level of radiation, they will be moved to work in a 'safer' area for a while.

FIGURE 5: A radiation badge.

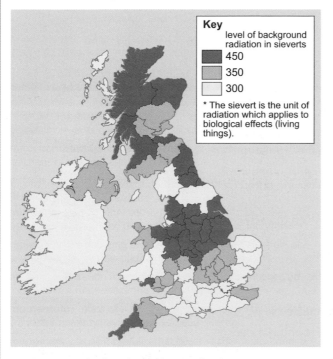

Key

level of background radiation in sieverts

450

350

300

* The sievert is the unit of radiation which applies to biological effects (living things).

FIGURE 4: How high is the background radiation in your area?

QUESTIONS

3 Radon gas builds up is a greater risk in new houses than old houses. Suggest why.

4 Suggest why radioactivity from food depends on the food source.

5 Some people worry about radiation from nuclear power stations. Do you agree that they are right to worry? Give your reasons.

6 During experimental work, why should you take background radiation into account? How would you do this?

Dangerous levels – or not?

Scientists do not know whether or not low doses of radiation, at the level of the background radiation, are harmful, beneficial or have no effect at all.

One of the problems is that it is impossible to establish a control group who have not been exposed to any radiation. Also, there are far too many other factors to be considered.

QUESTIONS

7 Why is it hard to establish whether background radiation is potentially harmful?

Half-life

You will find out:
> every radioactive isotope has its own unique half-life
> how the half-life of a radioactive isotope is determined

How old?

Is the Turin Shroud really 2000 years old? If so, it may be genuine. Radiocarbon dating has been used to gather evidence about the Shroud. The method is based on work that Willard Libby carried out – and for which he received the 1960 Nobel Prize. Radiocarbon dating now allows scientists and archaeologists to detect frauds as well as the exact age of genuine artefacts.

Did you know?

The decay of radioactive uranium has been used to date rocks billions of years old, accurate to within two million years.

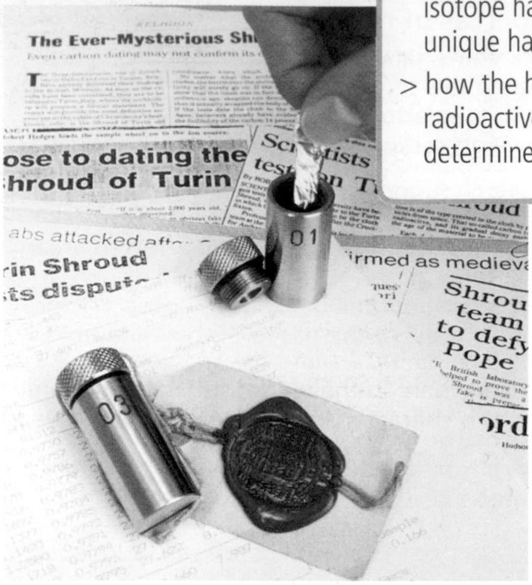

FIGURE 1: Fragments of the Turin Shroud being prepared for radiocarbon dating.

Half-life of radioactive isotopes

Radioactive isotopes (sometimes called **radioisotopes**) are unstable isotopes of any element that decay by emitting alpha, beta or gamma radiation.

The decay of any individual nucleus is random. As the original number of unstable nuclei decreases, the activity rate (the rate of clicks measured by a Geiger counter) decreases.

Scientists have worked out how to describe the rate at which the activity of an isotope changes. It is called **half-life**.

The half-life of a radioactive isotope, in a sample, is:

> the average time it takes for the number of nuclei of the isotope to halve, or

> the time it takes for the count rate to fall to half its initial level.

Figure 2 shows how the amount of a radioactive isotope changes. However much you start with:

> after one half-life, half the original atoms will have decayed and half the original amount will be left

> after two half-lives, one quarter of the original amount will be left

> after three half-lives, one eighth will be left and so on.

Remember

The half-life of a radioactive isotope is the average time it takes for the number of nuclei of the isotope in a sample to halve.

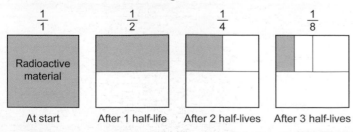

$\frac{1}{1}$ $\frac{1}{2}$ $\frac{1}{4}$ $\frac{1}{8}$

Radioactive material

At start   After 1 half-life   After 2 half-lives   After 3 half-lives

FIGURE 2: The amount of radioactive isotope left after each half-life.

QUESTIONS

1 Write down the definition of half-life.

2 A sample of radioactive material has 800 atoms.
(i) How many atoms are left after two half-lives?

(ii) How many half-lives does it take to reduce it to 50 atoms?

(iii) If the material has a half-life of 30 minutes, how long does it take for the number of atoms to reduce to 100?

Q radiocarbon dating ... half-life isotopes

Measuring half-life

The activity rate, or count rate, for a radioactive isotope is the mean (average) number of emissions (recorded as clicks of a Geiger counter) in a given time.

In one half-life, the activity rate will halve.

Each radioactive material has its own unique, constant rate of decay – its own unique half-life in which the count rate will halve. Examples are shown in Table 1.

Half-life values vary enormously, from millions of years to milliseconds. The half-life is also unique to each individual type of isotope. For example, barium-143 has a half-life of 12 seconds. Any sample of barium-143, anywhere in the world, will have the same half-life. Scientists use this characteristic to identify radioactive isotopes.

Remember

The half-life of a radioactive isotope is the time for the count rate from a sample of the isotope to fall to half its initial level.

Isotope	Radiation	Half-life
uranium-238	alpha	$4\frac{1}{2}$ billion years
plutonium-239	alpha and beta	24 000 years
carbon-14	beta	5700 years
cobalt-60	gamma	5 years
americium-241	alpha	460 years
iodine-131	beta	8 days
sodium-24	beta	15 hours
strontium-93	beta and gamma	8 minutes
barium-143	beta	12 seconds
polonium-213	alpha	0.000 004 second

TABLE 1: The half-life of some radioactive isotopes.

QUESTIONS

3 Iodine-131 has a half-life of 8 days. A sample has a count rate of 128 counts/minute.

(i) What will the count rate be after 3 half-lives?

(ii) Calculate the time it will take for the activity rate to drop to 4 counts/minute.

Graphs to identify radioactive isotopes

Plotting the activity (or count rate) of a radioactive isotope against time gives a graph similar to Figure 3. Several values of the time for the activity to fall by half can be read from the graph. This will give an accurate value of half-life and thus identify an unknown radioactive isotope.

QUESTIONS

4 Figure 3 shows the decay curve for strontium-93. Determine its half-life.

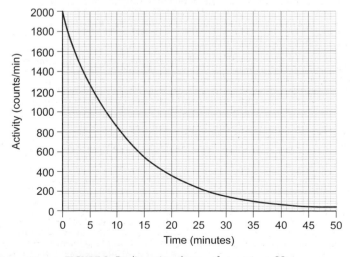

FIGURE 3: Radioactive decay of strontium-93.

Preparing for assessment: Analysing and interpreting data

To achieve a good grade in science, you not only have to know and understand scientific ideas, but you need to be able to apply them to other situations and investigations. These tasks will support you in developing these skills.

✳ Investigating radioactivity

Nuclei in a sample of a radioactive material constantly change into nuclei of a different element. The count rate measures the rate that these nuclei are changing. The table below, shows readings taken by a teacher during a lesson on radioactive decay.

Time (s)	0	25	50	75	100	125	150	175	200	225	250
Count rate (/s)	80	62	50	42	31	24	18	14	11	9	7

To collect the data shown in the table, the teacher took readings at regular intervals using a Geiger counter. A data logger would be better for this experiment as it can take readings at much closer intervals.

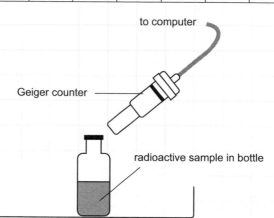

to computer

Geiger counter

radioactive sample in bottle

✳ Analysing data

1. Use the data, in the table, table to draw a graph of the teacher's results.

> Time is placed on the *x*-axis. The scale on each axis must be equally spaced. Include a smooth curve through the points.

2. Describe the pattern shown by your graph.

> Describe how the dependent variable changes when the independent variable changes.

3. The half-life is the time taken for the original count rate to halve. The teacher told the class that the half-life of this sample was 80 s.

Use your graph to decide if she was right.

> Work out what half the original count rate is, and use the graph find out when this occurred.

4. The teacher could have collected extra data. Describe what extra data you think should be collected, and explain how you would use it.

> Extra data could be used to improve the reliability of results, or to identify anomalous results, or to see if a trend continues.

Interpreting data

5. The teacher could have also have repeated the measurements a few times. Explain how these measurements would allow her to check the repeatability of the experiment.

> The reliability of data is shown by its repeatability and reproducibility.

6. The graph shows results from an experiment carried out by another scientist using the same equipment. The results were published in a science journal.

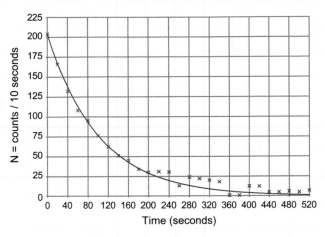

(a) Calculate the half-life of the sample using these data.

(b) Compare your answer with the half-life calculated using your graph of the teacher's data.

(c) Explain which set of data you think gives a more reliable idea of half-life.

> Compare the quantity of data collected. Bear in mind that the scale of the published graph makes it difficult to estimate the half-life with confidence.

7. Background radiation varies from place to place. Scientists themselves and their other activities may also contribute to background radiation.

Neither set of data (the teacher's and the other scientist's) allowed for background radiation. Suggest whether or not this would affect the conclusions of the investigations.

> Consider the validity of results as well as their reproducibility. Remember that background radiation is the same for all measurements at the place where they are made.

Connections

How Science Works

• Select and process primary and secondary data

• Analyse and interpret primary and secondary data

• Use scientific models and evidence to develop hypotheses, arguments and explanations

Science ideas

2.5.2 Atoms and radiation

Using nuclear radiation

You will find out:

> some of the uses of the three types of radiation

Thyroid cancer

Your thyroid is a gland that helps control how quickly your body transfers energy. Nuclear radiation is used to both detect and treat thyroid cancer. The thyroid strongly absorbs iodine. Small doses of radioactive iodine enable doctors to examine the thyroid. If the thyroid is surgically removed, larger doses of radioactive iodine can destroy any remaining cancerous cells.

Did you know?

Radiotherapy – using nuclear radiation to destroy cancer cells – is an important part of the treatment for many cancers today.

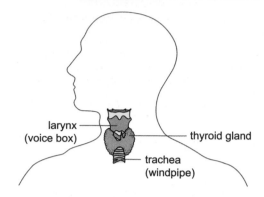

FIGURE 1: Your thyroid is in your neck.

Radiation in medicine and industry

Gamma radiation

Although exposure to radiation can cause cancer, radiation is also use in treating cancer. A fine beam of gamma rays, focused accurately on just the cancerous tumour, can kill the cancer cells without damaging too many healthy cells.

Gamma radiation is also used to sterilise surgical instruments and some foods. The washed instruments or fresh foods are:

> sealed in airtight packaging, and

> irradiated with a low dose of gamma rays.

The gamma rays kill bacteria and the sealed packaging prevents more entering.

Beta radiation

Beta particles are used to control paper thickness in paper mills. A beta source and a detector are put on opposite sides of the paper. The number of beta particles reaching the detector depends on the thickness – too many beta particles means the paper is too thin. Beta is the only type of radiation suitable, as:

> gamma rays would not be stopped by the paper

> all alpha particles would be stopped.

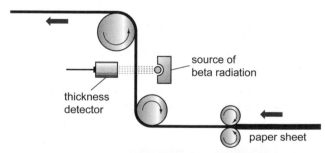

FIGURE 3: Controlling paper thickness at a paper mill.

QUESTIONS

1 How can gamma rays cause cancer and also help to kill the cancer?

2 How can gamma rays help with the sterilisation of a surgeon's instruments?

3 Explain why beta particles are used in a paper thickness detector.

FIGURE 2: These strawberries are the same age, but those on the left have been irradiated.

Q gamma rays uses ... beta particles paper

Using radioisotopes

Tracers

Finding a leak in an underground pipe would be tricky without radiation. An isotope, called a **tracer**, is put into the pipe and the radiation above the pipe measured. The count rate is higher where the tracer is leaking from the pipe.

An isotope emitting gamma rays is used; alpha and beta particles would not penetrate through the ground to the surface. A short half-life isotope is used so that the radiation levels in the area rapidly return to normal.

Tracers are also used in medicine, to examine how well different organs are functioning. For example:

> Radioactive iodine-123 is used to examine the thyroid and kidneys. This is because iodine tends to concentrate in the thyroid gland. The doctor gives the patient a drink containing a solution of iodine-123. The doctor then scans both the thyroid and then the kidneys as the iodine passes through the body.

> Barium-143 is used to examine the stomach and intestines.

> Sometimes, radioactive isotopes are injected into the bloodstream to examine other organs.

Smoke detectors

Smoke alarms contain small amounts of americium-241 which emits alpha particles. The alpha particles ionise the air, releasing free electrons which then maintain a tiny current in the detector circuit. If smoke particles block the beam of alpha particles, the ionisation stops and the current ceases. This sets off the alarm.

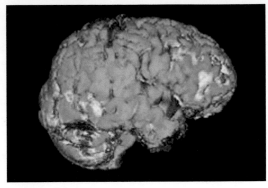

FIGURE 4: This brain scan used a medical tracer to highlight brain activity.

3. Current detector: when smoke prevents alpha particles from ionising air, the current stops and the detector triggers the alarm.

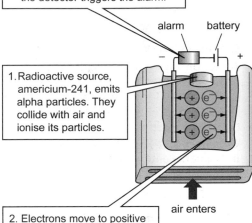

1. Radioactive source, americium-241, emits alpha particles. They collide with air and ionise its particles.

2. Electrons move to positive electrode. Positive ions move to negative electrode.

FIGURE 5: How a smoke detector works.

QUESTIONS

4 Look at Table 1 on page 263. Select one isotope as a suitable tracer and explain your choice.

5 Medical tracers emit beta particles or gamma rays. Suggest why these are better than alpha particles.

6 Suggest some factors that would affect the choice of half-life for a medical tracer.

Plant growth

Most hydrogen atoms have a nucleus containing just one proton. Tritium (sometimes called hydrogen-3) is a radioactive isotope of hydrogen with one proton and two neutrons. In a water molecule, if one of the hydrogen atoms is replaced with a tritium atom, the water becomes radioactive. The water emits beta particles.

Tritium can be used to investigate plant growth, by using radiation detectors to measure the uptake of water to various parts of the plant.

QUESTIONS

7 Suggest why tritium is a good radioisotope to use when investigating plant growth.

8 What does the use of tritium in plant growth studies, tell you about the half-life of tritium?

 radioactive tracers ... tritium half life

Nuclear fission

You will find out:

> how the energy contained within an atom is released in nuclear fission

The quest for energy

The world's industrialised nations use vast amounts of energy, most of it coming from burning fossil fuels that are non-renewable and believed to cause climate change. Developing nations are rapidly increasing their energy use, much of it coming from fossil fuels. There is a real need for energy supplies that are clean, reliable and affordable.

Did you know?

About 55 000 tonnes of uranium are mined each year, most of it from Kazakhstan, Canada and Australia.

FIGURE 1: Energy is released from these nuclear fuel rods in a controlled reaction.

 ## Energy in atoms

Energy and mass

When fuels are burned, energy is released as atoms recombine in new ways, and less energy is stored in the bonds holding the atoms together. Einstein showed that, if the nucleus of the atom itself is split, some of the mass of the nucleus changes directly into energy. His famous equation is:

$E = mc^2$

where
E is energy
m is mass
c is speed of light - about 300 000 000 metres per second

The equation shows that the energy released in this way is huge. Just a few grams could release as much as a power station transfers in a year.

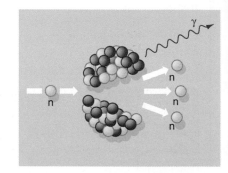

FIGURE 2: Splitting a single atom.

Splitting the atom

The atoms that split most easily to release energy are large isotopes. The ones most commonly used are uranium-235 and plutonium-239.

This is **nuclear fission**:

> A slow-moving neutron is absorbed by the nucleus.

> The nucleus splits into two almost equal parts.

> This releases large amounts of energy.

> Two or three new neutrons are also released.

Nuclear fission in one kilogram of plutonium supplies enough energy to generate electricity for a small town for a year.

QUESTIONS

1 Use Einstein's equation to calculate how much energy would be released if 1 g of mass turned entirely to energy.

2 Describe what happens in nuclear fission.

3 What must happen to the total mass, to release so much energy?

🔍 e=mc² ... nuclear fission GCSE

Critical mass and chain reaction

Each of the extra neutrons released, when an atom of uranium-235 or plutonium-239 is split, could go on to collide with another nucleus and split that. Whether or not they do, depends on the amount of nuclear fuel present.

If the amount of uranium is small (smaller than tennis ball size):

> most of the neutrons will escape from the uranium

> there will not be enough left to keep splitting atoms

> the reaction will stop.

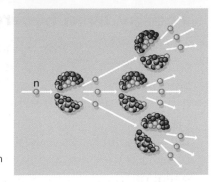

FIGURE 3: Chain reaction.

If the amount of uranium is larger:

> the extra neutrons split more atoms

> this releases more neutrons

> which split more atoms

> and so on.

It is like an avalanche, beginning slowly, then becoming faster and faster until it becomes unstoppable. This is a **chain reaction**.

The **critical mass** of a nuclear fuel is the minimum amount needed to sustain the chain reaction (to keep it going).

Uncontrolled reactions

An atomic bomb is an example of nuclear fission undergoing an uncontrolled chain reaction. The avalanche of neutrons being released, and atoms being split, becomes so great, that there is a huge release of destructive energy. Much of the damage done by atomic bombs is caused by the huge temperatures (millions of degrees Celsius at the centre of the explosion) and very high speed winds (hundreds of kilometres per hour).

Controlled reactions

Chain reactions can be controlled. This is how nuclear power stations work. Control rods absorb some of the neutrons released, so controlling the rate at which further atoms are split, and also the rate at which energy is released.

FIGURE 4: An atomic bomb detonation.

QUESTIONS

4 Describe how splitting an atom of uranium can lead to a chain reaction.

5 Describe how a fission reaction becomes unstoppable.

6 What makes an unstoppable fission reaction so damaging?

7 How can the energy from a nuclear fission reaction be harnessed safely?

Why is so much energy released?

The positively charged protons in an atomic nucleus do not repel each other, breaking the nucleus apart.

Protons are held together by a strong nuclear force that is much stronger than the electrical force of repulsion. The energy stored in the nucleus is the work done by this strong nuclear force to hold the protons together. When the atom is split, less work is done by this force to hold the new nuclei together. The difference is transferred as energy emitted as nuclear radiation.

QUESTIONS

8 Nuclei of neighbouring atoms do not stick to each other. What does this tell you about the range over which the strong nuclear force acts?

Nuclear fusion

You will find out:

> how energy can be released through the fusion of two small atoms

Energy from hydrogen?

In 2010, British scientists released plans to build Britain's first nuclear power station using nuclear fusion. They say that the power station could be generating electricity in 20 years time. So far though, world research has only succeeded in making a fusion reaction in a two millimetre pellet of frozen hydrogen – and they had to put in far more energy than it gave out.

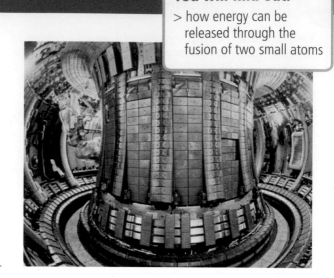

FIGURE 1: Inside the Tokamak nuclear fusion reactor at the research centre in Oxfordshire.

Fusing nuclei

What is nuclear fusion?

Nuclear fission is the process of very large nuclei splitting apart, releasing energy.

Nuclear fusion is the process of small nuclei *joining together*, releasing energy.

When small nuclei join together, the new nucleus has slightly less mass than the original nuclei. The mass changes to energy and is released. Although the difference in mass is tiny, the energy is large, according to Einstein's equation $E = mc^2$.

The word equation for the simplest nuclear fusion reaction is:

hydrogen → helium + energy

Nuclear fusion in stars

The energy given out by stars comes from nuclear fusion.

Stars are formed when the force due to gravity pulls together hydrogen atoms, gases and dust. If there is enough matter, the force of gravity pulling the cloud inwards compresses it into a dense ball. Work done by the force pulling the cloud inwards is transferred to kinetic energy of the particles – the cloud becomes hotter (temperature is a measure of the kinetic energy of particles). This is a **protostar**.

Clouds that are not big enough to become protostars stay cold and can become planets.

As gravity pulls more matter into the protostar, it heats up enough to start nuclear fusion reactions.

FIGURE 2: Protons join together in stages to make different elements.

energy

FIGURE 3: A star is born. How can dust and gas become a star?

⊙ QUESTIONS

1 Name the two ways in which energy can be released from inside a nucleus.

2 Write the equation which shows how energy is released by nuclear fusion.

Q nuclear fusion GCSE ... nuclear fusion energy source

Using nuclear fusion

Difficult conditions

Nuclear fusion can only take place in very unusual conditions.

> Positively charged protons in atomic nuclei usually repel each other, so nuclei cannot get close enough to join together. Protons or small nuclei need to travel at very high speeds to overcome the electrical force of repulsion. This only happens at temperatures above about 15 million degrees Celsius.

> Protons have to be densely packed to make collisions likely. It means that the pressure has to be very high.

These conditions have to be produced using forces, since no container is strong enough and materials vaporise at such high temperatures. Scientists are working to create the conditions for fusion reactions on Earth.

Nucleus size

Small nuclei give out energy when they join together (nuclear fusion), but large nuclei give out energy when they split apart (nuclear fission). You would have to put in a lot of energy to join nuclei together as a uranium nucleus.

Where do large nuclei come from?

> All the elements lighter than iron are made by nuclear fusion in stars, joining together smaller nuclei.

> Heavier nuclei are made when massive stars explode. Large nuclei in the star absorb huge amounts of energy, fusing them into even larger nuclei. These nuclei then give out this energy when they split apart again.

Did you know?

Nuclear fusion only happens at temperatures of millions of degrees Celsius, such as are found in the centre of stars.

QUESTIONS

3 Explain why the temperature has to be very high for fusion to happen.

4 Which elements 'lose' energy by nuclear fusion? Which lose it by nuclear fission?

5 Discuss, with reasons, which type of atomic nucleus is most stable.

The proton–proton chain

Figure 2 shows the three stages of the nuclear fusion of hydrogen into helium, called the proton–proton chain.

1. proton + proton = hydrogen-2 + electron + neutrino
(one of the protons changes to a neutron)

2. hydrogen-2 + proton = helium-3 + photon
(the photon transfers energy away from the reaction)

3. helium-3 + helium-3 = helium-4 + proton + proton
(the two protons can start another proton-proton chain)

QUESTIONS

6 Protons carry a positive charge, neutrons carry no charge. Why does it seem surprising that the nuclei of atoms do not break up spontaneously?

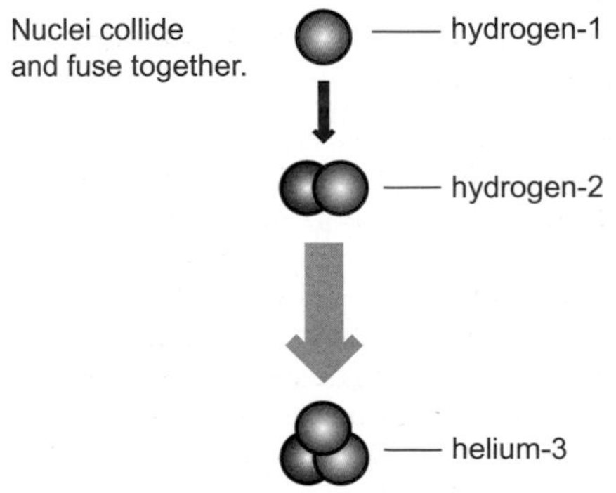

Nuclei collide and fuse together. —— hydrogen-1

—— hydrogen-2

—— helium-3

FIGURE 4: The proton–proton chain.

Life cycle of stars

You will find out:

> about the life cycle of a star

> about main sequence stars

Star struck

You may think of stars as permanent and unchanging, but nuclear fusion shows that all stars are continually changing. They are formed, have a life cycle that lasts billions of years, and then 'die'. Then, new stars may be formed from the leftover debris. Our solar system formed about 4.6 billion years ago from the remains of an earlier star.

Did you know?

Scientists estimate from the age of the universe, the age of our Sun, and from other stars in the Milky Way, that our Sun may be a third generation star.

FIGURE 1: A star may 'die' in a massive explosion. This is Cassiopeia – the youngest in the Milky Way. It exploded only 320 years ago.

The life cycle of a star

Main sequence stars

Stars form from protostars – clouds of gas and dust that are massive enough to collapse into a dense ball and heat up. The star then spends billions of years as a **main sequence star**, a star where nuclear fusion is changing hydrogen into helium and then into heavier nuclei.

Most of the fusion reactions in main sequence stars are hydrogen to helium reactions.

Rapid change

Eventually, when most of the hydrogen is 'used up', stars begin a relatively rapid sequence of changes. The changes that happen depend on the size of the main sequence star.

A In stars about the same mass as our Sun:

> As the hydrogen 'fuel' begins to run out, the star generates less heat by nuclear fusion. It becomes unstable.

> The core shrinks and the outer layers (still mostly hydrogen gas) expand outwards and cool. The star looks bigger and cooler. It is called a red giant star.

> The outer layers continue to expand out into space, becoming interstellar gas and dust. The inner core continues to collapse, getting hotter. It becomes a white dwarf star. More nuclear reactions take place, changing helium to the heavier elements, up to iron.

> Eventually the nuclear reactions run out, the star cools and becomes a black dwarf.

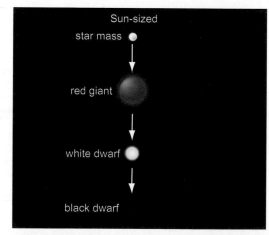

FIGURE 2A: The life cycle of a main sequence star depends on its size.

QUESTIONS

1 Why do stars not live forever?

2 What does the end of a star's life depend upon?

3 Describe the likely end of our own Sun. (Draw a diagram if it helps.)

Q star life cycle animation ... star life gases

B In stars much more massive than our Sun (ten times or more):

> As the hydrogen 'fuel' runs out, the star becomes unstable. The outer hydrogen gas layers expand to become a red supergiant star.

> The core collapses so quickly, and becomes so hot, that its outer layers explode in a supernova explosion. Elements heavier than iron are formed in supernovae explosions.

> If the dense remnant left from the explosion is small (less than three times the mass of our Sun), it collapses into a neutron star, giving off X-rays and gamma rays.

> If the dense remnant left from the explosion is large (more than three times the mass of our Sun), the force of gravity is so huge that it just keeps collapsing and drawing in surrounding matter. It becomes a black hole.

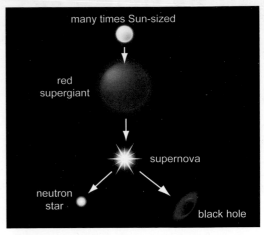

FIGURE 2B: The life cycle of a massive main sequence star.

Why does the main sequence last so long?

Stars spend most of their life in the stable, main sequence phase. There are two main forces acting inside any star:

> Gravity pulls all the mass of the star in towards the centre.

> The hot star would expand if it could, due to the fast moving particles 'bumping into each other' and pushing each other apart. The high temperature causes a 'force of expansion' that acts outwards.

During the main sequence phase of a star, these forces are balanced. The star is stable.

The main sequence phase can last up to ten billion years.

QUESTIONS

4 Why does the main sequence phase of a star last for so long?

5 What causes the heat that tries to make a star expand?

6 Write the word equation for the fusion of hydrogen.

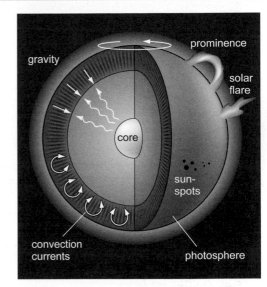

FIGURE 3: Inside a main sequence star.

A matter of size

Nuclear fusion continues in the core of a red supergiant until most of the core is iron. Then nuclear fusion reactions stop.

The core is no longer radiating energy, so the force of gravity takes over. The core collapses – sometimes in seconds.

Even a rise in temperature to over 100 billion degrees is not enough to produce an outward force – sufficient for a hot star trying to expand – that is big enough to balance the inward force of gravity.

The density increases until the force of repulsion between nuclei finally overcomes gravity and blows the star apart.

QUESTIONS

7 What difference would it make, if a supergiant was smaller to start with?

Checklist P2.4–2.6

To achieve your forecast grade in the exam you will need to revise

Use this checklist to see what you can do now. Refer back to the relevant topics in this book if you are not sure. Look across the three columns to see how you can progress. **Bold** text means Higher tier only.

Remember that you will need to be able to use these ideas in various ways, such as:

> interpreting pictures, diagrams and graphs

> applying ideas to new situations

> explaining ethical implications

> suggesting some benefits and risks to society

> drawing conclusions from evidence you are given.

Look at pages 278–299 for more information about exams and how you will be assessed.

To aim for a grade E	To aim for a grade C	To aim for a grade A
Know that electricity is supplied from the mains and from batteries.	Describe differences between alternating current (a.c.) and direct current (d.c.) supply.	Recall that mains electricity in the UK is supplied as 230 volts at 50 hertz.
Know that a plug and electric cable are used to connect equipment to the mains.	Recall the structure and wiring of a three-pin electric plug and be able to explain mistakes in their wiring. Recall the structure of three-core mains cable.	Evaluate the most suitable cable for an appliance.
Know that a fuse, circuit breaker and earthing are safety features. Recognise dangerous uses of mains electricity.	Describe how fuses, circuit breakers and earthing work as safety features. Describe ways to reduce the risks posed by mains electricity.	Explain how safety features protect the equipment and/or the user. Evaluate ways to reduce the risks posed by mains electricity.
Know that current has a heating effect.	Use equations to calculate power.	Rearrange and use equations to calculate power. **Rearrange and use equations to calculate energy transferred.**
Recall the structure of the atom.	Recall properties of sub-atomic particles. Explain what is meant by an ion. Describe the alpha scattering experiment.	Explain what is meant by an isotope. Explain how the alpha scattering experiment led to the nuclear model of the atom.

To aim for a grade E   To aim for a grade C   To aim for a grade A

Recall what is meant by radioactive materials. Recall what is meant by background radioactivity.	Describe some sources of background radioactivity.	Evaluate the effect of background radiation on people.
Recall that the radioactivity of a sample falls with time. Recall that exposure to ionising radiation can be harmful. Recall some ways to reduce the risks.	Describe the properties of alpha, beta and gamma radiation, their uses and effect on living cells. Explain what is meant by half-life. Explain why exposure to ionising radiation can be harmful. Describe uses of ionising radiation.	**Explain how the mass and charge of alpha and beta radiation affect their properties.** Evaluate the risks posed by ionising radiation and measures taken to reduce exposure. Explain how uses of ionising radiation are linked to their properties and half-lives.
Recall that there are different types of radioactivity.	Describe the structure of alpha and beta particles, and gamma rays.	**Complete nuclear equations that show alpha and beta decay.**
Know that nuclear fuels are used in nuclear power stations.	Recall that uranium and plutonium are nuclear fuels. Know that nuclear fission releases energy in a nuclear power station.	Describe the process of nuclear fission. Describe what happens during a chain reaction.
Know that nuclear fusion in stars releases energy and forms new elements. Know that stars have a life cycle.	Describe what happens during nuclear fusion. Describe the main stages in the life cycle of a star. Know that elements up to the atomic weight of iron form in stars and that elements heavier than iron form in supernovas.	Compare nuclear fusion with nuclear fission. Explain how the energy output from stars is maintained. Explain how supernovas distribute different elements throughout the Universe.

In the examination, equations will be given on a separate equation sheet.
Write down the equation that you will use. Show clearly how you work out your answer.

1. The diagram shows the inside of a plug that has been badly wired.

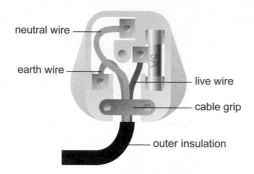

neutral wire

earth wire

live wire

cable grip

outer insulation

AO1 **(a)** Write down three mistakes in the wiring. [3]

AO2 **(b)** How does a fuse reduce the risk of fire? [2]

AO2 **(c)** A student plugged in her hair straightener and left it lying on the bathroom floor while she washed her hair.

Write down two reasons why this is not a good idea. [2]

2. A teacher showed her class some sources of ionising radiation. The teacher kept the radioactive sources in a lead box until she was ready to use them. She used tongs to lift them out of the box, and did not let the class touch the sources.

AO1 **(a)** People using radioactive sources should limit their exposure to the radioactivity. Explain why. [2]

AO3 **(b)** Explain whether these precautions were sensible in the lesson. [3]

3. Scientists have found traces of different elements in our Sun.

AO1 **(a)** Traces of elements such as helium and carbon were found. What is the name of the process that formed these elements in the Sun? [1]

AO2 **(b)** Traces of elements heavier than iron are found in the Sun. Explain how these elements provide evidence that there are stars older than our Sun. [3]

4. The diagram shows a chain reaction involving the fission of uranium nuclei.

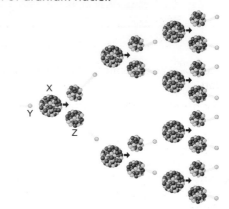

X
Y
Z

AO2 **(a)** State which part of the diagram represents the following:

(i) a neutron [1]

(ii) a uranium nucleus. [1]

AO1 **(b)** Describe one practical use of uranium fission. [2]

5. These pictures show two types of lightbulb. They both give out the same amount of light.

Filament bulb: power 100 W

Compact fluorescent lamp: power 23 W

AO2 **(a)** Explain why both bulbs give out the same amount of light even though their power is different. [2]

AO3 **(b)** Many homes still use filament bulbs. A letter from one electricity company told people to save electricity by changing their light bulbs.

Explain how people can save energy by choosing different light bulbs. [2]

AO1 recall the science   AO2 apply your knowledge   AO3 evaluate and analyse the evidence

✳ WORKED EXAMPLE – Foundation tier

The table gives some information about radon atoms.

Particle	Number in radon atom
protons	86
neutrons	136
electrons	86

(a) Write down the atomic number and mass number of radon. [2]

Atomic number = 86, mass number = 136

The atomic number is correct. The mass number should be 222 because it includes protons and neutrons.

Radon is a radioactive gas that constantly seeps out of the ground. Radon gas gives out alpha radiation all of the time. Traces of radon gas are found in buildings in some parts of the country. Having radon gas in your home increases the risk of getting lung cancer, but smoking tobacco is still the main cause.

(b) Why are people who live in places with high levels of radon more likely to get lung cancer? [3]

Radon gas gives out alpha particles, which damage cells.

This answer is only worth two marks because the candidate does not explain any link between radon gas and lung cancer.

(c) Radon causes half of all background radiation. Write down two other causes of background radiation. [2]

Cosmic rays, ground and buildings, medical uses, nuclear power.

Read the question carefully. You need to include only two types of background radiation. The candidate cannot receive more than two marks: one for each cause.

(d) A smoker lives in a house where the levels of radon gas are higher than normal. She is worried that she may get lung cancer. It costs £5000 to fit a pump to remove the gas.

Explain one way that this person could reduce her risk of lung cancer. [3]

She could give up smoking. She could fit a pump. She could move to another part of the country.

This answer receives one mark because more detail about one method only is needed. The candidate could have said: smoking is a bigger cause of lung cancer than radon gas so the person should stop smoking. This would not cost money and could happen straightaway.

*In the examination, equations will be given on a separate equation sheet.
Write down the equation that you will use. Show clearly how you work out your answer.*

1. This question is about the radioactive decay of strontium.

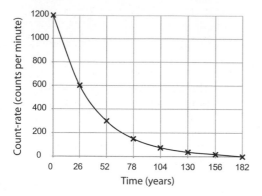

AO1 **(a)** Explain what is meant by half-life. [3]

AO2 **(b)** Use the graph to calculate the half-life of strontium. [2]

AO2 **(c)** A different material, X, has a longer half-life than strontium. The count rate for a sample of X is measured. The count rates of the samples of strontium and of X are the same at the start of the experiment.

Draw a graph the same as the one shown above. On your graph, draw another line to show how the count rate of X changes with time. [3]

AO3 **(d)** Doctors use radioactive sources as tracers in patients. Explain whether the most suitable sample should have a half-life of seconds, hours or years, and whether it should emit gamma or alpha radiation. [4]

AO3 **2.** *In this question you will be assessed on using good English, organising information clearly and using specialist terms where appropriate.*

This information was taken from a leaflet about electrical safety in the home:

"Most fires in homes in the UK are caused by faulty electrical wiring and equipment. Most accidents could be prevented if householders installed modern fuse boxes that include RCCBs (residual current circuit breakers). These are safer than relying on old-fashioned fuse boxes that use fuses."

Explain how changing to RCCBs would help to reduce electrical fires in homes. [6]

3. All stars go through a lifecycle.

AO1 **(a)** The Sun is at which stage of its life cycle, now? [1]

AO2 **(b)** Why do fusion processes in the Sun change during its lifecycle? [3]

AO3 **(c)** Silver has an atomic mass of 47. Will it form in the Sun? Explain your answer. [4]

4. One homeowner recently had a test carried out in his home to measure the radon levels. Radon is a radioactive gas that seeps out of rocks that contain uranium. Radon is responsible for about half of the background radiation in the UK.

AO1 **(a)** Explain what is meant by background radiation. [1]

AO2 **(b)** Radon gas emits alpha particles.

Explain why parts of the country with high levels of radon gas are linked with higher rates of lung cancer. [3]

AO3 **(c)** The homeowner could install a ventilation system to reduce the levels of radon gas. The system costs about £2000 to install. Explain whether you think the homeowner should install the vent. [4]

5. A current of 8.5 A flows through an electric fire when it is turned on. In the UK, mains electricity is supplied at 230 V.

AO1 **(a)** What is the frequency of mains electricity supply in the UK? [1]

AO2 **(b)** What is the power of the electric fire? Include the unit in your answer. [4]

AO1 **recall the science**   AO2 **apply your knowledge**   AO3 **evaluate and analyse the evidence**

✱ WORKED EXAMPLE – Higher tier

Cars use a battery to power the electrical equipment such as lights, heating and music system and to start the engine. The battery recharges while the engine is running.

(a) The power of the headlights used in one type of car is rated as 120 W for the pair.

If the car battery supplies 12 V, calculate the current in amps flowing through the headlights. [3]

power = potential difference x current
so current = power ÷ potential difference = 120 ÷ 12

(b) The owner of the car changes these headlights for LED headlights. LED headlights provide the same amount of light while using less power.

Describe one advantage for the driver of using LED headlights. [3]

Less power is taken from the battery while the lights are on so less fuel is used recharging the battery.

(c) The ratings of fuses in a car vary from 7.5 A to 40 A.

Explain what is meant by a fuse rating of 40 A. [2]

The fuse will melt if the current exceeds 40 A.

In part (d) of this question you will be assessed on using good English, organising information clearly and using specialist terms where appropriate.

(d) One car manufacturer has suggested redesigning the systems so that batteries recharge to 10 V instead of 12 V. Their press release said that this would benefit the environment as less fuel would be used charging the battery to 10 V, rather than 12 V.

Explain whether charging batteries to 10 V is a good idea or not. [6]

If the potential difference is 10 V but the lights remain the same power then the current in the cables will be larger which might cause fires unless the car is rewired for larger currents.

If the battery is only charged up to 10 V, it stores less energy for equipment. Energy = potential difference x charge and the total amount of charge in the battery stays the same.

How to raise your grade!
Take note of these comments – they will help you to raise your grade.

The answer was not completed, but the candidate receives two marks for substituting the correct values into the correct equation. Remember to read the question carefully.

This receives two marks as the candidate included some detail.

The candidate could also say that the driver will save money on fuel bills.

This answer receives two marks as it explains how a fuse works, and the rating value.

This answer receives five marks. The candidate applied physics to two points in detail but forgot to say whether this was a good idea. Examiners do not want long lists in answers, but they do want to see some physics used.

Carrying out practical investigations in GCSE Additional Science

Introduction

As part your GCSE Additional Science course, you will develop practical skills and have to carry out investigative work in science.

Your investigative work will be divided into several parts:

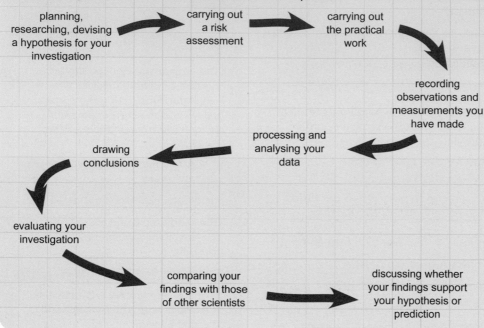

planning, researching, devising a hypothesis for your investigation → carrying out a risk assessment → carrying out the practical work → recording observations and measurements you have made → processing and analysing your data → drawing conclusions → evaluating your investigation → comparing your findings with those of other scientists → discussing whether your findings support your hypothesis or prediction

✳ Planning and researching your investigation

A scientific investigation usually begins with you thinking about an idea, answering a question, or trying to solve a problem.

Researching what other people know about the idea, question or problem may well suggest some variables that have an effect on what you decide to investigate.

From this you should develop a hypothesis. For example you might notice that plants grow faster in a heated greenhouse than an unheated greenhouse.

Your hypothesis would be that 'the rate of photosynthesis is increased by the temperature of the environment in which the plant is grown.'

You would then plan how you will carry out an investigation to test this hypothesis.

To formulate a hypothesis you are likely to need to research some of the background science.

First of all, use your lesson notes and your textbook. The topic you have been given to investigate will relate to the science that you have learned in class.

Also make use of the internet, but make sure that your internet search is closely focused on the topic that you are investigating.

Definition

A **hypothesis** is a possible explanation that someone suggests to explain some scientific observations.

Assessment tip

When devising your hypothesis, it is important that it is testable. In other words, you must be able to test the hypothesis in the school lab.

Assessment tip

You need to use your research to explain why you made your hypothesis.

- The search terms you use on the internet are very important. 'Investigating temperature and photosynthesis' is a better search term than just 'photosynthesis', as it is more likely to provide links to websites that are more relevant to your investigation.

- The information on websites also varies in its reliability. Free encyclopaedias often contain information that has not been written by experts. Some question and answer websites might appear to give you the exact answer to your question, but be aware that they may sometimes be incorrect.

- Most GCSE Science websites are more reliable but, if in doubt, use other information sources to verify the information.

As a result of your research, you may be able to extend your hypothesis and make a **prediction** based on science.

Example 1

Investigation: Plan and research an investigation into the effect of temperature on the change in height of a plant over two weeks.

Your hypothesis might be 'when I increase the temperature, the percentage increase in the height of the plant will be greater.'

You should be able to justify the hypothesis by some facts that you find. For example 'growing lettuces in greenhouses halves the time it takes for them to be ready to sell.'

Assessment tip

Make sure that you make a detailed note of which sources you have used: for a book, the author's name and the title; for a website, the name of it.

Assessment tip

Higher tier
You are expected to be able to balance chemical equations. So, for example, if the enzyme is being used to decompose hydrogen peroxide to water and oxygen you should be able to balance the equation: $H_2O_2 \rightarrow H_2O + O_2$ to give $2H_2O_2 \rightarrow 2H_2O + O_2$

Choosing a method and suitable apparatus

As part of your planning, you must choose a suitable way of carrying out the investigation.

You will have to choose suitable techniques, equipment and technology, if this is appropriate. How do you make this choice?

You will have already carried out the techniques you need to use during the course of practical work in class (although you may need to modify these to fit in with the context of your investigation). For most of the experimental work you do, there will be a choice of techniques available. You must select the technique:

- that is most appropriate to the context of your investigation, and

- that will enable you to collect valid data, for example if you are measuring the effects of light intensity on photosynthesis, you may decide to use an LED (light-emitting diode) at different distances from the plant, rather than a light bulb. The light bulb produces more heat, and temperature is another independent variable in photosynthesis.

Your choice of equipment, too, will be influenced by the measurements that you need to make. For example:

- you might use a one-mark or graduated pipette to measure out the volume of liquid for a titration, but

- you may use a measuring cylinder or beaker when adding a volume of acid to a reaction mixture, so that the volume of acid is in excess to that required to dissolve, for example, the calcium carbonate.

Assessment tip

Technology, such as data-logging and other measuring and monitoring techniques, for example heart sensors, may help you to carry out your experiment.

Definition

The **resolution** of the equipment refers to the smallest change in a value that can be detected using a particular technique.

Assessment tip

Carrying out a preliminary investigation, along with the necessary research, may help you to select the appropriate technique to use.

Variables

In your investigation, you will work with independent and dependent variables.

The factors you choose, or are given, to investigate the effect of are called **independent variables**.

What you choose to measure, as affected by the independent variable, is called the **dependent variable**.

 Independent variables

In your practical work, you will be provided with an independent variable to test, or will have to choose one – or more – of these to test. Some examples are given in the table.

Investigation	Possible independent variables to test
activity of amylase enzyme	> temperature > sugar concentration
rate of a chemical reaction	> temperature > concentration of reactants
stopping distance of a moving object	> speed of the object > the surface on which it's moving

Independent variables can be **categoric** or **continuous**.

> When you are testing the effect of different disinfectants on bacteria you are looking at categoric variables.
> When you are testing the effect of a range of concentrations of the same disinfectant on the growth of bacteria you are looking at continuous variables.

Range

When working with an independent variable, you need to choose an appropriate **range** over which to investigate the variable.

You need to decide:

✔ which treatments you will test, and/or
✔ the upper and lower limits of the independent variables to investigate, if the variable is continuous.

Once you have defined the range to be tested, you also need to decide the appropriate intervals at which you will make measurements.

The range you would test depends on:

✔ the nature of the test
✔ the context in which it is given
✔ practical considerations
✔ common sense.

Definition

Variables that fall into a range of separate types are called **categoric variables**.

Definition

Variables that have a continuous range, or are numeric, are called **continuous variables**.

Definition

The **range** defines the extent of the independent variables being tested, for example, from 15 cm to 35 cm.

Example 2

1 Investigation: Investigating the factors that affect how quickly household limescale removers work in removing limescale from an appliance

You may have to decide on which acids to use from a range that you are provided with. You would choose a weak acid, or weak acids, to test, rather than a strong acid, such as concentrated sulfuric acid. This is because of safety reasons, but also because the acid might damage the appliance you were trying to clean. You would then have to select a range of concentrations of your chosen weak acid to test.

2 Investigation: How speed affects the stopping distance of a trolley in the lab

The range of speeds you would choose would clearly depend on the speeds that you could produce in the lab.

Temperature

You might be trying to find out the best temperature at which to grow tomatoes.

The 'best' temperature is dependent on a number of variables that, taken together, would produce tomatoes as quickly as possible whilst not being too costly. You might use a high temperature, but the cost of the fuel may outweigh the advantage of growing the tomatoes more quickly.

You should limit your investigation to just one variable, temperature, and keep the other factors constant (such as watering and feeding the tomatoes). Later you can consider other variables, including fuel costs for heating.

Dependent variables

The dependent variable may be clear from the problem that you are investigating, for example the stopping distance of moving objects. You may have to make a choice.

Example 3

1 Investigation: Measuring the rate of photosynthesis in a plant

There are several ways in which you could measure the rate of photosynthesis in a plant. These include:

> counting the number of bubbles of oxygen produced in a minute by a water plant such as *Elodea* or *Cabomba*

> measuring the volume of oxygen produced over several days by a water plant such as *Elodea* or *Cabomba*

> monitoring the concentration of oxygen in a polythene bag enclosing a potted plant, using an oxygen sensor

> measuring the colour change of hydrogencarbonate indicator that contains algae embedded in gel.

2 Investigation: Measuring the rate of a chemical reaction

You could measure the rate of a chemical reaction in the following ways:

> the rate of formation of a product

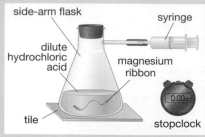

> the rate at which the reactant disappears

> a colour change

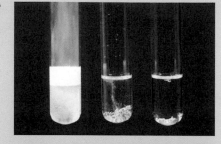

> a pH change.

✷ Control variables

The validity of your measurements depend on you measuring what you are supposed to be measuring.

Some of these variables may be difficult to control. For example, in an ecology investigation in the field, factors such as varying weather conditions are impossible to control.

Experimental controls

Experimental controls are often very important, particularly in biological investigations where you are testing the effect of a treatment.

Definition

Other variables that you are not investigating may also have an influence on your measurements. In most investigations, it is important that you investigate just one variable at a time. So other variables, apart from the one you are testing at the time, must be controlled, and kept constant, and not allowed to vary. These are called **control variables**.

Definition

An **experimental control** is used to find out whether the effect you obtain is from the treatment, or whether you get the same result in the absence of the treatment.

Example 4

Investigation: The effect of temperature on the growth of tomato plants.

The tomato plants grow well at 35 °C, but some plants grow just as well at lower temperatures. You need to be certain that the effect is caused by the temperature. There are many things that affect plant growth, so you should make sure that these variables are controlled. The volume of water they receive, the soil that the plants are grown in, the nutrients present in the soil, and the genetic status of the plants should all be as similar as possible. Farmers often use f1 hybrid seeds because the plants are virtually genetically identical and will be ready to harvest at the same time.

✳ Assessing and managing risk

Before you begin any practical work, you must assess and minimise the possible risks involved.

Before you carry out an investigation, you must identify the possible hazards. These can be grouped into biological hazards, chemical hazards and physical hazards.

Biological hazards include:	Chemical hazards can be grouped into:	Physical hazards include:
> microorganisms > body fluids > animals and plants.	> irritant and harmful > toxic > oxidising > corrosive > harmful to the environment.	> equipment > objects > radiation.

Scientists use an international series of symbols so that investigators can identify hazards.

Hazards pose risks to the person carrying out the investigation.

A risk posed by chlorine gas produced in the electrolysis of sodium chloride will be reduced if you devise a method to extract the gas from the process plant so that workers do not inhale it.

When you use hazardous materials, chemicals or equipment in the laboratory, you must use them in such a way as to keep the risks to absolute minimum. For example, one way is to wear eye protection when using hydrochloric acid.

Any action that you carry out to reduce the risk of a hazard happening is known as a 'control measure'.

> **Definition**
>
> A **hazard** is something that has the potential to cause harm. Even substances, organisms and equipment that we think of being harmless, used in the wrong way, may be hazardous.

Hazard symbols are used on chemical bottles so that hazards can be identified.

> **Definition**
>
> The **risk** is the likelihood of a hazard causing harm in the circumstances it's being used in.

> **Assessment tip**
>
> When assessing risk and suggesting control measures, these should be specific to the hazard and risk, and not general. Hydrochloric acid is dangerous as it is 'corrosive – skin and eye contact should be avoided' will be given credit, but 'wear eye protection' is too vague.

✹ Risk assessment

Before you begin an investigation, you must carry out a risk assessment. Your risk assessment must include:

✔ all relevant hazards (use the correct terms to describe each hazard, and make sure you include them all, even if you think they will pose minimal risk)

✔ risks associated with these hazards

✔ ways in which the risks can be minimised

✔ results of research into emergency procedures that you may have to take if something goes wrong.

You should also consider what to do at the end of the practical. For example, used agar plates should be left for a technician to sterilise; solutions of heavy metals should be collected in a bottle and disposed of safely.

Assessment tip

To make sure that your risk assessment is full and appropriate:

> Remember that, for a risk assessment for a chemical reaction, the risk assessment should be carried out for the products and the reactants.

> When using chemicals, make sure the hazard and ways of minimising risk match the concentration of the chemical you are using; many acids, for instance, while being corrosive in higher concentrations, are harmful or irritant at low concentrations.

✹ Collecting primary data

✔ You should make sure that any observations made are recorded in sufficient detail. For example, it is worth recording the appearance of a precipitate when making an insoluble salt, in addition to any other measurements that you make, such as its colour.

✔ Measurements should be recorded in tables. Have one ready so that you can record your readings as you carry out the practical work.

✔ Think about the dependent variable and define this carefully in your column headings.

✔ You should make sure that the table headings describe properly the type of measurements that you have made, for example 'time taken for magnesium ribbon to dissolve'.

✔ It is also essential that you include units – your results are meaningless without these.

✔ The units should appear in the column head, and not be repeated in each row of the table.

Definition

When you carry out an investigation, the data you collect are called **primary data.** The term 'data' is normally used to include your observations as well as measurements you might make.

Definition

One set of results from your investigation may not reflect what truly happens. Carrying out repeats enables you to identify any results that do not fit. These are called **outliers** or **anomalous results**.

Definition

If, when you carry out the same experiment several times, and get the same, or very similar results, the results are **repeatable**.

✹ Repeatability and reproducibility of results

When making measurements, in most instances, it is essential that you carry out repeats.

These repeats are one way of checking your results.

Results will not be repeatable of course, if you allow the conditions the investigation is carried out in to change.

You need to make sure that you carry out sufficient repeats, but not too many. In a titration, for example, if you obtain two values that are within 0.1 cm³ of each other, carrying out any more will not improve the reliability of your results.

This is particularly important when scientists are carrying out scientific research and make new discoveries.

Definition

Taking more than one set of results will improve the **reliability** of your data.

Definition

The **reproducibility** of data is the ability of the results of an investigation to be reproduced by:

> using a different method and reaching the same conclusion

> someone else, who may be in a different lab, carrying out the same work.

Processing data

Calculating the mean

Using your repeat measurements you can calculate the arithmetical mean (or just 'mean') of these data. Often, the mean is called the 'average.'

Here are the results of an investigation into the energy requirements of three different mp3 players. The students measured the energy using a joulemeter for ten seconds.

mp3 player	Energy used in joules (J)			
	trial 1	trial 2	trial 3	mean
Viking	5.5	5.3	5.7	5.5
Anglo	4.5	4.6	4.9	4.7
Saxon	3.2	4.5	4.7	4.6

Significant figures

When calculating the mean, you should be aware of significant figures.

For example, for the set of data below:

18	13	17	15	14	16	15	14	13	18

The total for the data set is 153, and ten measurements have been made. The mean is 15, and not 15.3.

This is because each of the recorded values has two significant figures. The answer must therefore have two significant figures. An answer cannot have more significant figures than the number being multiplied or divided.

Using your data

When calculating means (and displaying data), you should be careful to look out for any data that do not fit in with the general pattern. These are called anomalous data.

It might be the consequence of an error made in measurement, but sometimes it may be a genuine result. If you think that an anomalous result has been introduced by careless practical work, you should ignore it when calculating the mean. To check, the measurements may be repeated (if this is possible). You should examine possible reasons carefully before just ignoring data.

Definition

The **mean** is calculated by adding together all the measurements, and dividing by the number of measurements.

Assessment tip

You may also be required to use equations when processing data.

Sometimes, you will need to rearrange an equation in order to make the calculation you need. Practise using and rearranging equations as part of your preparation for assessment.

Definition

Significant figures are the number of digits in a number based on the precision of your measurements.

Displaying your data

Displaying your data – usually the means – makes it easy to pick out and show any patterns. It also helps you to pick out any anomalous data.

It is likely that you will have recorded your results in tables, and you could also use additional tables to summarise your results. The most usual way of displaying data is to use graphs. The table will help you decide which type to use.

Type of graph	When you would use the graph	Example
bar chart or bar graph	where one of the variables is categoric	'the energy requirements of different mp3 players'
line graph	where independent and dependent variables are both continuous	'the volume of carbon dioxide produced by a range of different concentrations of hydrochloric acid'
scatter graph	to show an association between two (or more) variables	'the association between length and breadth of a number of privet leaves'

It should be possible, from the data, to either join the points of a line graph using a single straight line, or using a curve. In scatter graphs, the points are plotted, but not usually joined. In this way, graphs can also help you to understand the relationship between the independent variable and the dependent variable.

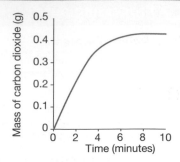

You can calculate the rate of production of carbon dioxide from the gradient of the graph.

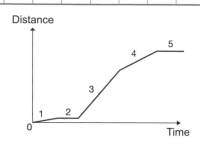

You can calculate the speed of the object from the gradient of the graph.

Assessment tip

Remember, when drawing graphs, plot the independent variable on the x-axis, and the dependent variable on the y-axis.

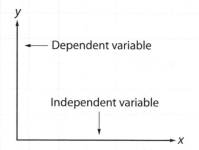

Conclusions from differences in data sets

When comparing two (or more) sets of data, you can often compare the values of two sets of means.

Example 5

Investigation: Comparing the braking distance of two tyres

Two groups of students compared the braking distance of two tyres, labelled A and B. Their results are shown in the table.

Tyre	Braking distance in metres (m)										Mean (m)
	1	2	3	4	5	6	7	8	9	10	
A	15	13	17	15	14	16	15	14	13	18	15
B	25	23	24	23	26	27	25	24	23	22	24

When the means are compared, it appears that tyre A will bring a vehicle to a stop in less distance than tyre B. The difference might have resulted from some other factor or could be purely by chance.

Scientists use statistics to find the probability of any differences having occurred by chance. The lower this probability is, which is found out by statistical calculations, the more likely it is that tyre A is better at stopping a vehicle than tyre B.

Statistical analysis can help to increase the confidence you have in your conclusions.

Definition

If there is a relationship between dependent and independent variables that can be defined, there is a **correlation** between the variables.

 Drawing conclusions

Observing trends in data or graphs will help you to draw conclusions. You may obtain a linear relationship between two sets of variables, or the relationship might be more complex.

Example 6

Conclusion: The higher the concentration of acid, the shorter the time taken for the magnesium ribbon to dissolve.

Conclusion: The higher the concentration of acid, the faster the rate of reaction.

When drawing conclusions, you should try to relate your findings to the science involved.

> In the first point in Example 6, your discussion should focus on the greater possibility/increased frequency of collisions between reacting particles as the concentration of the acid is increased.

> In the second point in Example 6, there is a clear scientific mechanism to link the rate of reaction to the concentration of acid.

Sometimes, you can see correlations between data which are coincidental, where the independent variable is not the cause of the trend in the data.

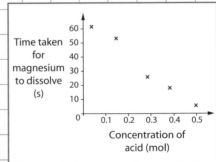

This graph shows **negative correlation.**

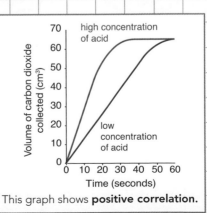

This graph shows **positive correlation.**

Evaluating your investigation

Your conclusion will be based on your findings, but must take into consideration any uncertainty in these introduced by any possible sources of error. You should discuss where these have come from in your evaluation.

The two types of errors are:

✔ random error
✔ systematic error.

This can occur when the instrument that you are using to measure lacks sufficient sensitivity to indicate differences in readings. It can also occur when it is difficult to make a measurement. If two investigators measure the height of a plant, for example, they might choose different points on the compost, and the tip of the growing point to make their measurements.

They are either consistently too high or too low. One reason could be down to the way you are making a reading, for example taking a burette reading at the wrong point on the meniscus. Another could be the result of an instrument being incorrectly calibrated, or not being calibrated.

The volume of liquid in a burette must be read to the bottom of the meniscus.

Definition

Error is a difference between a measurement you make and its true value.

Definition

With **random error**, measurements vary in an unpredictable way.

Definition

With **systematic error**, readings vary in a controlled way.

Assessment tip

A pH meter must be calibrated before use using buffers of known pH.

Assessment tip

Make sure that you relate your conclusions to the hypothesis you are investigating. Do the results confirm or reject the hypothesis. Quote some results to back up your statement. For example, 'My results at 35 °C and 65 °C show that over a 30 °C change in temperature the time taken to produce 50 cm^3 of carbon dioxide halved'

Accuracy and precision

When evaluating your investigation, you should mention accuracy and precision. If you use these terms, it is important that you understand what they mean, and that you use them correctly.

Precise but not accurate.

Precise and accurate.

Not precise and not accurate.

The terms accuracy and precision can be illustrated using shots at a dartboard.

Definition

When making measurements:
> the **accuracy** of the measurement is how close it is to the true value
> **precision** is how closely a series of measurements agree with each other.

✹ Improving your investigation

When evaluating your investigation, you should discuss how your investigation could be improved. This could be by improving:

✔ the reliability of your data. For example, you could make more repeats, or more frequent readings, or 'fine-tune' the range you chose to investigate, or refine your technique in some other way.

✔ the accuracy and precision of your data, by using measuring equipment with a higher resolution.

In science, the measurements that you make as part of your investigation should be as precise as you can, or need to, make them. To achieve this, you should use:

✔ the most appropriate measuring instrument

✔ the measuring instrument with the most appropriate size of divisions.

The smaller the divisions you work with, the more precise your measurements. For example:

✔ In an investigation on how your heart rate is affected by exercise, you might decide to investigate this after a 100 m run. You might measure out the 100 m distance using a trundle wheel, which is sufficiently precise for your investigation.

✔ In an investigation on how light intensity is affected by distance, you would make your measurements of distance using a metre rule with millimetre divisions; clearly a trundle wheel would be too imprecise.

✔ In an investigation on plant growth, in which you measure the thickness of a plant stem, you would use a micrometer or Vernier callipers. In this instance, a metre rule would be too imprecise.

✹ Using secondary data

As part of controlled assessment, you will be expected to compare your data – primary data – with **secondary data** that others have collected.

One of the simplest ways of doing this is to compare your data with other groups in your class who have carried out an identical practical investigation.

In your controlled assessment, you will be provided with a data sheet of secondary data.

It may be that not all the investigations provided are relevant: different independent or dependent variables may have been investigated. It may be that some of the data has errors or uncertainties.

You should identify and explain, with reasons, whether each set of results can be used to support the hypothesis, and also any features of the data that might allow the hypothesis to be further developed, such as 'the effect is reduced with slopes of angle greater than 25°'.

Remember to use some figures from the results to support your answers.

Definition

Secondary data are measurements/observations made by anyone other than you.

Assessment tip

In Section 2 of your ISA you will be required to critically evaluate this data to see how far it supports or does not support the hypothesis being investigated.

You should review secondary data and evaluate it. Scientific studies are sometimes influenced by the **bias** of the experimenter.

✔ One kind of bias is having a strong opinion related to the investigation, and perhaps selecting only the results that fit with a hypothesis or prediction.

✔ The bias could be unintentional. In fields of science that are not yet fully understood, experimenters may try to fit their findings to current knowledge and thinking.

There have been other instances where the 'findings' of experimenters have been influenced by organisations that supplied the funding for the research.

You must fully reference any secondary data that you have used, using one of the accepted referencing methods.

✳ Referencing methods

The two main conventions for writing a reference are the:

✔ Harvard system
✔ Vancouver system.

In your text, the Harvard system refers to the authors of the reference, for example 'Smith and Jones (1978)'.

The Vancouver system refers to the number of the numbered reference in your text, for example '... the reason for this hypothesis is unknown[1]'.

Though the Harvard system is usually preferred by scientists, it is more straightforward for you to use the Vancouver system.

Harvard system

In your references list a book reference should be written:

> Author(s) (year of publication). *Title of Book*, publisher, publisher location.

The references are listed in alphabetical order according to the authors.

Vancouver system

In your references list a book reference should be written:

> 1 Author(s). *Title of Book.* Publisher, publisher location: year of publication.

The references are numbered in the order in which they are cited in the text.

Assessment tip

Remember to write out the URL of a website in full. You should also quote the date when you looked at the website.

Do the data support your hypothesis?

You need to discuss, in detail, whether all, or some, of the primary data that you have collected, and the secondary data, support your original hypothesis. They may or may not.

You should communicate your points clearly, using the appropriate scientific terms.

If your data do not match your hypothesis completely, it may be possible to modify the hypothesis or suggest an alternative one.

You may be asked to suggest further investigations that can be carried out to support your original hypothesis or the modified version.

It is important to remember, however, that if your investigation does support your hypothesis, it can improve the confidence you have in your conclusions and scientific explanations, but it cannot prove that your explanations are correct. One result can never prove a theory, but it can disprove it.

Your controlled assessment

The **Controlled Assessment** is worth 25% of the marks for your GCSE Science. It is worth doing it well!

Controlled Assessment is a two-part (sections 1 and 2) *Investigative Skills Assignment* (ISA) test. Before you start, you will discuss a problem in a given context. For example, how does a bungee jumping company know how much rope to use to avoid jumpers hitting the ground?

> You will then research variables that affect the stretch of a bungee rope, and devise a method to test the stretch of a bungee rope.

> You will also suggest a hypothesis about the stretch of a bungee rope as the result of your research.

You are allowed to make brief notes about your research on one side of A4. You can use the notes to help answer Sections 1 and 2 of the ISA paper.

Section 1 (45 minutes, 20 marks) consists of questions relating to your research. This includes your planned investigation. Then you will carry out your investigation and record and analyse your results.

Section 2 of the ISA test (50 minutes, 30 marks) consists of questions related to your investigation. You will be given a sheet of secondary data by AQA. You should use it to select data to analyse and compare with the hypothesis. Finally, you will be asked to suggest how ideas from your investigation and research could be used in a specific context.

Assessment tip

Write your plan clearly, using the appropriate scientific terms, and checking carefully your use of spelling, punctuation and grammar. You will be assessed on this written communication as well as your science.

How to be successful in your GCSE Additional Science assessment

Introduction

AQA uses assessments to test how good your understanding of scientific ideas is, how well you can apply your understanding to new situations and how well you can analyse and interpret information you have been given. The assessments are opportunities to show how well you can do these.

To be successful in exams you need to:

✔ have a good knowledge and understanding of science
✔ be able to apply this knowledge and understanding to familiar and new
✔ situations
✔ be able to interpret and evaluate evidence that you have just been given.

You need to be able to do these things under exam conditions.

✳ The language of the assessment paper

When working through an assessment paper, make sure that you:

✔ re-read a question enough times until you understand exactly what the examiner is looking for
✔ make sure that you highlight key words in a question. In some instances, you will be given key words to include in your answer.
✔ look at how many marks are allocated for each part of a question. In general, you need to write at least as many separate points in your answer as there are marks.

✳ What verbs are used in the question?

A good technique is to see which verbs are used in the wording of the question and to use these to gauge the type of response you need to give. The table lists some of the common verbs found in questions, the types of responses expected and then gives an example.

Verb used in question	Response expected in answer	Example question
write down state give identify	these are usually more straightforward types of question in which you are asked to give a definition, make a list of examples, or the best answer from a series of options	'Write down three types of microorganism that cause disease.' 'State one difference and one similarity between radio waves and gamma rays.'
calculate	use maths to solve a numerical problem	'Calculate the percentage of carbon in copper carbonate ($CuCO_3$).'

estimate	use maths to solve a numerical problem, but you do not have to work out the exact answer	'Estimate from the graph the speed of the vehicle after 3 minutes.'
describe	use words (or diagrams) to show the characteristics, properties or features of, or build an image of something	'Describe how meiosis halves the number of chromosomes in a cell to make egg or sperm cells.'
suggest	usually in a new or unfamiliar context	'Suggest why tyres with different tread patterns will have different braking distances.'
demonstrate/ show how	use words to make something evident using reasoning	'Show how temperature can affect the rate of a chemical reaction.'
compare	look for similarities and differences	'Compare aerobic and anaerobic respiration.'
explain	to offer a reason for, or make understandable, information that you are given	'Explain why alpha and beta radiations can be deflected by a magnetic field, but gamma rays are not.'
evaluate	discuss different points of view or opinions, then decide which is best, usually giving a reason	'Evaluate the benefits of using a circuit breaker instead of a fuse in an electrical circuit.'

What is the style of the question?

Try to get used to answering questions that have been written in lots of different styles before you sit the exam. Work through past papers, or specimen papers, to get a feel for these. The types of questions in your assessment fit the three assessment objectives shown in the table.

Assessment objective	Your answer should show that you can...
AO1 recall the science	Recall, select and communicate your knowledge and understanding of science.
AO2 apply your knowledge	Apply skills, knowledge and understanding of science in practical and other contexts.
AO3 evaluate and analyse the evidence	Analyse and evaluate evidence, make reasoned judgements and draw conclusions based on evidence.

Assessment tip

Of course you must revise the subject material adequately. It is as important that you are familiar with the different question styles used in the exam paper, as well as the question content.

How to answer questions on: AO1 Recall the science

These questions, or parts of questions, test your ability to recall your knowledge of a topic. There are several types of this style of question:

✔ Fill in the spaces (you may be given words to choose from)
✔ Tick the correct statements
✔ Use lines to link a term with its definition or correct statement
✔ Add labels to a diagram
✔ Complete a table
✔ Describe a process

Example 7

a What is meant by the term *exothermic reaction*?
Tick (✓) **one** box.

☐ a reaction that gives out energy, heating the surroundings
☐ a reaction that takes in energy from the surroundings
☐ a reaction that can go in either direction.

How to answer questions on: AO1 Recall the science in practical techniques

You may be asked to recall how to carry out certain practical techniques, either ones that you have carried out before, or techniques that scientists use.

To revise for these types of questions, make sure that you have learned definitions and scientific terms. Produce a glossary of these, or key facts cards, to make them easier to remember. Make sure that your key facts cards also cover important practical techniques, including equipment, where appropriate.

Example 8

Describe how to find the work done when an object of 15 newtons is moved 1 metre.

Assessment tip

Don't forget that mind maps – either drawn by you or by using a computer program – are very helpful when revising key points.

How to answer questions on: AO2 Apply skills, knowledge and understanding

Some questions require you to apply basic knowledge and understanding in your answers.

You may be presented with a topic that is familiar to you, but you should also expect questions in your Science exam to be set in an unfamiliar context.

Questions may be presented as:

✔ practical investigations
✔ data for you to interpret
✔ a short paragraph or article.

The information required for you to answer the question might be in the question itself, but for later stages of the question, you may be asked to draw on your knowledge and understanding of the subject material in the question.

Practice will help you to become familiar with contexts that examiners use and question styles. However, you will not be able to predict many of the contexts used. This is deliberate; being able to apply your knowledge and understanding to different and unfamiliar situations is a skill the examiner tests.

Practise doing questions where you are tested on being able to apply your scientific knowledge and your ability to understand new situations that may not be familiar. In this way, when this type of question comes up in your exam, you will be able to tackle it successfully.

Assessment tip

Work through the Preparing for assessment: Applying your knowledge tasks in this book as practice.

Example 9

The force of gravity acts on all objects falling towards the Earth. Two identical packages are dropped out of an aircraft at the same time. Both have parachutes. One package's parachute opens, the other does not.

Suggest which package will hit the ground first. Explain your answer.

✳ How to answer questions on: AO2 Apply skills, knowledge and understanding in practical investigations

Some opportunities to demonstrate your application of skills, knowledge and understanding will be based on practical investigations. You may have carried out some of these investigations, but others will be new to you and based on data obtained by scientists. You will be expected to describe patterns in data from graphs you are given or that you will have to draw from given data.

Again, you will have to apply your scientific knowledge and understanding to answer the question.

Example 10

Look at the graph showing the volume of gas collected when 10 g of calcium carbonate is reacted with three different concentrations of hydrochloric acid.

a What is the maximum volume of gas that can be produced using 1 mol/dm³ of hydrochloric acid?

b Explain why this volume of gas is produced quicker when using 2 mol/dm³ of hydrochloric acid.

c Suggest why 0.5 mol/dm³ of hydrochloric acid does not produce this volume of gas.

You will also need to analyse scientific evidence or data given to you in the question. It is likely that you will not be familiar with the material.

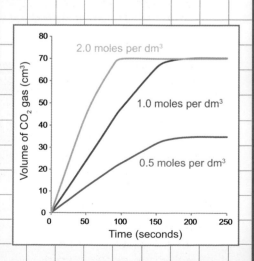

Analysing data may involve drawing graphs and interpreting them, and carrying out calculations. Practise drawing and interpreting graphs from data.

When drawing a graph, make sure that you:

✔ choose and label the axes fully and correctly

✔ include units, if this has not been done already

✔ plot points on the graph carefully – the examiner will check individual points to make sure that they are accurate

✔ join the points correctly; usually this will be by a line of best fit.

When reading values off a graph that you have drawn or one given in the question, make sure that you:

✔ do it carefully, reading the values as accurately as you can

✔ double-check the values.

When describing patterns and trends in the data, make sure that you:

✔ write about a pattern or trend in as much detail as you can

✔ mention anomalies where appropriate

✔ recognise there may be one general trend in the graph, where the variables show positive or negative correlation

✔ recognise the data may show a more complex relationship. The graph may demonstrate different trends in several sections. You should describe what is happening in each.

✔ describe fully what the data show.

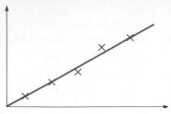

What type of line is drawn on this graph?

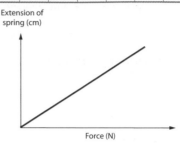

Make sure you know what type of relationship is shown in this graph.

✳ How to answer questions needing calculations

✔ The calculations that you are asked to do may be straightforward, for example the calculation of the mean from a set of data.

✔ They may be more complex, for example calculating the yield of a chemical reaction.

✔ Other questions will require the use of formulae.

You will be given an equation sheet with the question paper.

On page 299, there is a list of the maths skills that you will need. Remember, these are the same skills that you have learned in maths lessons.

Example 11

A parachutist with a mass of 75 kg is taken up 3000 metres in an aircraft. Calculate the parachutist's gravitational potential energy on leaving the plane. Assume gravitational field strength is 9.8 N/kg.

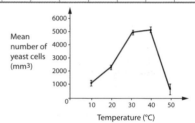

What type of relationship does this graph show?

Assessment tip

When completing your calculation, make sure that you include the correct units.

Assessment tip

Check the specification, or with your teacher, to make sure that you know the formulae that you have to learn and remember.

Assessment tip

Remember, when carrying out any calculations, you should include your working at each stage. You may get credit for getting the process correct, even if your final answer is wrong.

How to answer questions on: AO3 Analysing and evaluating evidence

For these types of questions, in addition to analysing data, you must also be able to evaluate information that you are given. This is one of the hardest skills. Think about the validity of the scientific data: did the technique(s) used in any practical investigation allow the collection of accurate and precise data?

Your critical evaluation of scientific data in class, along with the practical work and controlled assessment work, will help you to develop the evaluation skills required for these types of questions.

> **Example 12**
>
> Explain why it is cheaper to produce chlorine gas by the electrolysis of a sodium chloride solution, rather than the electrolysis of molten sodium chloride.

You may be expected to compare data with other data, or come to a conclusion about its reliability, its usefulness or its implications. Again, it is possible that you will not be familiar with the context. You may be asked to make a judgement about the evidence or to give an opinion with reasons.

> **Example 13**
>
> When investigating the extension of a spring using different forces, explain why overloading the spring may make it hard to achieve repeatable results.

Assessment tip

Work through the Preparing for assessment: Analysing and evaluating data tasks, in this book, as practice.

Assessment tip

Wherever possible, use as much data as you can in your answer, particularly when explaining trends or conclusions, so you can gain full marks. Try to use numbers and values rather than just trends in data or graphs. 'At 45 °C…' is always better than 'as the temperature rises it gets greater'.

The quality of your written communication

Scientists need good communication skills to present and discuss their findings. You will be expected to demonstrate these skills in the exam. You will be assessed in the longer-response exam questions that you answer. These questions are clearly indicated in each question paper. The quality of your written communication will also be assessed in your controlled assessment.

You will not be able to obtain full marks unless you:

✔ make sure that the text you write is legible
✔ make sure that spelling, punctuation and grammar are accurate so that the meaning of what you write is clear
✔ use a form and style of writing appropriate for its purpose and for the complexity of the subject matter
✔ organise information clearly and coherently
✔ use the scientific language correctly.

You will also need to remember the writing and communication skills that you have developed in English lessons. For example, make sure that you understand how to construct a good sentence using connectives.

Assessment tip

You will be assessed on the way in which you communicate science ideas.

Assessment tip

When answering questions, you must make sure that your writing is legible. An examiner cannot award marks for answers that he or she cannot read.

Revising for your Additional Science exam

You should revise in the way that suits you best. It is important that you plan your revision carefully, and it is best to start well before the date of the exams. Take the time to prepare a revision timetable and try to stick to it. Use this during the lead up to the exams and between each exam.

When revising:

✔ Find a quiet and comfortable space in the house where you will not be disturbed. It is best if it is well ventilated and has plenty of light.

✔ Take regular breaks. Some evidence suggests that revision is most effective when you revise in 30 to 40 minute slots. If you get bogged down at any point, take a break and go back to it later when you're feeling fresh. Try not to revise when you are feeling tired. If you do feel tired, take a break.

✔ Use your school notes, textbook and possibly a revision guide. But also make sure that you spend some time using past papers to familiarise yourself with the exam format.

✔ Produce summaries of each topic.

✔ Draw mind maps covering the key information on a topic.

✔ Set up revision cards containing condensed versions of your notes.

✔ Ask yourself questions, and try to predict questions, as you are revising a topic.

✔ Test yourself as you go along. Try to draw key labelled diagrams, and try some questions under timed conditions.

✔ Prioritise your revision of topics. You might want to allocate more time to revising the topics you find most difficult.

Assessment tip

Try to make your revision timetable as specific as possible – don't just say 'science on Monday, and Thursday', but list the topics that you will cover on those days.

Assessment tip

Start your revision well before the date of the exams, produce a revision timetable, and use the revision strategies that suit your style of learning. Above all, revision should be an active process.

How do I use my time effectively in the exam?

Timing is important when you sit an exam. Do not spend so long on some questions that you leave insufficient time to answer others. For example, in a 60-mark question paper, lasting one hour, you will have, on average, one minute per question.

If you are unsure about certain questions, complete the ones you are able to do first, then go back to the ones you're less sure of.

If you have time, go back and check your answers at the end of the exam.

On exam day...

A little bit of nervousness before your exam can be a good thing, but try not to let it affect your performance in the exam. When you turn over the exam paper keep calm. Look at the paper and get it clear in your head exactly what is required from each question. Read each question carefully. Do not rush.

If you read a question and think that you have not covered the topic, keep calm – it could be that the information needed to answer the question is in the question itself or the examiner may be asking you to apply your knowledge to a new situation.

Finally, good luck!

✹ Mathematical skills

You will be allowed to use a calculator in all assessments.

These are the maths skills that you need, to complete all the assessments successfully.

You should understand:

✔ the relationship between units, for example, between a gram, kilogram and tonne
✔ compound measures such as speed
✔ when and how to use estimation
✔ the symbols = < > ~
✔ direct proportion and simple ratios
✔ the idea of probability.

You should be able to:

✔ give answers to an appropriate number of significant figures
✔ substitute values into formulae and equations using appropriate units
✔ select suitable scales for the axes of graphs

✔ plot and draw line graphs, bar charts, pie charts, scatter graphs and histograms
✔ extract and interpret information from charts, graphs and tables.

You should be able to calculate:

✔ using decimals, fractions, percentages and number powers, such as 10^3
✔ arithmetic means
✔ areas, perimeters and volumes of simple shapes

In addition, if you are a higher tier candidate, you should be able to:

✔ **change the subject of an equation**

and should be able to use:

✔ **numbers written in standard form**
✔ **calculations involving negative powers, such as 10^{-1}**
✔ **inverse proportion**
✔ **percentiles and deciles.**

✹ Some key physics equations

With the written papers, there will be an equation sheet. In order to make best use of the sheet, it will help if you practise using the following equations.

Equation	Meaning of symbol and its unit	Equation	Meaning of symbol and its unit
$F = m \times a$ or $a = \dfrac{F}{a}$	F is resultant force in newtons, N m is mass in kilograms, kg a is acceleration in metres per second squared, m/s²	$E_k = \dfrac{1}{2} \times m \times v^2$	E_k is kinetic energy in joules, J m is mass in kilograms, kg v is speed in metres per second, m/s
$a = \dfrac{v - u}{t}$	a is acceleration in metres per second squared, m/s² v is final velocity in metres per second, m/s u is initial velocity in metres per second, m/s t is time taken in seconds, s	$p = m \times v$	p is momentum in kilogram metre per second, kg m/s m is mass in kilograms, kg v is velocity in metres per second, m/s
$W = m \times g$	W is weight in newtons, N m is mass in kilograms, kg g is gravitational field strength in newtons per kilogram, N/kg	$I = \dfrac{Q}{t}$	I is current in amperes (amps), A Q is charge in coulombs, C t is time in seconds, s
$F = k \times e$	F is force in newtons, N k is spring constant in newtons per metre, N/m e is extension in metres, m	$v = \dfrac{W}{Q}$	V is potential difference in volts, V W is work done in joules, J Q is charge in coulombs, C
$W = F \times d$	W is work done in joules, J F is force applied in newtons, N d is distance moved in the direction of the force in metres, m	$V = I \times R$	V is potential difference in volts, V I is current in amperes (amps), A R is resistance in ohms, Ω
$P = \dfrac{E}{t}$	P is power in watts, W E is energy transferred in joules, J t is time taken in seconds, s	$P = I \times V$	P is power in watts, W I is current in amperes (amps), A V is potential difference in volts, V
$E_p = m \times g \times h$	E_p is change in gravitational potential energy in joules, J m is mass in kilograms, kg g is gravitational field strength in newtons per kilogram, N/kg h is change in height in metres, m	$E = V \times Q$	E is energy in joules, J V is potential difference in volts, V Q is charge in coulombs, C

Bad Science for Schools

When the evidence doesn't add up

Sometimes people use what sound like scientific words and ideas to sell you things or persuade you to think in a certain way. Some of these claims are valid, and some are not. The activities on these pages are based on the work of Dr Ben Goldacre and will help you to question some of the scientific claims you meet. Read more about the work of Ben at www.badscience.net.

Brown goo

You may have seen adverts for a foot spa that can remove toxins from your body. They are sometimes used in beauty salons or you might even buy one to use at home. The basin is filled with water, a sachet of special salts is added and then it is plugged in. You put your feet in to soak and the water turns brown!

It looks impressive, but is that because toxins have left your body through your feet?

Now, the advertisers of these products would tell us that we are being 'detoxed' and that horrible chemicals, toxins, which have accumulated in our bodies are at long last being released. It's perhaps not surprising that people are keen to be cleansed. However the talk doesn't match the facts. The chemicals in the water didn't come from your body which (as you know) is quite capable of getting rid of substances it doesn't need without using special equipment.

We are learning to:

> use primary and secondary evidence to investigate scientific claims

> apply scientific concepts to evaluate 'health products'

> explore the implications of these evaluations

✴ CAN YOU DETOX VIA YOUR FEET?

Read the leaflet – it sounds scientific but is it? Think about what you have learnt in science.

> Human metabolism is complex with the 'building blocks' of molecules being reshaped into new arrangements. The same molecule can be a waste product or a valued ingredient, depending on when and where it is in the body. There is no such thing as a 'detox system' in any medical textbook. Sometimes the body does need to dispose of waste but it does so by well-known ways.

> Electrolysis occurs when a direct electric current is passed through a liquid containing mobile ions, resulting in a chemical reaction at the electrodes.

Can you come up with a hypothesis about what's going on? How would you prove it? Ben came up with a good idea and gave his Barbie™ a foot bath – you might get a chance to replicate his experiment. Can you predict what might happen?

What would happen if you gave this toy a detoxifying foot spa?

Collins Detox Foot Bath

Before

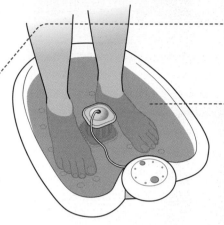

After 30 minutes

This looks like a serious piece of equipment.

This brown water looks horrible but is it brown because of toxins from the body?

This explanation sounds scientific, but is it?

The patented Collins Detox Foot Bath stimulates the active release of tingling ions that surge back and forth around your feet generating a flow of both negative and positive energy. This refreshes and renews the tissues, cleansing your body of accumulated toxins, readjusting the balance of energy at a bio-cellular level and removing excretory residues.

The centrally located micro-voltaic electrodes cause the flow of bi-polar ions producing an energy field that carries essential nutrients and life-giving oxygen. The release of toxins takes places through the myriad of microscopic pores in the soles of your feet. Graduated colour changes in the water present conclusive evidence of the beneficial effects.

The many enthusiastic users report a range of exhilarating effects including a heightened sense of awareness, improved circulation and relief of arthritic pain. The results are personal to each user as their toxin levels and combinations vary, but all report positive outcomes. One recent example of enthusiastic feedback said "The colour of the water shocked me in the realisation of what had accumulated in my body but the lightness I felt lasted for days!"

The people who tested it were impressed but did they enjoy the effects of detox or a relaxing foot bath?

✳ DETOX SELLS!

Words like 'toxins' and 'detoxification' (the removal of toxins) are sometimes used to promote products and techniques. Nobody likes to think of toxins accumulating in their body but we must consider whether there's any scientific basis for these ideas.

> Can you think of other products that claim to 'detox'?

> Why do you think that 'detoxing' can be used to sell these products?

> These treatments could all be said to be a little theatrical. How does this help to convince people that they're effective?

Bad Science for Schools

When the evidence doesn't add up

Sometimes people use what sound like scientific words and ideas to sell you things or persuade you to think in a certain way. Some of these claims are valid, and some are not. The activities on these pages are based on the work of Dr Ben Goldacre and will help you to question some of the scientific claims you meet. Read more about the work of Ben at www.badscience.net.

Bad news

In science you learn about ideas that scientists have developed by collecting evidence from experiments; you are also learning to collect and evaluate evidence yourself. You can use this outside of the laboratory to weigh up information you come across every day. Let's look at this example about how data can be used to support a story for a newspaper.

When data are produced you might think that there's only one way they can be used, and only one meaning that can be supported. This isn't always true.

We are learning to:

> understand how data can be used to make a good news story

> understand how science reports may be distorted to make headlines

> consider why science reports may be represented in various different ways

✱ GOOD ADVICE?

If a woman wants to be sexually active but doesn't want to get pregnant, one of the contraceptive methods available to her is to use a contraceptive implant. There are a number of factors to take into account; one of the most important ones is, of course, 'how well does it work?' Think about the headline on the right. What kind of questions might you ask that would reveal whether this method is, in fact, a failure?

600 pregnancies despite contraceptive implant

✱ STICKING TO THE NUMBERS

One of the questions we might want to consider is 'over what timescale?' Is this 600 over the last month, last year or since records began? In fact, the contraceptive implant had been available for ten years when this data was released, so it's 60 unintended pregnancies per year, on average. Still not ideal, but maybe not as disastrous as at first thought.

We might also want to know how widespread the use of the implant was. If the 60 pregnancies a year was out of say, 1,000 people, then that's not very good: it would mean that 6 out of every hundred women with an implant had got pregnant over a year.

If it was out of 100,000 then that means 6 out of every 10,000 women got pregnant over a year, so this method of contraception would compare well with other methods.

In fact, around 1.3 million implants have been used over the last ten years, and each lasts for three years. This works out as 1.4 unwanted pregnancies for every 10,000 women using the method per year if we assume that each implant lasts for the full 3 years.

Making the headlines

Four students are talking about this story.

Jo says

I think the journalists were doing a good job here to tell people about the fact that 600 women who thought that they couldn't get pregnant, then did. They got hold of the facts and then reported them.

Adam says

The journalists didn't write this up very well. Most of the people reading this story would be women who would be wondering if this method of contraception was one that they should use. The headline suggests that it's not safe and it is. Well, most of the time.

Will says

Journalists have to be responsible. If this story frightens women off one of the safest methods of contraception they've let people down.

Emma says

The main job of journalists is to be entertaining. Boring stories don't get read. '600 women using contraceptive get pregnant' makes you read the story. '0.014% of women using contraceptive get pregnant' looks boring.

- Look at these comments. Who do you think is right?

- Do you think the main purpose of a journalist is:
  - To be informative, even if it's sometimes boring?
  - To be engaging, even if it may sometimes give a false impression?

- If you had been the journalist assigned to this story, what headline would you have used?

✳ NUMBERS IN THE REAL WORLD

A useful way of presenting data like this is to use what's called the natural frequency. Out of a set number, this indicates how many will have a changed outcome as a result of this. In this case it's 1.4 out of 10,000. The figure of 600 isn't wrong, neither is the 0.014%, but 1.4 in 10,000 puts it in a simple form that people can make sense of and use to assess the likely impact on them.

When the evidence doesn't add up

Sometimes people use what sound like scientific words and ideas to sell you things or persuade you to think in a certain way. Some of these claims are valid, and some are not. The activities on these pages are based on the work of Dr Ben Goldacre and will help you to question some of the scientific claims you meet. Read more about the work of Ben at www.badscience.net.

MMR – don't die of ignorance

Autism is a condition which affects between one and two people in every thousand, affecting neural development and causing restricted and repetitive behaviour. It affects social behaviour and language. It is usually diagnosed from the age of three onwards.

In Britain, as in many countries, the majority children are vaccinated against measles, mumps and rubella using a combined vaccine (MMR) between the ages of one and two. In 1998 a British doctor wrote a report on twelve children who had been vaccinated with the MMR vaccine and were subsequently diagnosed as autistic. The result of this report was that media interest was raised, many anti-MMR stories appeared and there was a significant fall in the number of children who were given the MMR vaccine.

Consider these questions:

> Does the fact that the children in the report were diagnosed with autism after being given the MMR vaccine prove that the vaccine caused the autism?

> At the time of the report being written well over 90% of children had the MMR vaccine. Why should it not be a surprise if some of those children are diagnosed with autism?

> What kind of survey would have helped to identify whether the MMR vaccine caused autism?

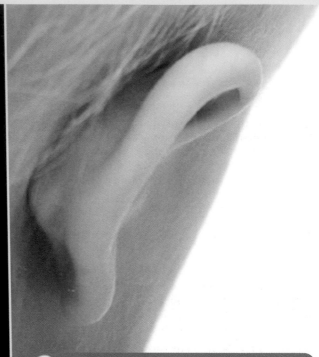

✳ THE RISE OF MEASLES

As the number of MMR vaccinations fell, the number of measles cases rose. Measles is a very dangerous disease that even in developed countries kills one in every 3000 people and causes pneumonia in one in 20.

It was subsequently established beyond reasonable doubt that there is no causal link between MMR vaccination and autism. The doctor had a commercial interest in the alleged link and was subsequently struck off. The scare affected no other countries. MMR vaccination rates in Britain are rising again. Doctors are still not sure why some children develop autism; its causes are unknown.

Consider these points of view:

"The doctor who wrote the report was right to alert people to his concerns and suggest that more research should be carried out."

"The media got hold of the story and turned it into a huge scare. It's their fault."

"There was never any evidence to prove a link. Thousands of children have caught diseases that could otherwise have been avoided."

We are learning to:

> Understand the difference between correlation and cause and effect

> To apply this understanding to a variety of contexts

> To explore the professional dilemma facing scientists who have concerns and whose actions have serious consequences

✳ THE RISE OF MEASLES

One of the things this story illustrates is what can happen when you look at only a very small sample and the importance of working with large-scale surveys wherever possible. Such a study was carried out in Denmark: the Madsen study. Because Denmark tracks patients and the care they receive on a central system they have been able to study the correlation between vaccination and illness. The data clearly shows that there is no correlation between MMR vaccination and the incidence of autism.

The study was based on data from over half a million children: over 440,000 had been vaccinated and there was no greater incidence of autism in children vaccinated than in those not vaccinated.

> Identify the features of this study that make its findings reliable.

> What might you say to someone who still wasn't convinced by this study and decided to 'play it safe' by not having their child vaccinated for MMR?

Cause and effect?

Sometimes it looks like something causes something else, perhaps because they both happen at the same time. But scientists need to be very careful before saying that one thing causes another.

• Often you need to use common sense and extra information to help decide if there is true causation. For example, cocks crow in the morning, but nobody thinks that cocks crowing causes the sun to come up, because there's no conceivable mechanism for that, and it conflicts with everything we know about the sun and the earth. On the other hand, we can observe that when it gets warmer, people wear fewer clothes, and it seems reasonable to say that the warm weather causes people to wear less.

• Sometimes two things are correlated, but it's harder to say what causes what, and there might be a third factor causing both of the things that we are observing. Let's say, for example, that a study finds that there is a strong correlation between a child's IQ and their height: perhaps both height and IQ are themselves related, through a complex causal pathway, to something else, like general health, or diet, or social deprivation.

• Often, although things happen at the same time, there is no link at all. For example, Halley's Comet appears once every 76 years. Previous appearances have coincided with King Harold's defeat at the Battle of Hastings, Genghis Khan's invasion of Europe and both the birth and death of great American novelist Mark Twain, author of Tom Sawyer.

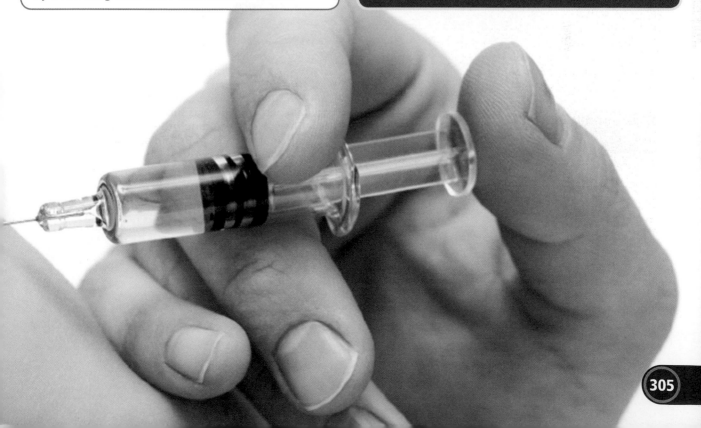

Glossary

acceleration rate at which an object speeds up, calculated from change in velocity divided by time

acid hydrogen compound that can produce hydrogen ions, H⁺

acidic solution aqueous solution with pH less than 7 – the concentration of $H^+(aq)$ is higher than that of $OH^-(aq)$

activation energy minimum energy needed to break bonds in reactant molecules to allow a reaction to occur

active site a depression in an enzyme molecule into which its substrate fits

actual yield mass of product actually obtained from a reaction

aerobic respiration a process in which energy is released from glucose, using oxygen

aerosol tiny particles of liquid or solid dispersed in the air

air resistance force resisting the movement of an object travelling through air

algae simple, plant-like organisms

alkali metals very reactive metals in Group 1 of the periodic table, for example, sodium

alkaline solution aqueous solution with pH more than 7 – the concentration of $OH^-(aq)$ is higher than that of $H^+(aq)$

alkali compound that contains hydroxide ions, OH^-

allele a particular form of a gene

alloy a mixture of two or more metals, with useful properties different from the individual metals

alpha particle particle emitted from the nuclei of radioactive atoms consisting of two protons and two neutrons

alternating current (a.c.) electric current where the direction of the flow of current constantly reverses, as in mains electricity

amino acids small molecules from which proteins are built

amplitude size of wave oscillations – for a mechanical wave, how far the particles vibrate around their central position

amylase an enzyme that breaks down starch molecules to maltose (sugar) molecules

anaerobic respiration process in which energy is released form glucose, without using oxygen

anode positive electrode

antibody protein normally present in the body or produced in response to an antigen which it neutralises, thus producing an immune response

aqueous solution substance dissolved in water

atomic number number of protons in the nucleus of an atom

average speed total distance of a journey divided by total time taken

background radiation low level nuclear radiation that exists everywhere, from rocks and other environmetal sources

bacterium (pl: bacteria) single-celled microorganisms; they do not have a nucleus

balanced equation equation where the number of atoms of each element is the same in the reactants as in the products

balanced forces system of forces with resultant force

base substance that neutralises an acid to form a salt and water

beta particle type of nuclear radiation emitted as an electron by a radioactive nucleus

bile a liquid produced by the liver, which flows into the small intestine where it neutralises the acidic juices from the stomach

biological detergent washing powder or liquid that contains enzymes

biosensor sensitive material that detects very low levels of chemical or biological agents in the surroundings

biotic factor something that influences a living organism and is caused by other living organisms, such as competition

braking distance distance a vehicle travels before stopping, once the brakes are applied

breathing movements that move air into and out of the lungs

buckminsterfullerene carbon molecule C_{60} – 60 carbon atoms arranged in the form of a hollow sphere

buckyballs common name for fullerenes, such as C_{60}, buckminsterfullerene

burette item of glassware used to measure the volume of liquid added during a titration

carbohydrase an enzyme that breaks down carbohydrates; examples include amylase and isomerase

carbon dating method of determining the age of archaeological specimens by measuring the proportion of the carbon-14 isotope present

carrier organism that carries a recessive allele for a genetic disorder but does not itself have the disorder

catalyst substance added to a chemical reaction to alter the reaction rate, without being used up in the process

cathode negative electrode

cell membrane the outer covering of every cell, which controls the passage of substances into the cell

cell wall a strong covering made of cellulose, found on the outside of plant cells

cellulose a carbohydrate; a polysaccharide used to make plant cell walls

chain reaction a series of nuclear fission reactions where neutrons released from one reaction cause another nuclear fission reaction and so on

chemical analysis process of performing tests to determine what chemical substances are present in a sample, and/or the amount of each substance present

chemical formula shows the elements present in a compound and the number of atoms of each, such as H_2SO_4

chlorophyll green pigment inside chloroplasts in some plant cells, which absorbs energy from sunlight

chromatogram pattern of spots produced by paper or thin-layer chromatography

chromatography analytical technique for separating and identifying the components of a mixture, using paper or a thin layer or column of absorbent

collision frequency   number of collisions per second between the particles involved in a chemical reaction

collision theory   relates reaction rates to the frequency and energy of collisions between the reacting particles

column chromatography   chromatography method using a solvent to carry substances down a column of absorbent

compact fluorescent lamp (CFL)   light bulb that is efficient at transferring the energy in an electric current as light

compound   substance composed of two or more elements joined together by chemical bonds, for example, H_2O

concentration   amount of chemical present in a given volume of a solution – usually measured as g/dm^3 or mol/dm^3

conduction (electrical)   flow of electrons through a solid, or ions through a liquid

conduction (thermal)   heat passing through a material by transmitting vibrations from one particle to another

contact process   industrial process for making sulfuric acid

conventional current flow   direction of flow of electric current around a circuit, from positive to negative – the opposite direction to the flow of electrons

covalent bond   bond between atoms in which some of their outer electrons are shared

critical mass   the minimum mass of nuclear fuel needed to make a chain reaction happen

crumple zone   part of a vehicle designed to absorb energy in an accident, so reducing injuries to passengers

crystallise   form crystals from a liquid – for example, by partly evaporating a solution and leaving to cool

current   flow of electricity around a circuit – carried by electrons through solids and by ions through liquids

cystic fibrosis   a genetic disorder caused by a recessive allele, where lungs become clogged with mucus

cytoplasm   the jelly-like material inside a cell, in which metabolic reactions take place

deceleration   see negative acceleration

delocalised electrons   electrons not attached to any particular atom, so free to move through the structure, allowing electrical conduction – present in metals and graphite

denatured   the shape of an enzyme molecule has changed so much that it can longer bind with its substrate

diatomic molecule   molecule consisting of just two atoms

differentiation   change of a cell from general-purpose to one specialised to carry out a particular function

diffusion   spreading of particles of a gas, or of any substance in solution, resulting in a net movement from a region where they are in a high concentration to a region where they are in a lower concentration

digestive juices   liquids secreted within the digestive system and containing enzymes that help to digest food molecules

diode   semiconductor device that allows an electric current to flow through it in only one direction

direct current (d.c.)   electric current where the direction of the flow of current stays constant, as in cells and batteries

distance–time graph   graph showing how the distance an object travels varies with time: its gradient shows speed

DNA   deoxyribonucleic acid – the chemical from which chromosomes are made: its sequence determines genetic characteristics, such as eye colour

dominant   a dominant allele has an effect even when another allele is present

double covalent bond   two covalent bonds between the same pair of atoms – each atom shares two of its own electrons plus two from the other atom

ductile   can be drawn out into thin wires

earth wire   wire connecting the case of an electrical appliance, through the earth pin on a three-pin plug, to earth

earthed   safety feature where part of an appliance is connected to earth to protect users from electrocution if there is a fault

efficiency   a measure of how effectively an appliance transfers the energy that flows into the appliance into useful effects

egg cell   female gamete

elastic   material that returns to its original shape when the force deforming it is removed

elastic collision   collision where colliding particles or objects bounce apart after collision

elastic potential energy   energy stored in an object because it is stretched, compressed or deformed, and released when the object returns to its original shape

electrode   solid electrical conductor through which the current passes into and out of the liquid during electrolysis – and at which the electrolysis reactions take place

electrolysis   decomposing an ionic compound by passing a d.c. electric current through it while molten or in solution

electrolyte   solution or molten substance that conducts electricity

electron   small negatively-charged particle within an atom and orbiting the nucleus

electronic structure (or configuration)   arrangement of electrons in shells, or energy levels, in an atom

electrostatic induction   electric charge induced on an object made of an electrical insulator, by another electrically charged object nearby

empirical formula   ratio of elements in a compound, as determined by analysis – for example, CH_2O for glucose (molecular formula $C_6H_{12}O_6$)

endothermic reaction   chemical reaction which takes in heat, or energy from other sources

energy levels   electrons in shells around the nucleus – the further from the nucleus, the higher the electron's energy level

enzyme   biological catalyst that increases the speed of a chemical reaction but not used up in the process

epidermis   a tissue covering the outer surface of a plant's leaf, stem or root

epithelial tissue   tissue forming a covering over a part of an animal's body

equal and opposite forces   balanced forces equal in size but acting in opposite directions

exothermic reaction   chemical reaction which gives out heat

extinct   no longer existing

filament bulb   lightbulb giving out light by current flowing through a fine wire and heating the wire until it glows white hot

flexible (material)   can be bent without the material breaking

force meter   device measuring the size of a force, by measuring how much the force stretches a spring

forward reaction   reaction from left to right in an equation for a reversible reaction

fossil   preserved remains of a long-dead organism

'free' electrons   electrons that move readily from one atom to another to transmit an electric current through a conductor

frequency   number of waves passing a set point per second

friction   force acting at points of contact between objects moving over each other, to resist the movement

fullerenes   cage-like carbon molecules containing many carbon atoms, for example, C_{60}, buckminsterfullerene

fuse   a fine wire that melts if too much current flows through it, breaking the circuit and so switching off the current

gamete   sex cell – a cell containing the haploid number of chromosomes, such as an egg or sperm

gamma rays   ionising electromagnetic radiation – radioactive and dangerous to human health

gas chromatography   method that uses a gas to carry the substances through a long, thin tube of absorbent

gene   a length of DNA that codes for a particular characteristic

genetic diagram   a format used to describe and explain the probable results of a genetic cross

genotype   the pair of alleles that an organism possesses for a particular gene

geographical isolation   the separation of two populations of a species by a geographical barrier, such as a mountain chain

giant covalent structure   solid structure made up of a regular arrangement of covalently bonded atoms – may be an element or a compound

giant ionic structure   solid structure made up of a regular arrangement of ions in rows and layers

gland   organ that secretes a useful substance

glandular tissue   tissue made up of cells that are specialised to secrete a particular substance

glucose   a simple sugar, made by plants in photosynthesis, and broken down in respiration to release energy inside all living cells

glycogen   carbohydrate used for energy storage in animal cells

gravitational potential energy   energy that an object has because of its position, for example, increasing the height of an object above the ground increases its gravitational potential energy

gravity   the attractive force acting between all objects with mass – on Earth the attractive force due to gravity pulls objects downwards

Haber process   industrial process for making ammonia

haemoglobin   chemical in red blood cells which carries oxygen

haemophilia   disease where blood lacks the ability to clot

halide ion   ion of a halogen – halide ions have a 1- charge

halogens   reactive non-metals in Group 7 of the periodic table

hard water   water supply containing dissolved calcium or magnesium salts – these react with soap, making it hard to form a lather

heterozygous   possessing two different alleles of a gene

homozygous   possessing two identical alleles of a gene

Hooke's Law   for an elastic object, the extension is proportional to the force applied, provided the limit of proportionality is not exceeded

hormone   chemicals that act on target organs in the body (hormones are made by the body in special glands)

hydrogen ion H^+ ion – hydrogen ions in solution, $H^+(aq)$, make the solution acidic

hydroxide ion OH^- ion – hydroxide ions in solution, $OH^-(aq)$, make the solution alkaline

inelastic collision   collision where the colliding particles or objects stick together after collision

inhibitor   substance used to decrease a reaction rate – also called a 'negative catalyst'

initial reaction rate   reaction rate at the start of the reaction

insoluble salt   salt which is not soluble in water, so forms a precipitate

instantaneous reaction rate   reaction rate at a particular instant during the reaction

intermolecular forces   forces between molecules

ion   atom (or group of atoms) with a positive or negative charge, caused by losing or gaining electrons

ionic bonding   chemical bond formed by attractions between ions of opposite charges

ionic compound   compound composed of positive and negative ions held together in a regular lattice by ionic bonding, for example, sodium chloride

ionic equation   shows only the ions that actually react – anything that does not change during the reaction is omitted

isomerase   enzyme that changes glucose to fructose

isotopes   forms of element where their atoms have the same number of protons but different numbers of neutrons

kinetic energy   energy an object has because of its movement – it is greater for objects with greater mass or higher speed

lactic acid   a waste product of anaerobic respiration in muscle cells

lattice   regular arrangement of ions or atoms in a solid – may be covalent or ionic

Law of Conservation of Energy   energy can be transferred but cannot be created or destroyed

Law of Conservation of Momentum   total momentum before a collision is equal to total momentum after the collision, if no outside forces are acting

LDR (light dependent resistor)   resistor with a resistance that decreases when light is shone on it

LED (light emitting diode)   diode that gives off light when a current flows through it

limit of proportionality (for Hooke's Law) the point for an elastic material when the extension stops being proportional to the force: materials break or are permanently damaged when stretched beyond this point

limiting factor anything that is in short supply and therefore stops a process from happening faster

lipase enzyme that breaks down fat molecules to fatty acids and glycerol molecules

macromolecule very large molecule made up of hundreds of thousands, or millions, of atoms, for example, a polymer or crystal with a giant covalent structure

main sequence star a star in which nuclear fusion reactions combine small atomic nuclei into elements with larger nuclei

mains supply domestic electricity supply – in the UK, mains supply is 230 V at 50 Hz

malleable can be hammered into shape without breaking

mass a measure of the amount of 'stuff' in an object

mass number total number of protons and neutrons in the nucleus of an atom – always a whole number

mass spectrometer instrument for identifying chemicals by measuring their relative formula mass very accurately

meiosis type of cell division producing four genetically different daughter cells, each with half the normal number of chromosomes

metallic bonding type of bonding in metals – a regular lattice of metal ions is held together by delocalised electrons

mitochondrion (pl: mitochondria) organelle in which the reactions of aerobic respiration take place

mitosis type of cell division producing two genetically identical daughter cells

mixtures two or more substances mixed together – they can usually be separated by physical methods such as filtration

mole unit for counting atoms and molecules – one mole of any substances contains the same number of particles

molecular ion ion formed when an electron is knocked off a molecule in a mass spectrometer

momentum mass of a moving object multiplied by its velocity – a vector quantity having both size and direction

monomers small molecules that become chemically bonded to each other to form a polymer chain

muscular tissue a tissue that is specialised for contraction, causing movement

mutation an unpredictable change in an organism's DNA

nanometre unit used to measure very small length (1 nm = 0.000 000 001 m, or one-billionth of a metre)

nanoparticles very small particles (1–100 nanometres in size)

nanoscience the study of nanoparticles

nanotube carbon molecule in the form of a cylinder

native (relating to metals such as gold) occurs in rocks as the element – not combined in compounds

natural selection the increased chance of survival of individual organisms that have phenotypes that adapt them successfully to their environment

negative acceleration rate at which an object slows down, or decelerates, calculated from change in velocity divided by time

neutral solution aqueous solution with pH 7 – the concentrations of $H^+(aq)$ and $OH^-(aq)$ ions are equal

neutralisation reaction between an acid and a base to make a salt and water (H^+ ions react with OH^- or O_2^- ions)

neutron small particle that does not have a charge – found in the nucleus of an atom

newton standard unit of force – one newton is about the same as the weight of a small apple

Newton's Third Law when two objects are in contact with each other, they exert equal and opposite forces on one another

noble gas structure stable arrangement of electrons achieved by gaining, losing or sharing electrons to obtain an outer shell of eight (two in the case of hydrogen)

non-ohmic device device that does not obey Ohm's law, so the current through it is not directly proportional to the potential difference across it

nuclear fission nuclear reaction in which large atomic nuclei split into smaller ones, giving off large amounts of energy

nuclear fusion nuclear reaction in which small atomic nuclei join together into larger ones, giving off large amounts of energy

nuclear radiation radiation given out by nuclear reactions – three types: alpha particles, which are helium nuclei, beta particles, which are electrons and gamma rays, which are electromagnetic radiation

nucleus (cells) a structure found in most animal cell and plant cells, which contains the chromosomes made of DNA, and which controls the activities of the cell

Ohm's Law for a device with a constant value of resistance, the current through it is always directly proportional to the potential difference across it

organ structure within an organism's body, made up of different types of tissues, that carries out a particular function

organelle structure within a cell

organic compound compound containing carbon and hydrogen and possibly oxygen, nitrogen or other elements – living organisms are made up of organic compounds

oscilloscope device with screen to show how amplitude and frequency of an input wave varies – also called a cathode ray oscilloscope.

ovary organ in a female in which eggs are made

oxidation process that increases the amount of oxygen in a compound – opposite of reduction

oxygen debt the extra oxygen that has to be taken into the body after anaerobic respiration has taken place

parallel circuit electrical circuit with more than one possible path for the current to flow around

pathogen harmful microorganism that causes disease

pepsin protease enzyme found in the stomach

percentage yield theoretical yield actually obtained (percentage yield = actual yield ÷ theoretical yield x 100)

pH scale scale from 0 to 14 which shows how acidic or alkaline a substance is

phenotype appearance or characteristics or an organism, affected by its genes and its environment

phloem tissue made up long tubes that transports sugars from the leaves to all other parts of a plant

photochromic photochromic materials change colour in response to changes in light level

photosynthesis process in green plants where sunlight, carbon dioxide and water produce glucose and oxygen

physical factor something that influences a living organism that is caused by non-living aspects of their environment, such as temperature or light intensity

pipette used to measure out an exact volume of liquid

plastics compounds produced by polymerisation, capable of being moulded into various shapes or drawn into filaments and used as textile fibres

plum pudding model model that said the atom was like a positively charged 'jelly' with negatively charge electrons dotted through it – later shown to be incorrect

polydactyly having more than five fingers or toes on a hand or foot

polymer large molecule made up of a chain of monomers

polymerisation chemical process that links monomers together to form a polymer chain

power amount of energy that something transfers each second and measured in watts (or joules per second)

power rating measure of how fast an electrical appliance transfers energy supplied as an electrical current

precipitate solid product formed by reacting two solutions

precipitation reaction between two solutions to form a solid product (a precipitate)

protease enzyme that breaks down protein molecule to amino acid molecules

proton small positive particle found in the nucleus of an atom

protostar dense cloud of dust and gas that can form a new star, if it contains enough matter

quadrat a square area within which type and numbers of living organisms can be counted or estimated

radioactive materials giving off nuclear radiation

radioisotope a radioactive isotope of an element

RCCB (residual current circuit breaker) measures the current flowing into and out of an appliance, and switches the current off if they are not equal

reaction conditions physical conditions under which a reaction is performed, for example, temperature and pressure

reaction rate the speed at which a chemical reaction takes place – measured as the amount of reaction per unit time

reaction rate (average) total amount of reaction ÷ total time

reaction rate (initial) reaction rate at the start of the reaction

reaction time time between when a driver sees a hazard and when they begin to respond to it – increased by tiredness, drugs, or distractions

reactivity series list of metals in order of their reactivity with oxygen, water and acids

recessive a recessive allele only has an effect when a dominant allele is not present

reduction process that reduces the amount of oxygen in a compound, or removes all the oxygen from it – opposite of oxidation

regenerative braking type of braking which transfers some of the kinetic energy wasted by the slowing car into electricity which is used to operate the car brakes

relative atomic mass average mass of all the atoms in an element, taking into account the presence of different isotopes – often rounded to the nearest whole number

relative formula mass total mass of all atoms in a formula = each relative atomic mass × number of atoms present

relative molecular mass same as relative formula mass, but limited to elements or compounds that have separate molecules

reproducibility how likely it is that measurements made again under similar conditions, give the same results

resistance (electrical) measure of how hard or how easy it is for an electric current to flow through a component

respiration process occurring in all living cells, in which energy is released from glucose

resultant force the single force that would have the same effect on an object as all the forces that are acting on the object

retention factor (R_f) used to help identify individual spots in a chromatogram (R_f = distance moved by the spot ÷ distance moved by the solvent)

retention time (R_t) time taken for a component to travel through the tube of absorbent during gas chromatography

reverse reaction reaction from right to left in the equation for a reversible reaction

reversible reaction a reaction that can also occur in the opposite direction – that is, the products can react to form the original reactants again

ribosomes tiny structures within a cell, where protein synthesis takes place

salt compound composed of metal ions and non-metal ions – formed by acid–base neutralisation

sample take measurements, or make counts, in a small area rather than over the entire area in question

scalar quantity quantity that only has size, but not direction, for example, energy is a scalar quantity

series circuit electrical circuit with only one possible path for the current to flow around

sex chromosomes the X and Y chromosomes

shape memory alloy alloy that 'remembers' its original shape and returns to it when heated

shells electrons are arranged in shells (or orbits) around the nucleus of an atom – also known as 'energy levels'

smart material material which changes in response to changes in its surroundings, such as light levels or temperature

soften (water) treat water so as to remove the calcium ions that cause hardness

soluble salt salt which dissolves in water

solute substance which dissolves in a liquid to form a solution

solvent   liquid in which solutes dissolve to form a solution

speciation   formation of a new species

species   group of organisms that share similar characteristics and that can breed together to produce fertile offspring

speed   how quickly an object is moving, usually measured in metres per second (m/s)

sperm cell   male gamete

starch   a carbohydrate; a polysaccharide that is used for storing energy in plant cells, but not in animal cells

static electricity   an electric charge on an insulating material, caused by electrons flowing onto or away from the object

stem cell   a cell that has not yet differentiated – it can divide to form cells that form various kinds of specialised cell

stopping distance   total distance it takes a vehicle to stop – the sum of thinking distance and braking distance

sub-atomic particle   particle that make up an atom, such as proton, neutron or electron

subcutaneous   just under the skin

sublime   turn directly from solid into a gas without melting

substrate   molecule on which an enzyme acts – the enzyme catalyses the reaction that changes the substrate into a product

successful collisions   collisions with enough energy to break bonds in the reactant particles, and thus cause a reaction

surface area (of a solid reactant)   measure of the area of an object that is in direct contact with its surroundings

synthetic   artificial or made by people

syrup   concentrated solution of sugar

tarnish   go dull and discoloured by reacting with oxygen, moisture or other gases in the air

terminal velocity   maximum velocity an object can travel at – at terminal velocity, forward and backward forces are the same

testes   organs in a male in which sperms are made

tetrahedral structure   structure in which atoms have four covalent bonds to other atoms positioned at the four corners of a tetrahedron, for example, diamond

theoretical yield   mass of product that a given mass of reactant should produce according to calculations from the equation – the actual yield is always less than this

thermistor   resistor made from semiconductor material: its resistance decreases as temperature increases

thermochromic   thermochromic materials change colour in response to changes in temperature

thermosetting polymer   plastic polymer that sets hard when heated and moulded for the first time – it will not soften or melt when heated again

thermosoftening polymer   plastic polymer that softens and melts when heated and reheated

thinking distance   distance a vehicle travels while a signal travels from the driver's eye to brain and then to foot on the brake pedal: thinking distance increases with vehicle speed

thin layer chromatography (TLC)   chromatography using a plate coated with a thin layer of powdered absorbent

three core cable   electrical cable containing three wires, live, neutral and earth

three-pin plug   type of plug used for connecting to the mains supply in the UK: it has three pins, live, neutral and earth

tissue   group of similar cells that work together to carry out a particular function

titration   procedure to determine the volume of one solution needed to react with a known volume of another solution

tonne   1 tonne = 1000 kg (1 million grams)

toxic   poisonous

tracer   radioactive element used to track the movement of materials, such as water through a pipe or blood through organs of the body

transect   line along which organisms are sampled – transects are often used to investigate how the distribution of organisms changes when one type of habitat merges into another

triple covalent bond   three covalent bonds between the same two atoms – each atom shares three of its own electrons plus three from the other atom

upthrust   upward force on an object in water – for a floating object, the upthrust is equal to the weight of the object

vacuole   liquid-filled space inside a cell – many plant cells contain vacuoles full of cell sap

validity   how well a measurement really measures what it is supposed to be measuring

Van der Graaff generator   device for investigating static electricity: a large static electricity charge builds up on a metal dome insulated from earth

vector quantity   a quantity that has both size and direction: velocity is a vector quantity, having size in a particular direction

velocity   measure of how fast an object is moving in a particular direction

velocity–time graph   graph showing how the velocity of an object varies with time: its gradient shows acceleration

virus   very small structure made of a protein coat surrounding DNA (or RNA); viruses can only reproduce inside a living cell

VO$_2$max   the maximum volume of oxygen the body can use per minute

'wasted' energy   energy that is transferred by a device or appliance in ways that are not useful

water of crystallisation   water molecules present in crystals of some metal salts – shown separately in the formula, for example, hydrated copper sulfate, $CuSO_4.5H_2O$

watt   unit of energy transfer – one watt is a rate of energy transfer of one joule per second

weight   the downward force on a mass due to gravity, measured in newtons (N)

work   amount of energy transferred to an object by a force moving the object through a distance: work done = force × distance moved in the direction of the force

xylem   tissue made up of long, empty, dead cells that transports water from the roots to the leaves of a plant

yeast   a single-celled fungus

yield   mass of product made from a chemical reaction

Index

Internet research

The internet is a great resource to use when you are working through your GCSE Science course.

Below are some tips to make the most of it.

1 Make sure that you get information at the right level for you by typing in the following words and phrases after your search: 'GCSE', 'KS4', 'KS3', 'for kids', 'easy', or 'simple'.

2 Use OR, AND, NOT and NEAR to narrow down your search.

 > Use the word OR between two words to search for one or the other word.

 > Use the word AND between two words to search for both words.

 > Use the word NOT, for example, 'York NOT New York' to make sure that you do not get unwanted results (hits).

 > Use the word NEAR, for example, 'London NEAR Art' to bring up pages where the two words appear very close to each other.

3 Be careful when you search for phrases. If you search for a whole phrase, for example, A Room with a View, you may get a lot of search results matching some or all of the words. If you put the phrase in quote marks, 'A Room with a View' it will only bring search results that have that whole phrase and so bring you more pages about the book or film and less about flats to rent!

4 For keyword searches, use several words and try to be specific. A search for 'asthma' will bring up thousands of results. But, a search for 'causes of asthma' or 'treatment of asthma' will bring more specific and fewer returns. Similarly, if you are looking for information on cats, for example, be as specific as you can by using the breed name.

5 Most search engines list their hits in a ranked order so that results that contain all your listed words (and so most closely match your request) will appear first. This means the first few pages of results will always be the most relevant.

6 Avoid using lots of smaller words such as A or THE unless it is particularly relevant to your search. Choose your words carefully and leave out any unnecessary extras.

7 If your request is country-specific, you can narrow your search by adding the country. For example, if you want to visit some historic houses and you live in the UK, search 'historic houses UK' otherwise it will search the world. With some search engines you can click on a 'web' or 'pages from the UK only' option.

8 Use a plus sign (+) before a word to force it into the search. That way only hits with that word will come up.

Periodic Table

	1	1 H hydrogen

Group	1	2												Group 3	4	5	6	7	0
																			4 2 He helium
	7 3 Li lithium	9 4 Be beryllium												11 5 B boron	12 6 C carbon	14 7 N nitrogen	16 8 O oxygen	19 9 F fluorine	20 10 Ne neon
	23 11 Na sodium	24 12 Mg magnesium												27 13 Al aluminium	28 14 Si silicon	31 15 P phosphorus	32 16 S sulfur	35 17 Cl chlorine	40 18 Ar argon
	39 19 K potassium	40 20 Ca calcium	45 21 Sc scandium	48 22 Ti titanium	51 23 V vanadium	52 24 Cr chromium	55 25 Mn manganese	56 26 Fe iron	59 27 Co cobalt	59 28 Ni nickel	64 29 Cu copper	65 30 Zn zinc		70 31 Ga gallium	73 32 Ge germanium	75 33 As arsenic	79 34 Se selenium	80 35 Br bromine	84 36 Kr krypton
	85 37 Rb rubidium	88 38 Sr strontium	89 39 Y yttrium	91 40 Zr zirconium	93 41 Nb niobium	96 42 Mo molybdenum	99 43 Tc technetium	101 44 Ru ruthenium	103 45 Rh rhodium	106 46 Pd palladium	108 47 Ag silver	112 48 Cd cadmium		115 49 In indium	119 50 Sn tin	122 51 Sb antimony	128 52 Te tellurium	127 53 I iodine	131 54 Xe xenon
	133 55 Cs caesium	137 56 Ba barium	139 57 La lanthanum	178 72 Hf hafnium	181 73 Ta tantalum	184 74 W tungsten	186 75 Re rhenium	190 76 Os osmium	192 77 Ir iridium	195 78 Pt platinum	197 79 Au gold	201 80 Hg mercury		204 81 Tl thallium	207 82 Pb lead	209 83 Bi bismuth	210 84 Po polonium	210 85 At astatine	222 86 Rn radon
	223 87 Fr francium	226 88 Ra radium	227 89 Ac actinium																

Acknowledgements

The publishers wish to thank the following for permission to reproduce photographs. Every effort has been made to trace copyright holders and to obtain their permission for the use of copyright materials. The publishers will gladly receive any information enabling them to rectify any error or omission at the first opportunity.

p. cover D. Roberts/Science Photo Library, p. 8t Biophoto Associates/Science Photo Library, p. 8c Roca/Shutterstock, p. 8b Seleznev Oleg/Shutterstock, p. 9t blickwinkel/Alamy, p. 9u Sebastian Kaulitzki/Shutterstock, p. 9l Juan Estey/iStockphoto, p. 9b Chris Howes/Wild Places Photography/. Alamy, p. 10t Russell Kightley/Science Photo Library, p. 10c Biophoto Associates/Science Photo Library, p. 10b Michael Eichelberger, Visuals Unlimited/Science Photo Library, p. 11 Scott Camazine/Alamy, p. 12t Wim Van Egmond, Visuals Unlimted/Science Photo Library, p. 12r blickwinkel/Alamy, p. 12l gardeningpix/Alamy, p. 14 Henrik Larsson/iStockphoto, p. 15 Glow Wellness/Alamy, p. 16 Steve Gschmeissner/Science Photo Library, p. 17 Biodisc, Visuals Unlimited/Science Photo Library, p. 18 Mauro Fermariello/Science Photo Library, p. 19 Peter Arnold, Inc./Alamy, p. 20 Science Photo Library, p. 22t MikeE/Shutterstock, p. 22b Gustom Life Science Images/Alamy, p. 23t Geoff Jones, p. 23c Geoff Jones, p. 23b Geoff Jones, p. 24 Andrejs Pidjass/Shutterstock, p. 26 Yellowj/Shutterstock, p. 27 Gavin Hellier/Alamy, p. 28 Photofusion Picture Library/Alamy, p. 29 Igor Dutina/Shutterstock, p. 30 Photoshot Holdings Ltd/Alamy, p. 31 Juniors Bildarchiv/Alamy, p. 32t Photodynamic/Shutterstock, p. 32c Holmes Garden Photos/Alamy, p. 32b Marketa Mark/Shutterstock, p. 33 Photoshot/Alamy, p. 34 Paul Glendell/Alamy, p. 36 Robert Read Thames Portfolio/Alamy, p. 43 Grant Heilman Photography/Alamy, p. 44t Pete Saloutos/iStockphoto, p. 44c Tony Campbell/Shutterstock, p. 44b Volodymyr Goinyk/Shutterstock, p. 45t Deco/Alamy, p. 45u deepblue-photographer/Shutterstock, p. 45l Steve Gschmeissner/Science Photo Library/Alamy, p. 45b Heiko Grossmann/iStockphoto, p. 46 Aleksandar Todorovic/Shutterstock, p. 47 UgputuLf SS/Shutterstock, p. 48 imagebroker/Alamy, p. 50 SHOUT/Alamy, p. 52t Giovanna Tondelli/Shutterstock, p. 52b Sinisa Botas/Shutterstock, p. 53 Saivann/Shutterstock, p. 54t outis/Alamy, p. 54l Monkey Business Images/Shutterstock, p. 54r deniss09/Shutterstock, p. 55 Danicek/Shutterstock, p. 58t Robert Estall photo agency/Alamy, p. 58b Imageshop/Alamy, p. 59 Steve Gschmeissner/Science Photo Library, p. 60 INTERFOTO/Alamy, p. 61t T-Service/Science Photo Library, p. 61b Steve Gschmeissner/Science Photo Library, p. 62 Aflo Foto Agency/Alamy, p. 64t Eye of Science/Science Photo Library, p. 64c Steve Gschmeissner/Science Photo Library/Alamy, p. 64b Medical-on-Line/Alamy, p. 65 PR. G Gimenez-Martin/Science Photo Library, p. 66t CNRI/Science Photo Library, p. 66b Look at Sciences/Science Photo Library, p. 68t Pictorial Press Ltd/Alamy, p. 68b Steve Gschmeissner/Science Photo Library/Alamy, p. 70 Rex Features, p. 71 Leah-Anne Thompson/Shutterstock, p. 72t haak78/Shutterstock, p. 72b Wally Eberhart/Visuals Unlimited/Corbis, p. 73 Pictorial Press Ltd/Alamy, p. 74 omkar.a.v/Shutterstock, p. 75 Sidney Moulds/Science Photo Library, p. 76 Julie Fairman/iStockphoto, p. 78 Science Source/Science Photo Library, p. 79 Omikron/Science Photo Library, p. 80t Dr.Margorius/Shutterstock, p. 80b CNRI/Science Photo Library, p. 81t Angela Hampton Picture Library/Alamy, p. 81b Doug Steley A/Alamy, p. 82 Jason Stitt/iStockphoto, p. 84t Volodymyr Goinyk/Shutterstock, p. 84c Bob Ainsworth/iStockphoto, p. 84b Paul Rackham/Alamy, p. 86t Mauricio Anton/Science Photo Library, p. 86b Duncan Walker/iStockphoto, p. 87 Rebecca Cairns-Wicks, p. 88t Toenne/Shutterstock, p. 88l neelsky/Shutterstock, p. 88r Anke van Wyk/Shutterstock, p. 89t Purestock/Photolibrary, p. 89b John Cancalosi/Photolibrary, p. 95l gallofoto/Shutterstock, p. 95r worldswildlifewonders/Shutterstock, p. 96t MrJafari/Shutterstock, p. 96c nikolpetr/Shutterstock, p. 96b Martyn F. Chillmaid/Oxford Scientific/Photolibrary, p. 97t Jaap Hart/iStockphoto, p. 97c Vasilyev/Shutterstock, p. 97b tanewpix/Shutterstock, p. 100 kevin bampton/Shutterstock, p. 101 James King-Holmes/Science Photo Library, p. 102t Aaron Amat/Shutterstock, p. 102b johanna goodyear/iStockphoto, p. 103 Andrew Lambert Photography/Science Photo Library, p. 104 Rob Sylvan/iStockphoto, p. 105t Dr Ajay Kumar Singh/Shutterstock, p. 105b RF Company/Alamy, p. 106t Charles D. Winters/Science Photo Library, p. 106b Andrew Lambert Photography/Science Photo Library, p. 108t Chris Howes/Wild Places Photography/Alamy, p. 108b Martyn F. Chillmaid/Science Photo Library, p. 109 Phil Degginger/Alamy, p. 110t Peter Arnold, Inc./Alamy, p. 110b Andrew Lambert Photography/Science Photo Library, p. 112t Pecold/Shutterstock, p. 112b Yoav Levy/Phototake Science/Photolibrary, p. 113 Health Protection Agency/Science Photo Library, p. 114t Jarno Gonzalez Zarraonandia/Shutterstock, p. 114b stanislaff/Shutterstock, p. 116 Efired/Shutterstock, p. 117 Scott Camazine/Alamy, p. 118t Dinodia Photos/Alamy, p. 118c Vasilyev/Shutterstock, p. 119 Rob Lavinsky/Wikimedia Commons, p. 120t Mike Tingle, used with kind permission of Dry-Planet.com, p. 120c Aleksandra Novakovic/Shutterstock, p. 120b John Kasawa/Shutterstock, p. 121 Mike Tingle, p. 122t Diego Cervo/Shutterstock, p. 122c Laguna Design/Science Photo Library, p. 122b Studio Araminta/Shutterstock, p. 124t Mark Lorch/Shutterstock, p. 124u R-O-M-A/Shutterstock, p. 124c Cordelia Molloy/Science Photo Library, p. 124b Tyler Boyes/iStockphoto, p. 125t mangostock/Shutterstock, p. 125b Laguna Design/Science Photo Library, p. 126t Paul Cowan/Shutterstock, p. 126b Andrew Lambert Photography/Science Photo Library, p. 128t Hedser van Brug/Shutterstock, p. 128b Luigi Chiesa/Wikimedia Commons, p. 130t Juburg/Shutterstock, p. 130b Bernard Bisson/Sygma/Corbis, p. 131 E.R.Degginger/Science Photo Library, p. 132t david olah/iStockphoto, p. 132b Stock Connection Blue/Alamy, p. 134t DocCheck Medical Services GmbH/Alamy, p. 134b Christian Darkin/Science Photo Library, p. 136 Adrian Sherratt/The Sunday Times/NI Syndication, p. 138 sciencephotos/Alamy, p. 139 Paul Marcus/Shutterstock, p. 140t Susumu Hishinaga/Science Photo Library, p. 140b Steve Lupton/Corbis, p. 141t Martyn F. Chillmaid/Science Photo Library, p. 141b Andrew Lambert Photography/Science Photo Library, p. 142 sciencephotos/Alamy, p. 150t Sasha Radosavljevic/iStockphoto, p. 150c Jacob Hamblin/Shutterstock, p. 150b Shawn Hempel/Shutterstock, p. 151t Alexander Gitlits/Shutterstock, p. 151u Solent News & Photo Agency/Rex Features, p. 151l sciencephotos/Alamy, p. 151b Ria Novosti/Science Photo Library, p. 152t Martyn F. Chillmaid/Science Photo Library, p. 152b Charles D. Winters/Science Photo Library, p. 154t marilyn barbone/Shutterstock, p. 154b Martyn F. Chillmaid/Science Photo Library, p. 156t Richard Peterson/Shutterstock, p. 156b Stu49/Shutterstock, p. 160 Peter Arnold, Inc./Alamy, p. 162 Martyn F. Chillmaid/Science Photo Library, p. 163 Agphotographer/Shutterstock, p. 164 Document General Motors/Reuter R/Corbis Sygma, p. 165 Martyn F. Chillmaid/Science Photo Library, p. 166t Ching-Yeh Ching-Yeh Lu/iStockphoto, p. 166b sciencephotos/Alamy, p. 167t James L. Amos/Corbis, p. 167b Nigel Cattlin/Alamy, p. 168 Galyna Andrushko/Shutterstock, p. 169 Thermit Welding (GB) Limited; used with kind permission, p. 170 Dr Jeremy Burgess/Science Photo Library, p. 171 Victor I. Makhankov/Shutterstock, p. 172 LianeM/Shutterstock, p. 174t Wikimedia Commons, p. 174b Leslie Garland Picture Library/Alamy, p. 175 Andrew Lambert Photography/Science Photo Library, p. 176 Zurbagan/Shutterstock, p. 178 tunart/iStockphoto, p. 180t David W. Leindecker/Shutterstock, p. 180b Jacqueline Abromeit/Shutterstock, p. 188t Martyn F. Chillmaid/Science Photo Library, p. 188c teekid/iStockphoto, p. 188b Ivaschenko Roman/Shutterstock, p. 188t luigi nifosi/Shutterstock, p. 188b X.D.Luo/Shutterstock, p. 190 Alex Segre/Alamy, p. 192 NASA, p. 194 Studio 1One/Shutterstock, p. 195 Ilja Mašík/Shutterstock, p. 196t Michael Shake/Shutterstock, p. 196b Vuk Vukmirovic/Shutterstock, p. 197 NASA, Cornell Univ., JPL and M. Di Lorenzo et al., p. 198t testing/Shutterstock, p. 198b Mandy Godbehear/Shutterstock, p. 199t photomak/Shutterstock, p. 199b Justin Kase ztwoz/Alamy, p. 200t Edward Shaw/iStockphoto, p. 200b Pakhnyushcha/Shutterstock, p. 201 idp oulton park collection/Alamy, p. 202 Hazan/Shutterstock, p. 203 Jacom Stephens/iStockphoto, p. 204t Kathy deWitt/Alamy, p. 204b Valentyn Hontovyy/Shutterstock, p. 206 Fribus Ekaterina/Shutterstock, p. 208t Marcel Jancovic/Shutterstock, p. 208b The Garden Picture Library/Alamy, p. 209t Photo Dassault-Breguet/Science Photo Library, p. 209c Art Directors & TRIP/Alamy, p. 209b llukee/Alamy, p. 210t Knud Nielsen/Shutterstock, p. 210u RTimages/Shutterstock, p. 210c Glue Stock/Shutterstock, p. 210b Stuart Miles/Shutterstock, p. 211t Kapu/Shutterstock, p. 211c Ljupco Smokovski/Shutterstock, p. 211b mathom/Shutterstock, p. 212t Bryan Busovicki/Shutterstock, p. 212b Mervyn Rees/Alamy, p. 214t PCN Photography/Alamy, p. 214b William Caram/Alamy, p. 215t Dmitriy Shironosov/Shutterstock, p. 215b ClassicStock/Alamy, p. 216 Duncan Shaw/Alamy, p. 218 H. Powers/Wikimedia Commons, p. 220 pakowacz/Shutterstock, p. 222 sciencephotos/Alamy, p. 224 Przemyslaw Rzeszutko/iStockphoto, p. 226 Jiri Pavlik/Shutterstock, p. 227 dean bertoncelj/Shutterstock, p. 228 Kevin Foy/Alamy, p. 230 Viktor Gladkov/Shutterstock, p. 232 Oleksandr Chub/iStockphoto, p. 238 MariusdeGraf/Shutterstock, p. 240t Jeff Banke/Shutterstock, p. 240c Stasys Eidiejus/Shutterstock, p. 240b Johan Ramberg/iStockphoto, p. 241t TVR/Shutterstock, p. 241c Pola36/Shutterstock, p. 241b Bill Frische/Shutterstock, p. 242 Tatiana Popova/Shutterstock, p. 243 Gordon Heeley/Shutterstock, p. 244tl Chris leachman/Shutterstock, p. 244tr Chris leachman/Shutterstock, p. 244b Photoseeker/Shutterstock, p. 245t Ingvar Bjork/Shutterstock, p. 245c Four Oaks/Shutterstock, p. 245b tankist276/Shutterstock, p. 246t Master3D/Shutterstock, p. 246b StockImages/Alamy, p. 247t Vydrin/Shutterstock, p. 247b Leslie Garland Picture Library/Alamy, p. 248t Blackout Concepts/Alamy, p. 248b mathieukor/iStockphoto, p. 249 Cico/Shutterstock, p. 250 Mike Kemp/Rubberball/Corbis, p. 251t Osipava Alena/Shutterstock, p. 251b StudioSource/Alamy, p. 252 Archive Pics/Alamy, p. 254 Mark Kostich/iStockphoto, p. 256 Science Source/Science Photo Library, p. 258t Jorg Hackemann/Shutterstock, p. 258b Hank Morgan/Science Photo Library, p. 259 Martyn F. Chillmaid/Science Photo Library, p. 260 James King-Holmes/Science Photo Library, p. 264 Cordelia Molloy/Science Photo Library, p. 265 medicalpicture/Alamy, p. 266 Corbis RF/Alamy, p. 267 Rex Features , p. 268t EFDA-JET/Science Photo Library, p. 268b Bill Frische/Shutterstock, p. 270 NASA, p. 282t Andrew Lambert Photography/Science Photo Library, p. 282c Pedro Salaverría/Shutterstock, p. 282b Shawn Hempel/Shutterstock, p. 288 Martyn F. Chillmaid/Science Photo Library, p. 300-301 Tischenko Irina/Shutterstock, p. 300 Kletr/Shutterstock, p. 302 Diego Cervo/Shutterstock, p. 303 Dmitry Naumov/Shutterstock.